Viswanathan S. Saji
Abdelkader A. Meroufel • Ahmad A. Sorour

Editors

Corrosion and Fouling Control in Desalination Industry

 Springer

Editors
Viswanathan S. Saji
King Fahd University of Petroleum
and Minerals
Center of Research Excellence in Corrosion
Dhahran, Saudi Arabia

Abdelkader A. Meroufel
Corrosion Department
Saline Water Conversion Corporation
Jubail, Saudi Arabia

Ahmad A. Sorour
King Fahd University of Petroleum
and Minerals
Center of Research Excellence
in Corrosion & Mechanical Engineering
Department
Dhahran, Saudi Arabia

ISBN 978-3-030-34286-9   ISBN 978-3-030-34284-5   (eBook)
https://doi.org/10.1007/978-3-030-34284-5

This Springer imprint is published by the registered company Springer Nature Switzerland AG
The registered company address is: Gewerbestrasse 11, 6330 Cham, Switzerland

Preface

Corrosion and fouling continue to be severe issues for modern society affecting all life aspects, including the water desalination industry. Many authors from different perspectives discussed the understanding of these two phenomena. However, within the context of sustainable and cost-effective freshwater production through desalination, it is necessary in our opinion to address these challenges by assessing the progress of phenomena understanding, best engineering practices, and research and development efforts. We have made an effort in this direction to collectively present important aspects of corrosion and fouling and their control methods that are specific to the desalination industry. The term fouling here is used in a broader perspective that also includes inorganic scaling.

This book consists of three parts: (I) desalination processes, (II) corrosion in desalination, and (III) fouling in desalination. In each part, the authors recall fundamentals of the topic before discussing practical solutions and pending challenges for future efforts.

Part I (Chaps. 1, 2, 3 and 4) covers the key design components, performance indicators, economics, and challenges of conventional and unconventional desalination systems. Chapter 1 focuses on the concepts and system components of industrial desalination systems and provides a detailed account of the different desalination techniques, their energy consumption, and environmental sustainability. Chapters 2 and 3 respectively deal with the thermal and reverse osmosis desalination systems and provide interesting accounts of process components, performance, and challenges of the two widely employed industrial desalination systems. Chapter 4 describes in detail the unconventional desalination technologies, including membrane distillation, forward osmosis, adsorption desalination, and freeze desalination.

Part II (Chaps. 5, 6, 7, 8, 9, 10 and 11) deals with corrosion challenges and mitigation practices within desalination plants. Chapters 5 and 6 respectively detail forms and mitigation practices of corrosion in thermal and reverse osmosis desalination plants. Chapter 7 focuses explicitly on corrosion and environment-assisted cracking of stainless steels, the most preferred corrosion-resistant alloy in desalination systems. Chapter 8 explains various corrosion monitoring techniques that are currently in use in industrial desalination systems, namely, both direct and indirect corrosion monitoring techniques. Chapter 9 focuses on corrosion control in desalination plants utilizing chemical additives, particularly on corrosion inhibitors, and explains the inhibitors for MIC and oxygen scavengers. Chapter 10 deals with corrosion control strategies during acid cleaning in heat exchangers, which are a vital part of thermal desalination industry. Chapter 11 describes two cutting-edge corrosion control strategies that can be applied in desalination industry, namely smart coating and photoelectrochemical cathodic protection.

Part III (Chaps. 12, 13, 14, 15, 16 and 17) concentrates on fouling and its control methods in desalination industries. Chapter 12 focuses on inorganic scaling in desalination systems, while Chapter 13 provides an interesting account on biofouling of membranes in reverse osmosis desalination plants. Chapters 14 and 15, which both deal with control approaches to scaling, gives a general overview of various scale control strategies and provides a specific account of antiscalant chemical additives, respectively. Chapter 16 provides an interesting description of the practical ways of biofouling control in desalination systems and an account of biofouling monitoring techniques, while Chapter 17, the last chapter, explains strategies in fabricating antifouling desalination membranes.

The book encompassed a multidisciplinary group of authors, including academicians and industrial professionals, to combine knowledge and practical experience for a broad spectrum of interested readers. We hope that the present book will be a handy reference tool for scientists and engineers who are working in water desalination.

Dhahran, Saudi Arabia   Viswanathan S. Saji
Jubail, Saudi Arabia   Abdelkader A. Meroufel
Dhahran, Saudi Arabia   Ahmad A. Sorour

Acknowledgments

We are thankful to all the authors for their valued contribution to this book. We would also like to express our gratitude to all those granting us the copyright permissions for reproducing illustrations. We acknowledge King Fahd University of Petroleum and Minerals (KFUPM), and Saline Water Conversion Corporation (SWCC), Saudi Arabia for the moral support provided. Our sincere thanks for the Springer team in evolving this book into its final shape.

Contents

List of Abbreviations

AD	Adsorption Desalination
AGMD	Air Gap Membrane Distillation
AOC	Assimilable Organic Carbon
CAPEX	Capital Expenditure
CP	Cathodic Protection
CFD	Computational Fluid Dynamics
CF	Corrosion Fatigue
CRA	Corrosion Resistant Alloy
CUI	Corrosion Under Insulation
DEEP	Desalination Economic Evaluation Program
DCMD	Direct Contact Membrane Distillation
EDI	Electrodeionization
ED	Electrodialysis
EDR	Electrodialysis Reversal
ERDs	Energy Recovery Devices
EAC	Environment-Assisted Cracking
EFD	Eutectic Freeze Desalination
EPS	Extracellular Polymeric Substance
FRP	Fiber-Reinforced Polymer
FO	Forward Osmosis
FRC	Free Residual Chlorine
FD	Freeze Desalination
GOR	Gain Output Ratio
GRP	Glass-Reinforced Plastic
GDP	Gross Domestic Product
GCC	Gulf Cooperation Council
HIC	Hydrogen-Induced Cracking
MVC	Mechanical Vapor Compression
MD	Membrane Distillation
MIC	Microbial Induced Corrosion
MF	Microfiltration

MGD	Million Gallons per Day
MED	Multiple Effect Distillation
MSF	Multistage Flash
NF	Nanofiltration
NOM	Natural Organic Matter
NDT	Nondestructive Test
OPEX	Operational Expenditure
PR	Performance Ratio
PECP	Photoelectrochemical Cathodic Protection
PV	Photovoltaic
PREN	Pitting Resistance Equivalent Number
RO	Reverse Osmosis
SWRO	Seawater Reverse Osmosis
SCC	Stress Corrosion Cracking
SRB	Sulfate-Reducing Bacteria
SGMD	Sweeping Gas Membrane Distillation
TVC	Thermal Vapor Compression
TFC	Thin Film Composite
TDS	Total Dissolved Solids
TOC	Total Organic Carbon
TRC	Total Residual Chlorine
TSS	Total Suspended Solids
TMP	Transmembrane Pressure
VMD	Vacuum Membrane Distillation
VMDC	Vacuum Membrane Distillation and Crystallization

Contributors

Hasan Al Abdulgader Research & Development Center, Saudi Aramco, Dhahran, Saudi Arabia

Mohamed Afizal Mohamed Amin Advanced Membrane Technology Research Centre, School of Chemical and Energy Engineering, Faculty of Engineering, Universiti Teknologi Malaysia, Johor, Johor Bahru, Malaysia

Department of Chemical Engineering and Sustainability Energy, Faculty of Engineering, Universiti Malaysia Sarawak, Kota Samarahan, Sarawak, Malaysia

Nawrin Anwar Building, Civil and Environmental Engineering Department, Concordia University, Montreal, QC, Canada

M. C. M. Bruijs Pecten Aquatic, Lent, The Netherlands

Dayang Norafizan Awang Chee Advanced Membrane Technology Research Centre, School of Chemical and Energy Engineering, Faculty of Engineering, Universiti Teknologi Malaysia, Johor, Johor Bahru, Malaysia

Faculty of Resource Science and Technology, Universiti Malaysia Sarawak, Kota Samarahan, Sarawak, Malaysia

Tiantian Chen Department of Building, Civil and Environmental Engineering, Concordia University, Montreal, QC, Canada

Mahbuboor Rahman Choudhury Department of Building, Civil and Environmental Engineering Department, Concordia University, Montreal, QC, Canada

Civil and Environmental Engineering Department, Manhattan College, Bronx, NY, USA

Konstantinos D. Demadis Crystal Engineering, Growth and Design Laboratory, Department of Chemistry, University of Crete, Crete, Greece

A. Mohammed Farooque Desalination Technologies Research Institute, Saline Water Conversion Corporation, Jubail, Saudi Arabia

Pei Sean Goh Advanced Membrane Technology Research Centre, School of Chemical and Energy Engineering, Faculty of Engineering, Universiti Teknologi Malaysia, Johor, Johor Bahru, Malaysia

Osman Ahmed Hamed Desalination Technologies Research Institute, Saline Water Conversion Corporation, Jubail, Saudi Arabia

Ahmad Fauzi Ismail Advanced Membrane Technology Research Centre, School of Chemical and Energy Engineering, Faculty of Engineering, Universiti Teknologi Malaysia, Johor, Johor Bahru, Malaysia

H. A. Jenner Aquator, IJsselstein, The Netherlands

Ashish Kapoor Department of Chemical Engineering, SRM Institute of Science and Technology, Kattankulathur, TN, India

Juneseok Lee Department of Civil and Environmental Engineering, Manhattan College, Riverdale, NY, USA

Wen Ma Building, Civil and Environmental Engineering Department, Concordia University, Montreal, QC, USA

Chemical and Environmental Engineering Department, Yale University, New Haven, CT, USA

Wesley Meertens Building, Civil and Environmental Engineering Department, Concordia University, Montreal, QC, Canada

Abdelkader A. Meroufel Desalination Technologies Research Institute, Saline Water Conversion Corporation, Jubail, Saudi Arabia

H. J. G. Polman H2O Biofouling Solutions B.V., Bemmel, The Netherlands

Sivaraman Prabhakar Department of Chemical Engineering, SRM Institute of Science and Technology, Kattankulathur, TN, India

Md. Saifur Rahaman Building, Civil and Environmental Engineering Department, Concordia University, Montreal, QC, Canada

Sayeed Rushd Department of Chemical Engineering, College of Engineering, King Faisal University, Al Ahsa, Saudi Arabia

Viswanathan S. Saji King Fahd University of Petroleum and Minerals, Center of Research Excellence in Corrosion, Dhahran, Saudi Arabia

Moses M. Solomon Center of Research Excellence in Corrosion, King Fahd University of Petroleum & Minerals, Dhahran, Saudi Arabia

Argyro Spinthaki Crystal Engineering, Growth and Design Laboratory, Department of Chemistry, University of Crete, Crete, Greece

Khaled Touati Building, Civil and Environmental Engineering Department, Concordia University, Montreal, QC, Canada

Saviour A. Umoren Center of Research Excellence in Corrosion, King Fahd University of Petroleum & Minerals, Dhahran, Saudi Arabia

Haamid Sani Usman Building, Civil and Environmental Engineering Department, Concordia University, Montreal, QC, Canada

Liuqing Yang Building, Civil and Environmental Engineering Department, Concordia University, Montreal, QC, Canada

Tamim Younos Green Water-Infrastructure Academy, Washington, DC, USA

Part I
Desalination Processess

Chapter 1
Desalination: Concept and System Components

Tamim Younos and Juneseok Lee

1.1 Introduction

About 70% of the world's population is likely to be dealing with problems linked to water scarcity by 2025 [1]. The primary factor driving water scarcity is the high potable water demand in densely populated urban areas. Water scarcity issues are most critical in coastal areas within 100 km of the ocean, where approximately 40% of the world's population lives, although they are also a significant problem in arid/ semi-arid regions and island countries. The limited availability of freshwater resources and its high transportation cost from distant sources to high water demand areas have led to a renewed focus on developing seawater and brackish waters as alternative sources of water. Brackish water is available in estuarine/tidal surface waters, coastal aquifers, and some deep inland aquifers.

A broad definition of desalination includes the treatment of all non-potable water sources such as seawater, brackish water, wastewater, and stormwater runoff [2, 3]. In this chapter, the definition of desalination is limited to removing salts from seawater and brackish water. This chapter aims to recall the fundamental concept of desalination and to present an overview of modern desalination and system design components. Major system components considered include desalination techniques, energy consumption, environmental sustainability and the economics of desalination.

T. Younos (✉)
Green Water-Infrastructure Academy, Washington, DC, USA
e-mail: tamim.younos@gwiacademy.org

J. Lee
Department of Civil and Environmental Engineering, Manhattan College,
Riverdale, NY, USA

© Springer Nature Switzerland AG 2020
V. S. Saji et al. (eds.), *Corrosion and Fouling Control in Desalination Industry*,
https://doi.org/10.1007/978-3-030-34284-5_1

1.2 Desalination Concept

Since ancient times, desalination, i.e. separating salt and water via evaporation of seawater, has been practiced to produce freshwater for human consumption in small communities. In modern times, high water demand for municipalities and industrial complexes has necessitated developing of advanced and large-scale desalination systems. Desalinated water is also a vital water source for crop irrigation and power plant operation. Remote communities and ships depend on small-scale desalination systems as well.

For scientific and technical purposes, water quality in terms of salinity is best expressed by the concentration of total dissolved solids (TDS) which represents the sum of all minerals, salts, organic matter and metals that can dissolve in the water. As many as 50–70 dissolved elements can be found in seawater and brackish waters. More than 99% of the TDS in seawater or brackish water is comprised of the following six species: chloride (Cl^-), sodium (Na^+), sulfate (SO_4^{2-}), Magnesium (Mg^{2+}), calcium (Ca^{2+}), and potassium (K^+). Table 1.1 shows the possible range of TDS concentration for various categories of water sources. The ionic composition of seawater in TDS varies across different geographic locations (Table 1.2).

From a water use perspective, high TDS concentrations in drinking water can pose a health risk and may also convey an objectionable taste and odor issues. Other problems associated with high TDS concentration include but not limited to scaling in pipes, staining of bathroom fixtures, corrosion of piping and fixtures, and reduced soap lathering. Acceptable TDS concentrations vary depending on the intended use of water, but as a general rule of thumb, TDS of below 300 mg/L is considered excellent quality and levels above 1200 mg/L are considered unacceptable [5].

The World Health Organization (WHO) has published a document that highlights the principal health risks related to different desalination processes and provides guidance on appropriate risk assessment and management procedures that ensure the safety of desalinated drinking water [6]. The WHO report identified boron (B), borate (BO_3^{3-}), bromide (Br^-), sodium (Na^+), potassium (K^+), and magnesium (Mg^{2+}), as well as naturally occurring chemicals such as humic and fulvic acids and the by-products of algal and seaweed growth as chemicals of concern in source water. The U.S. Environmental Protection Agency (USEPA) has included TDS in its list of 15 nuisance chemicals and has set the Secondary Maximum Contaminant Level (SMCL) or aesthetic standard for TDS in potable water as being

Table 1.1 Water salinity based on TDS concentration [3]

Water source	TDS concentration (mg/L)
Freshwater (streams, rivers, lakes, aquifers)	< 500
Brackish water (estuarine/tidal surface waters, coastal aquifers, deep inland aquifers)	> 500–30,000
Seawater (saline water)	30,000-50,000
Brine water	> 50,000

Table 1.2 Ionic composition of seawater [4]. Reprinted with permission of Water Purification & Conditioning International

TDS composition	Typical seawater (mg/L)	Eastern mediterranean (mg/L)	Arabian Gulf at Kuwait (mg/L)	Red Sea at Jeddah (mg/L)
Chloride (Cl^-)	18,980	21,200	23,000	22,219
Sodium (Na+)	10,556	11,800	15,850	14,255
Sulfate (SO_4^{2-})	2649	2950	3200	3078
Magnesium (Mg^{2+})	1262	1403	1765	742
Calcium (Ca^{2+})	400	423	500	225
Potassium (K^+)	380	463	460	210
Bicarbonate (HCO_3^-)	140	–	142	146
Bromide (Br^-)	65	155	80	72
Borate (BO_3^{3-})	26	72	–	–
Strontium (Sr^{2+})	13	–	–	–
Fluoride (F^-)	1	–	–	–
Silicate (SiO_3^{2-})	1	–	1.5	–
Iodide (I^-)	<2	1	–	–
Others	–	–	–	–
Total TDS	34,483	38,600	45,000	41,000

below 500 mg/L, suggesting that a TDS concentration of less than 200 mg/L in drinking water is desirable [7]. Conventional water treatment processes – coagulation, sedimentation, and sand filtration technologies – are not effective in removing or lowering TDS from either seawater or brackish water; hence the need to develop techniques that can remove TDS from such sources to make them acceptable for potable water use and other intended uses is critical.

According to a 2019 report, there are 15,906 operational desalination plants around the world producing around 95 million m³/day of desalinated water of which 48% is produced in the Middle East and North Africa region [8]. Table 1.3 shows key components of desalination systems practiced around the world.

It should be noted that energy use is embedded within all components of a desalination system and significantly impact the economic efficiency of desalination plants.

1.3 Desalination Techniques

The two main types of modern desalination techniques adopted around the world are thermal (distillation) and membrane technologies. Many early desalination projects developed in the 1940s used thermal desalination and are still the dominant desalination technology in the Middle East. Membrane technologies were developed in the 1960s, and at present constitute a slightly larger portion of desalination

Table 1.3 Key components of desalination systems

Component	Description
Feed water intake	Site selection and structures built to extract seawater and brackish water for desalination purposes
Pretreatment	Removal of suspended solids and control of biological growth, to prepare the source water for further processing
Desalination techniques	Technologies used to remove ions, i.e., separate salt and water, to produce freshwater for the intended use
Post-treatment	The addition of chemicals to the desalinated water that makes water suitable for the intended use and to prevent corrosion of downstream infrastructure piping
Concentrate (brine) management	The treatment and/or reuse of rejected salt, residuals from the desalination process

plants around the world. Since these two technologies are well-described in Chaps. 2 and 3 of this book, a brief overview of these desalination techniques is provided below. Alternative desalination technologies, for example, membrane distillation (MD), that aim to enhance desalination system efficiency are discussed in Sect. 1.7.

1.3.1 Membrane Technologies

Membrane technologies have been extensively described in the literature [9]. A membrane is a thin film of porous material that allows water molecules to pass through while simultaneously preventing the passage of undesirable components such as salts, microorganisms and metallic elements [10]. Membranes can be made from a wide variety of materials, including polymeric materials such as cellulose, acetate, and nylon, and non-polymeric materials such as ceramics, metals and composites. Synthetic membranes are the most widely used for the desalination process. The American Water Works Association (AWWA) Manual M46 provides detailed information about applications of synthetic membranes for desalination [10].

Water membrane technologies include pressure-driven membranes and electrical-driven membranes. Pressure-driven membranes, namely reverse osmosis (RO), are applied to desalination of both seawater and brackish water. Table 1.4 shows various types of pressure-driven membrane technologies and characteristics. Electrical-driven membranes are mainly used for desalination of brackish water and sometimes as a pre-treatment step for the RO process, which are further discussed later[1].

Pressure-driven membranes can also be characterized by their Molecular Weight Cut-Off (MWCO). For example, the MWCO for RO is 50–200 daltons compared to MWCO of 100,000 daltons for microfiltration. The lower dalton value allows removal of very fine particles and dissolved solids in water.

---

[1] Natural organic matter

Table 1.4 Characteristics of pressure-driven membrane processes [9, 11]

Membrane process	Membrane pore size (Å)	Pressure applied Psi (kPa)	Contaminant removal efficiency	Application
Microfiltration	> 1000	4–70 (30–500)	Suspended particles, algae, protozoa, bacteria	Pre-treatment of feedwater
Ultrafiltration	100–1000	4–70 (30–500)	Large macromolecules, small colloids, viruses	Pre-treatment of feedwater
Nanofiltration	10–100	70–140 (500–1000)	NOM, Hardness (Ca^{2+}, Mg^{2+})	Partial desalination
Reverse Osmosis	1–10	140–700 (1000–5000)	Dissolved contaminants, Salt (Na^+, Cl^-)	Desalination

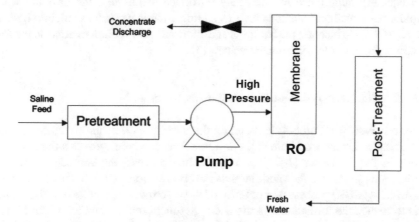

Fig 1.1 RO schematic of the overall operation. (Source: Sandia National Laboratories) [13]

1.3.1.1   Reverse Osmosis

Reverse Osmosis (RO) is a pressure-driven membrane technique where hydraulic pressure greater than the osmotic pressure is applied to saltwater (known as the feedwater) to reverse the natural flow direction through the membrane [12, 13]. The RO process is using the solution/diffusion mechanism whereby the applied pressure forces water molecules to diffuse through the tiny pore of the membrane leaving the majority of salts behind in a high salt solution called concentrate (reject salt or brine) (Fig 1.1).

The RO membrane characteristics are shown in Table 1.4. The RO process is effective for removing TDS concentrations of up to 45,000 mg/L with TDS removal efficiency of >99% [9, 11]. The management of concentrate is a critical environmental problem and is discussed later under environmental issues of desalination.

Pre-treatment of the feedwater is one of the most critical factors in the successful operation of the RO plant. Then, the RO feedwater should be free of large particles, suspended and colloidal particle, NOM, bacteria and viruses, and oil and grease. All these components will contribute to the fouling build up causing the blockage of membrane pores. This fouling affects water productivity and quality. The primary mechanisms of fouling include scaling, plugging, adsorption and bio-fouling caused by biological growth [13]. It's essential to protect the RO membrane from fouling, reduce energy use and cost, and increase water recovery rate. To achieve these objectives, typical pre-treatment steps involve multimedia, cartridge and sand filtration, as well as the addition of chemicals. Depending on feedwater quality, microfiltration and ultrafiltration membranes (Table 1.4) are used for pre-treatment purposes.

Post-treatment of desalinated water is also a required component. Depending on the number of RO stages, the RO desalination process results in near total removal of TDS (>99%), and low hardness and alkalinity in the produced water, which is consequently quite corrosive, and may introduce metals into the drinking water. Typical post-treatment methods involve adding chemicals such as calcium hydroxide ($Ca(OH)_2$) to increase the hardness and alkalinity and sodium hydroxide (NaOH) to adjust the pH of the desalinated water [13].

1.3.1.2 Electrodialysis and Electrodialysis Reversal

Electrodialysis (ED) process is based on the use of an electromotive force applied to electrodes adjacent on both sides of a membrane, which separates the dissolved solids in the feedwater [14] (Fig. 1.2). In this process, the cathode attracts the sodium ions (Na^+), and the anode attracts the chloride ions (Cl^-). In the electrodialysis reversal (EDR) process, the polarity of the electrodes is switched at fixed intervals to reduce the formation of scale and subsequent fouling and allow the EDR to achieve higher water recoveries [15]. The required pressure for these desalination processes is between 500–640 kPa (70 and 90 psi) [15].

ED and EDR processes can remove 75–98% of TDS from feedwater but are only effective for treating water with TDS concentration of up to 4000 mg/L (brackish water) and are not applicable to the desalination of seawater [14]. For example, the City of Suffolk, Virginia (U.S.) is operating a 17,100 m³/day (3.75 MGD) EDR plant to treat brackish water [9]. However, ED/EDR can be used for pre-treatment of seawater since the process can remove or reduce a host of contaminants and is less sensitive to pH or hardness levels in the feedwater. Furthermore, EDR membranes can treat waters that have a high scaling potential from elevated levels of contaminants such as barium (Ba) and strontium (Sr); is effective for treating high silica (SiO_2) feedwater; and is resistant to chlorine, making them more robust for processing feedwaters with higher levels of organic matter that would typically foul RO membranes [16].

Pre-treatment requirements for the EDR feedwater include the removal of particles that are greater than about 10 μm in diameter to prevent membrane pore clogging, as well as the removal of substances such as large organic anions, colloids, iron oxides (Fe_2O_3) and manganese oxide (MnO_2) [17]. Pre-treatment methods applied to ED/EDR processes include active carbon filtration (for organic matter removal), flocculation (for colloids) and standard filtration techniques.

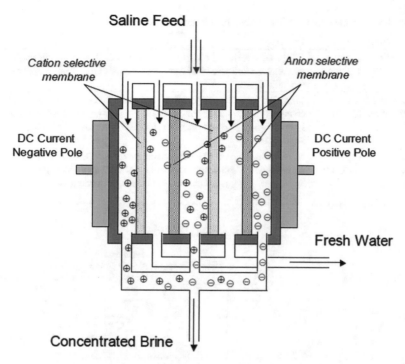

Fig. 1.2 Schematic diagram of electrodialysis desalination process (Source: Sandia National Laboratories) [13]

1.3.2   Thermal Technologies

Thermal technologies, which are based on the evaporation mechanism and distillation processes, were developed in the 1940s. Thermal technologies to desalinate seawater on a commercial basis are mature technologies and continue to be a logical regional choice for desalination particularly in the Middle East where fossil fuels as an energy source are readily available. Table 1.5 shows a summary of thermal technologies most commonly applied to desalination of seawater. Thermal processes require pre-treatment to avoid scaling and to control corrosive constituents of the source water. Removal of sand and suspended solids may also be necessary to prevent pipe erosion.

In MSF process (Fig. 1.3), saltwater travelling through tubes is cooler than the vapor surrounding the tubes. Then this vapor preheat the saltwater and condense to form distillate by heat transfer across the MSF heat exchanger. The vapor is condensed to form potable water, and the brine becomes the feed water for the next stage. The MSF process is energy intensive but can be operated using waste thermal energy [9, 13, 16].

In MED (Fig. 1.4), saltwater is sprayed overtop of hot tubes. It evaporates, and the vapor is collected to run through the tubes in the next effect. As the cool saltwa-

Table 1.5 Dominant thermal desalination technologies [9, 13, 16]

Technology	Description
Multi-Stage Flash (MSF) (Fig. 1.3)	MSF uses a series of chambers, or stages, each with successively lower temperature and pressure, to rapidly vaporize (or "flash") water from the bulk liquid.
Multiple Effect Distillation (MED) (Fig. 1.4)	MED is a thin-film evaporation approach. The vapor produced by one chamber subsequently condenses in the next chamber, which is at a lower temperature and pressure, providing additional heat for vaporization.
Mechanical Vapor Compression (MVC) (Fig. 1.5)	MVC utilizes an electrically driven mechanical device, powered by a compression turbine, to compress the water vapor. As vapor is generated, it is passed over a heat exchanging condenser that returns the vapor to water.

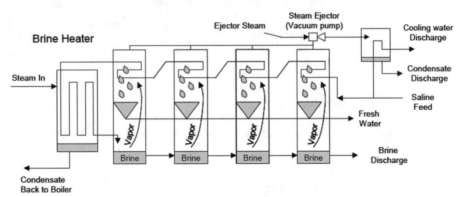

Fig. 1.3 Schematic of multi-stage flash (MSF) desalination process. (Source: Sandia National Laboratories) [13]

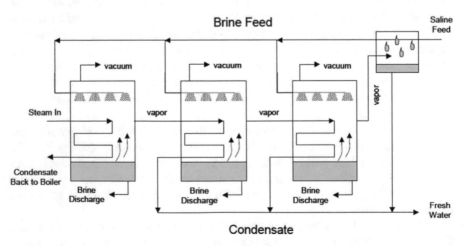

Fig. 1.4 Schematic of multi-effect distillation (MED) process. (Source: Sandia National Laboratories) [13]

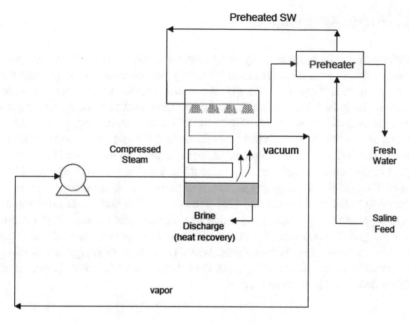

Fig. 1.5 Schematic of single stage mechanical vapor compression (MVC) desalination process. (Source: Sandia National Laboratories) [13]

ter is sprayed over the vapor filled tubes, the vapor condenses inside the tubes and is collected as distilled water. The resulting brine collects in the bottom of each effect and is either circulated to the next effect or exited from the system which requires less energy than MSF. MED technology is popular for applications where thermal evaporation is preferred or required due to its reduced pumping requirements and, thus, its lower power use compared to conventional MSF [9, 13, 16].

Mechanical vapor compression (MVC) (Fig. 1.5) is based on vapor compression mechanism, where vapor from the evaporator is compressed and the heat released is used for the subsequent evaporation of feedwater [9, 13, 16]. This method utilizes an electrically driven mechanical device, powered by a compression turbine to compress the water vapor. As the vapor is generated, it is passed over a heat exchanger (condenser) that converts the vapor into water. The resulting freshwater is moved to storage, while the heat removed during condensation is transmitted to the remaining feedstock. Another option is the thermal vapor compression (TVC) method, where an ejector system powered by steam under manometric pressure from an external source is used to recycle vapor from the desalination process. Large MED plants incorporate thermal vapor compression (TVC), where the pressure of the steam is used (in addition to the heat) to improve the efficiency of the process as discussed in Chap. 2 [9, 13, 16].

1.4   Energy Consumption

Desalination technologies are very energy intensive and it should be a high priority to employ the water and energy nexus for energy use efficiency [18]. Pumps, which require a significant amount of energy, are used in various stages of all desalination processes, including the feedwater intake, pretreatment and treatment processes, discharge of product water and concentrate management. Energy consumption depends on the type of desalination technique, the TDS and temperature of the feedwater, the capacity of the treatment plant, and the physical location of the plant with respect to the source of the intake water and the concentrate discharge site [19]. In general, the combined energy requirements of thermal technologies are greater than those of RO membrane processes [20]. However, MSF and MED are capable of using low-grade and/or waste heat, which can significantly improve their economic efficiency [16, 21]. Low-grade heat refers to heat energy that is available at relatively low (near-ambient) temperatures. Waste heat contains energy that is released to the environment without being used. Both have potential value for desalination which is described in the next section.

1.4.1   Energy Conservation Practices

Currently, various energy conservation measures in desalination plants are practiced around the world (Table 1.6).

1.4.1.1   Energy Recovery Devises

In the RO process, due to low net recoveries of the highly pressurized feedwater, typically 40–60% of the applied energy in the process can be lost to atmosphere without any attempt to recover that energy. In general, energy recovery devices (ERDs) can recover from 75 to 96% of the input energy from the brine stream in a seawater RO plant [17]. Two categories of ERDs exist:

1. devices that transfer the concentrate pressure directly to the feedstream (e.g., pressure exchanger, work exchanger), which have energy recovery efficiencies of about 95%; and

Table 1.6 Energy conservation practices

Energy conservation measures	Application
Energy recovery devices	RO membrane desalination
Dual operation of desalination and power generation plants	RO membrane and thermal desalination plants
Hybrid desalination plants	Integrated desalination technologies

2. devices that transfer concentrate pressure to mechanical power, which is then converted back to feed pressure (e.g., Pelton impulse turbines, hydraulic turbo-chargers, reverse-running pumps) which have recovery efficiency of about 74% [17, 22].

1.4.1.2 Dual Operation of Desalination and Power Generation Plants

Most of the large desalination facilities in the world are dual purpose facilities that produce both freshwater and electricity. Dual operation systems exploit the water and energy nexus in coastal environments to achieve energy use efficiency through the energy recovery concept. The dual operation of desalination and power plants exist in two approaches described below including cogeneration plants and co-located plants.

a. Cogeneration plants

Cogeneration plants integrate a power plant and a desalination plant and are operated jointly. A typical power plant produces high pressure and high temperature steam. A cogeneration plant uses this steam as an additional energy source (mechanical energy converted to electricity) during the desalination process to reduce fossil fuel costs [23, 24]. In the Middle East, the larger MSF and MED plants are built along with power plants and use the low temperature steam exhausted from the power plant steam turbines.

A cogeneration plant benefits both the power plant and the desalination plant. The power plant gains extra revenue by selling the waste steam to the desalination plant, while the desalination plant does not have to pay for the construction and operation of its own energy source, thus also reducing its costs. One disadvantage of cogeneration plants is that since a power plant's electricity generation depends on the electricity demand, it is not constant, and this can have an adverse impact on the power available to the desalination plant [23, 24].

b. Co-located plants

In this process, a desalination plant is co-located with a power plant and they function together as follows. A coastal power plant draws large volumes of cooling water directly from the ocean. A co-located RO desalination plant draws heated seawater from the power plant's cooling water loop and uses it for two purposes. This include as feedwater and as blend water to reduce the brine salinity before its discharge in the sea [22]. Because the desalination plant "piggybacks" on the existing cooling water loop, it can substantially reduce both the construction and operating costs. A co-located desalination plant shares the same advantages as a cogeneration plant but enjoys the additional benefit that the higher feedwater temperature requires less energy for the desalination process. The main disadvantage of a co-located plant is that it depends entirely on the power plant for its existence [25].

c. Hybrid plants

Hybrid plants take benefits of different water treatment and efficient energy use technologies. This enables the desalination system to reduce energy costs, and optimize its performance [26, 27]. The type and necessity for a hybrid plant can be considered on a case-by-case basis. The following case studies describe some real-world examples of how turbines, cogeneration and hybrid plants can reduce energy consumption in desalination plants.

Cape Hatteras, a resort area in North Carolina (U.S.A.), has operated a hybrid RO/Ion Exchange plant since 2000 [28]. This desalination plant withdraws water from two separate wells that have different water quality characteristics. The high salinity water from Well 1 is processed by the RO process and the water with high organic content from Well 2 is processed by the Ion Exchange process. The treated water from both processes is mixed as the final product water. This plant has also incorporated an energy recovery turbine into the RO treatment process [28].

Studies in Kuwait have also shown how different combinations of turbines and technologies affect energy consumption. Darwish and Al-Najem [23] compared two gas turbines with varying combinations of heat recovery systems and reported that for a simple gas turbine power plant operating in cogeneration with reverse osmosis, the fuel energy consumption was 92.78 kJ/kg. Adding a heat recovery steam generator (HRSG) to each gas turbine to supply MSF units with recovered steam lowered the energy consumption to 86.88 kJ/kg. If a condensing steam turbine and a HRSG were then added to each gas turbine, the energy consumption decreased even further, falling to 63.6 kJ/kg.

1.4.2 Nuclear Energy Use

Nuclear power plants generate power using nuclear fission, where the power comes from the energy released when a large atom splits into smaller atoms. The released energy is controlled and contained to heat a coolant material and ultimately generates steam that drives turbines, which rotate a coil in a magnetic field to produce electricity.

Combining nuclear power plants and desalination plants is considered economical because two-thirds of the thermal power generated by the fission process is waste heat, which is typically released to the surrounding water or air [29]. The International Atomic Energy Agency (IAEA) has worked with teams of researchers from several countries to study seawater desalination combined with nuclear reactors, publishing their findings as a document entitled 'New Technologies for Seawater Desalination Using Nuclear Energy' [30]. One of the main findings of this study revealed that the most efficient combination of desalination-nuclear power plant include high temperature MED and hybrid desalination systems.

1.4.3 Renewable Energy Use

Potential renewable energy resources for water treatment and desalination include solar energy, wind energy, ocean energy (tidal and wave) and geothermal energy [19]. The use of renewable energy for desalination has been reported since the mid-1990s [26–31], but new renewable energy technologies are now becoming available for desalination applications. For example, a pilot project utilizing wave power technology for seawater desalination using submerged buoys began operating in Perth, Australia in 2015 [32]. At present, various types of solar energy and wind energy (particularly solar energy) are the most commonly used and showing great promise as renewable energy sources for desalination projects around the world. The details of renewable energy use are elaborated in the following section focusing on solar and wind energy resources.

1.4.3.1 Solar Energy

Abou-Rayan and Djebedjian [33] discussed recent advances in desalination, focusing particularly on solar desalination. Solar energy can be used either directly or indirectly in desalination processes. Solar stills are a typical example of direct solar energy use, taking advantage of the greenhouse effect [24, 29]. In this process, a black-painted basin, sealed tightly with a transparent cover, stores the saline water. As the sun heats the water, the water in the basin evaporates, and the vapor comes into contact with the cool glass ceiling, where it condenses to form pure water. The water can then be drained away from the solar still for portable use. This technology is optimized when running at low production capacities of close to 0.757 m^3/d, although the use of heat recovery devices and hybrid systems may make solar stills more cost-competitive. Indirect solar technologies for desalination are based on using solar energy concentrators/collectors and solar photovoltaic (PV) arrays, described in more detail below.

a. Solar energy concentrators/collectors

MSF and MED desalination technologies can use solar collectors as an indirect source of solar energy to develop the thermal energy needed to drive the desalination process. A heliostat tracks the sun as it moves across the sky and collects parallel solar radiations using flat mirrors, directing them to fixed concave solar energy collectors. The collectors focus the energy collected on pipes filled with air or water to create steam or heated air that can then be used as a power source [25]. Parabolic trough radiation collector is another option. The collectors can withstand high temperatures without degrading the collector efficiency and are preferred for a solar steam generation [27]. Solar ponds can also be used as radiation collectors and some researchers consider solar pond-powered desalination to be one of the most cost-effective methods available in many parts of the world [28].

Table 1.7 Examples of solar energy use in desalination [19, 24, 27, 30]

Country/Location	Type of solar energy	Desalination technique	Plant capacity (Liter/Day)
El Paso, Texas (U.S.)	Solar pond	MSF	16,000
Margarita de Savoya, Italy	Solar pond	MSF	50,000–60,000
Yanbu, Saudi Arabia	Dish collectors	MSF	199,962
Arabian Gulf	Solar-parabolic trough	MED	5,999,225
Al-Ain, UAE	Solar-parabolic trough	MED-MSF	499,999
Cituis West, Jawa, Indonesia	Solar PV	BW-RO	35,995
Red Sea, Egypt	Solar PV	BW-RO	50,000
Lipari Island, Italy	Solar PV	SW-RO	47,994
Oshima Island, Japan	Solar PV	SW-ED	10,000
Fukue City, Japan	Solar PV	BW-ED	199,897

b. Solar PV

Solar PV arrays offer another way to generate electricity. In this process, PV arrays convert solar energy into electricity through the transfer of electrons. The arrays, made of silicon chips, facilitate the transfer of electrons and thus generate power. Table 1.7 shows some examples of how solar PV energy can be used in conjunction with desalination techniques around the world [19, 24, 27, 30].

1.4.3.2 Wind Energy

Wind energy creates mechanical energy by turning the blades of wind turbines that is then converted to electrical energy. Turbines utilizing wind energy for low power (10–100 kW), medium power (100 kW-0.5 MW), and high power (> 0.5 MW) applications are mature technologies [28].

Wind energy can be converted to shaft power and either directly powers the desalination process or is sent to the local grid. Electrodialysis and MVC systems are well suited to operate using direct wind energy [28]. Table 1.8 shows two examples of desalination plants powered by wind energy.

At present, similar to other applications, the major disadvantage of integrating renewable energy in desalination plants is the lack of continuity and consistency in energy supply. In most cases, battery storage and the requirement for a large number of batteries is cost prohibitive. To compensate, some control system or energy storage unit is required, especially if a backup energy source is unavailable. A common way to resolve this problem is to connect renewable energy sources to a conventional electricity grid or use diesel generators as a backup to power the desalination plant [28].

Table 1.8 Examples of wind energy use in desalination [26, 28]

Country/location	Wind power generated 10³ Btu/h (kW)	Desalination technique	Plant capacity (m³/day)
Shark Bay, Western Australia	109 (32)	RO	12,998–16,798
Ruegen Island, Germany	683 (200)	MVC	11,998–29,996

1.5 Environmental Sustainability

Desalination plants can have both direct and indirect impact on the environment. Developing environmentally friendly desalination system designs should be a high priority for twenty-first century water resource management and water infrastructure initiatives.

1.5.1 Environmental Issues

Site selection is the first step in planning and designing a desalination plant. The plant should not be placed in a densely populated area due to possible environmental and human health impacts, including noise pollution generated by pumps and potential gas emissions. Gaseous emissions from desalination plants using fossil fuels include carbon monoxide (CO), nitric oxide (NO), nitrogen dioxide (NO_2), and sulfur dioxide (SO_2). The large amounts of chemicals stored at the plants and the risk of chemical spills in populated areas may also be of concern [34].

Other site selection factors include the risks associated with the construction of what can be an extensive water intake infrastructure and network of pipes transporting the feedwater to the plant, as well as the location of the concentrate discharge, which may disturb environmentally sensitive areas. Feedwater (water source) intake structures are site specific which generally falls into one of two categories: surface intakes (open intakes) located above the seafloor and subsurface intakes located beneath the seafloor. The design of intake may impact quality of feedwater and desalination plant cost. At this point, few environmental regulations directly pertain to desalination plants. Younos [34, 35] discussed environmental issues of desalination and regulations in the U.S. applicable to desalination plants. The most critical environmental factor in desalination planning is the concentrate management, which should be a high priority during the planning phase.

1.5.2 Concentrate Management

Concentrate is the main byproduct of a desalination plant. The TDS concentration in the concentrate depends on the desalination technique involved. For example, RO plants usually produce concentrates with a TDS higher than 65,000 mg/L, while the

TDS for MSF plants will be at around 50,000 mg/L [16]. The temperature of this concentrate also depends on the desalination technique. The concentrate from a RO process remains at the ambient water temperature, while the concentrate from a thermal desalination process is typically 5.5–8.3 °C above ambient water temperature [36]. Desalination plants' concentrate may also contain some of the chemicals used for the feedwater pretreatment and post-treatment (or cleaning) processes.

Several critical factors should be considered when selecting the best concentrate management option [35, 36]. These factors include the volume or quantity of the concentrate to be produced, the quality of the concentrate, the location of the desalination plant, and the local environmental regulations. Other factors include the capital and operating costs incurred and the potential impact on future plant expansions. An overview of concentrate management options and practices reported in the literature are summarized below [32-35, 36–38].

a. Surface water disposal: the concentrate is discharged to receiving waters at a point that is adjacent to or near the desalination plant, which could include tidal rivers and streams, estuarine waters and the ocean. Concentrate disposal into freshwater systems is not recommended, However, the main risks associated with concentrate surface water disposal include a potentially adverse impact on the receiving waters' ecosystems, and the long term effect on the water quality of coastal aquifers.

b. Submerged disposal: the concentrate is transported away from the desalination plant via underwater pipes to an estuarine and/or ocean location. The creatures most at risk in this scenario are the benthic marine organisms living on the sea bottom.

c. Deep well injection: the concentrate is directly injected into deep groundwater aquifers that are not used as a source of drinking water. Injection well depths range from 0.32 km to 2.57 km below the ground surface. In many locations, deep well injection may not be feasible because of geologic conditions or regulatory constraints imposed to protect drinking water sources.

d. Evaporation ponds: evaporation ponds are constructed in a similar way to the ponds historically used for salt production. These ponds facilitate concentrate water content removal via evaporation and salt accumulation at the bottom of the pond. Evaporation ponds are especially useful in warm climates, where the evaporation rate is high. Evaporation ponds must be equipped with liners to prevent saltwater leaking into groundwater aquifers and regular maintenance to avoid the drying and cracking of liners. Although evaporation ponds can be a very cost-effective option, the practice is land intensive and can cause significant water source loss via evaporation.

e. Land application: another option is the land application of concentrate via methods such as spray irrigation, infiltration trenches, and percolation ponds. The feasibility of land application depends on land availability, climate, vegetation tolerance to salt, and depth of the groundwater table.

f. Integrated disposal with wastewater treatment plant: this includes concentrate disposal to the front or end products of a wastewater treatment plant. 'Front disposal' practices merge the concentrate with wastewater to be treated. This practice is not recommended due to the associated problems incurred: (1) the high TDS levels in the concentrate disrupt the biological wastewater treatment performance; and (2) conventional wastewater treatment processes do not remove TDS so that the treatment plant discharge water can pose a significant threat to the receiving waters. The concentrate 'end disposal' method involves mixing and dilution of the concentrate with treated wastewater, thus reducing the TDS load before it is discharged into the receiving waters. A major disadvantage of this practice is the requirement for a separate pipeline to transport the concentrate to the wastewater treatment plant and the consequent additional cost incurred.

g. Brine concentrators: brine concentrator process uses heat exchangers, deaerators, and vapor compression to convert liquid concentrate into a slurry. With a brine concentrator, 95% of the water can be recovered as a high purity distillate with less than 10 mg/L of TDS concentration. The remaining 5% of concentrated slurry can be reduced to dry solids in a crystallizer to create dry, solid cake, which is easy to handle for disposal.

h. Zero liquid discharge: the 'ZLD' technique originally developed for solid waste management is a promising new technology for concentrate management that brings significant environmental benefits. The ZLD technique uses evaporation mechanism to convert the liquid concentrate (brine) into a dry solid that can then be utilized for useful purposes [39]. Table 1.9 shows the energy consumption required to achieve ZLD using existing thermal technologies (MSF, MED, MCV).

1.6   The Economics of Desalination

Desalination cost is affected by several factors such as type of technology, energy availability, geographic location, plant capacity, and feedwater quality. Other important factors include costs associated with transporting water from source to desalination plant, distribution of treated water, and concentrate management. Financial factors such as financing options and subsidies also affect the product water cost [40]. Major cost factors associated with desalination plants are summarized in Table 1.10.

Table 1.9 Energy consumption for ZLD with various thermal technologies [39]. (Reprinted with permission of LENNTECH)

Brine treatment technology	Electrical energy (KWh/ m³)	Thermal energy (kWh/ m³)	Total energy equivalent[a] (kWh/ m³)	Typical capacity (m³/ day)	Max TDS (mg/L)
MSF	3.68	77.5	38.56	<75,000	250,000
MED	2.22	69.52	33.50	<28,000	250,000
MVC	14.86	0	14.86	<3000	250,000

[a]Total Energy Equivalent = Electric Energy + 0.45 × Thermal Energy

Table 1.10 Factors affecting desalination cost [40]

Cost factor	Description
Quality of feedwater	TDS concentration, contaminants in feedwater and pre-treatment requirement are major design factors.
Site characteristics	The proximity of plant location to water source and concentrate discharge point is a major factor. For example, pumping cost and costs of pipe installation will be substantially reduced if the plant is located near the water source.
Desalination plant capacity	Plant capacity affects the size of treatment units, pumping, water storage tank, and water distribution system. Large capacity plants require high initial capital investment compared to low capacity plants. However, due to the economy of scale, the unit production cost for large capacity plants can be lower.
Concentrate management	Concentrate management is a major economic factor and is affected by several factors that include site characteristics (geologic features, soil conditions, proximity to potential disposal site), regulatory requirements, public approval, and the type of concentrate management method.
Desalination implementation costs	Desalination plant implementation costs can be categorized as construction costs (starting costs) and operation and maintenance (O & M) costs. These costs are detailed in the next section.

1.6.1 Desalination Implementation Costs

Desalination plant implementation costs can be categorized as construction costs (starting costs) and operation and maintenance (O & M) costs.

1.6.1.1 Desalination Plant Construction Costs

Construction costs include direct and indirect capital costs. The indirect capital cost is usually estimated as percentages of the total direct capital cost. Descriptions of various direct and indirect costs associated with constructing a desalination plant are summarized in Tables 1.11a and 1.11b.

1.6.1.2 Desalination Plant Operating Maintenance Costs

The O & M costs consist of fixed costs and variable costs [40].

a. Fixed costs. Fixed costs include insurance and amortization costs. Usually, insurance cost is estimated as 0.5% of the total capital cost. Amortization compensates for the annual interest payments for direct and indirect costs and depends on the interest rate and the life-time of the plant. Typically, an amortization rate in the range of 5–10% is used.

Table 1.11a Direct costs associated with the construction of a desalination plant [40]

Direct Costs	Description
Land	It may vary considerably, from zero to a sum that depends on site characteristics and plant ownership (public vs. private).
Surface water intake structure	It depends on plant capacity and meeting environmental regulations. Also, see auxiliary equipment below.
Production wells	It depends on plant capacity and well depth. Also, see auxiliary equipment below.
Process equipment	The process equipment includes water treatment units (membranes), instrumentation and controls, pre- and post-treatment units and cleaning systems. Process equipment costs depend on plant capacity and feedwater quality.
Auxiliary equipment	Auxiliary equipment includes open water intakes, wells, storage tanks, generators, transformers, pumps, pipes, valves, electric wiring, etc.
Buildings	It include the construction of structures such as control room, laboratory, workshops, and offices. Construction cost is site-specific depending on site condition and type of building.
Concentrate management	It depends on the type of desalination technology, plant capacity, discharge location, and environmental regulations.

Table 1.11b Indirect costs associated with the construction of a desalination plant [40]

Indirect Costs	Description
Freight and insurance	It is typically estimated as 5% of total direct costs.
Construction overhead	It include labor costs, fringe benefits, field supervision, temporary facilities, construction equipment, small tools, contractor's profit and miscellaneous expenses. This cost is typically estimated as 15% of direct material and labor costs.
Owner's cost	It includes land acquisition, engineering design, contract administration, administrative expenses, commissioning and/or startup costs, and legal fees. It is estimated as approximately 10% of direct materials and labor costs.
Contingency cost	This cost is included for possible additional services. It is generally estimated at 10% of the total direct costs.

b. Variable costs. Major variable costs include the cost of labor, energy, chemicals, and maintenance. Labor costs can be site-specific and depends on plant ownership (public or private) or special arrangements such as outsourcing of plant operation. Energy cost depends on the availability of inexpensive electricity (or alternative power source). For example, energy cost can be reduced if the desalination plant is co-located with a power generation plant. Chemical use depends mainly on feedwater quality and degree of pre−/post-treatment and cleaning processes. The cost of chemicals depends on the type and quantity of such chemicals as well as global market prices and special arrangements with vendors. In the RO process, the major maintenance cost pertains to the frequency of membrane replacement, which is affected by the feedwater quality.

1.6.2   Desalination Cost Estimation Models

Several models are available for estimating desalination costs. Model applications are mostly limited to site-specific conditions and give approximate estimates. Nevertheless, cost models can be used as an indicator of potential costs for planning a desalination facility. A brief overview of two typical cost models is provided below. For details of these models and applications, readers are referred to reference citations [41, 42].

1.6.2.1   Desalination Economic Evaluation Program (DEEP-3.0)

DEEP is a Desalination Economic Evaluation Program developed by the International Atomic Energy Agency [41]. The program can be useful for evaluating desalination strategies by calculating estimates of technical performance and costs for various alternative energy and desalination technology configurations. Desalination technology options modelled include MSF, MED, RO and hybrid options (RO-MSF, RO-MED). Energy source options include nuclear, fossil, renewables and grid electricity (stand-alone RO) [41].

1.6.2.2   WTCost II Model

U.S. Department of the Interior Bureau of Reclamation [42] has developed a computer cost estimating program, WTCost II© that can be used for all commercial desalting processes involving membrane desalinations [RO and nanofiltration (NF)] and thermal desalination plants (MSF, MED and MVC). The WTCost© model provides estimates of capital costs, indirect costs and annual operating costs [42].

1.7   Futuristic Approaches

Hundreds of research and technical articles have been published on various aspects of desalination technologies. It's recognized that the cost-effectiveness of desalination technologies depends on energy use efficiency, water treatment technique, membrane performance, and environmental sustainability of desalination plants.

Elimelech and Phillip [43] reviewed the possible reductions in energy demand by state-of-the-art seawater desalination technologies. Specifically, they focused on the potential role of advanced materials and innovative technologies in improving performance and sustainability aspects of desalination. Basic research is underway on manufactured membranes to control membrane fouling and increase water recovery rate [16]. Examples include membrane modification to improve fouling resistance, and manufacturing carbon nanotube/graphene-based desalination membranes and various nanocomposite membranes. According to research cited in NAP report [16], modification of commercially available membranes to alter surface characteristics to reduce fouling while maintaining or improving flux and selectivity is an established research area that shows promising results for RO and NF membranes. Although many types of modification methods exist, graft polymerization is the method most commonly utilized in RO and NF membranes [16]. Also cited research indicates that theoretical studies and molecular dynamics simulations suggest that hydrophobic channels, like carbon nanotubes, increase water recovery rate; and nanocomposite RO membranes formed by the dispersion of nanoparticles or molecular sieves in polymers would yield enhanced membrane performance [16]. Antifouling membranes are discussed further in the Chap. 17.

As stated above, energy use efficiency remains a major research and development theme in the twenty-first century. Recently, the U.S. Department of Energy's [44] Advanced Manufacturing Office analyzed the range (or bandwidth) of potential energy savings for different unit operations within seawater desalination. The DOE report provides technology-based estimates of potential energy savings opportunities across the desalination system [44]. Also, the report presents a framework to evaluate and compare energy savings potential within and across different sectors of energy use. Several hybrid desalination techniques that incorporate combinations of existing water treatment technologies (e.g. RO and thermal technology) are being investigated in order to take advantage of the unique characteristics of different desalination techniques for implementing energy use efficiency. Table 1.12 summarizes some of futuristic desalination technologies reported in literature [9, 16].

Table 1.12 Examples of alternative and hybrid desalination techniques

Technology type	Description
Forward osmosis (FO)	FO process uses osmotic pressure difference between a concentrated "draw" solution and a feed stream to drive water flux across a semipermeable membrane.
Membrane distillation (MD)	In the MD technique, the temperature difference on opposing sides of the membrane creates different vapor pressures, and this is being utilized to drive the system, with only vapor passing through the membrane.
Freeze desalination (FD)	The basis of freeze desalination technologies is to change the phase of water from liquid to solid. Freezing the saltwater forms pure water ice crystals that can then be separated and melted to obtain potable water, requiring less energy than conventional evaporation techniques. This approach seeks to take advantage of the relatively low enthalpy of phase change—The freezing of water at atmospheric conditions (334 kJ/kg)—Whereas evaporation would require 2326 kJ/kg. The
Electrodeionization (EDI)	EDI uses a combination of ion exchange and electrodialysis, where an electrical charge is applied to plates mounted outside the membranes with resin beads between them. Saltwater ions take the place of ions on the resin and are then pulled through the membrane to the electrically charged plates. Water passes through the resin without hindrance as it contains no ions, thus producing purified water.

Environmental sustainability of desalination practices depends on two factors: (1) energy use efficiency as discussed above and/or shift toward increased integration of renewable energy technologies in desalination process; (2) enhanced concentrate management; and (3) enhanced regulatory requirements.

Advanced renewable energy technologies provide significant opportunities for environmental sustainable desalination projects. Abou-Rayan and Djebedjian [29] discussed several unique solar desalination case studies. For example, scientists at the Technological Institute of the Canary Islands (ITC) and the Aachen University of Applied Sciences installed a pilot plant called DESSOL (Desalination with Solar energy) in Pozo lzquierdo (Gran Canaria island) to demonstrate the technical feasibility of this technology. The RO plant, which has a nominal production capacity of 10 m³/day (specific energy consumption of 0.5 kWh/m³) is supplied by a 4.8 kWh PV generator and a 19 kWh battery back-up system. A second example discussed by Abou-Rayan and Djebedjian [29] is the High Concentration Photovoltaic Thermal (HCPVT) Project being developed by IBM in cooperation with the King Abdulaziz Research Center based on research conducted at the Massachusetts Institute of Technology (MIT). Their prototype HCPVT system uses a large parabolic dish made from a faceted mirror attached to a sun tracking system.

Ongoing research related to concentrate management is mostly focused on developing new zero liquid discharge (ZLD) techniques, as these support environmental preservation as well as identifying beneficial uses of the salts produced as a byproduct of the desalination process.

There is an urgent need to develop and implement more effective desalination permitting and monitoring programs in jurisdictions around the world that balance the cost-effectiveness of evolving technologies with environmental preservation. Younos [35] discussed desalination permits and regulatory requirements in the U.S. The State of California has published the California Desalination Planning Handbook [45]. It provides the most comprehensive guidelines for planning of desalination projects. Chap. 6 of the Handbook describes regulations and permitting requirements which could be a useful resource where regulations are lacking.

1.8   Conclusions and Outlook

The desalination concepts, basic categories of desalination system designs, and the status of pertinent technologies presented in this chapter, are only a brief introduction to exciting and rapidly growing technologies vital to providing clean water for human populations around the world. The worldwide availability of vast seawater and brackish water resources combined with evolving desalination techniques and system design offer significant opportunities to address global water scarcity in both urban and remote environments for the twenty-first century and beyond. Furthermore, evolving energy use efficiency and renewable energy technologies development and use will also facilitate the emergence of environmentally friendly large-scale desalination plants, as well as, small-scale and decentralized desalination infrastructure invaluable for densely populated urban areas and rural and remote communities. At present, few environmental regulations directly apply to desalination plant development and operation. It is critical to develop and implement effective desalination plant permitting and monitoring systems covering both water and air aspects of desalination plant impacts, and balancing the cost-effective aspects of evolving technologies with the need to protect vulnerable environments and meeting increasingly high potable water demand around the globe.

References

1. UNDESA, International decade for action 'Water for Life' 2005–2015, United Nations Department of Economic and Social Affairs (UNDESA, 2015) (http://www.un.org/waterfor-lifedecade/scarcity.shtml)
2. U.S. Bureau of Reclamation, *Desalting Handbook for Planners, Desalination and Water Purification Research and Development Program Report no* (U.S. Department of the Interior, Bureau of Reclamation, Technical service center, 2003), p. 72
3. T. Younos, Desalination: Supplementing freshwater supplies - approaches and challenges. J. Contemp. Water Res. Educ. **132**, 1–2 (2005)
4. J.A. Cotruvo, Water desalination processes and associated health and environmental issues, (Water Condition & Purification International, 2005) January 2005
5. WHO, Guidelines for drinking-water quality. Vol. 2. Health criteria and other supporting information, World Health Organization. Geneva **101**, 2 (1984)
6. WHO, Safe drinking-water from desalination. (World Health Organization, 2011) WHO/HSE/WSH/11.03, (https://apps.who.int/iris/handle/10665/70621)
7. USEPA ,Secondary drinking water standards: Guidance for nuisance chemicals. United States Environmental Protection Agency (EPA). (2017) (https://www.epa.gov/dwstandardsregulations/secondary-drinking-water-standards-guidance-nuisancechemicals)
8. E. Jones, M. Qadir, M.T.H. van Vliet, V. Smakhtin, S. Kang, The state of desalination and brine production: A global outlook. Sci. Total Environ. **657**, 1343–1356 (2019)
9. T. Younos, K.E. Tulou, Overview of desalination techniques. J. Contemp. Water Res. Educ. **132**, 3–10 (2005)
10. AWWA, *Manual of Water Supply Practices: Reverse Osmosis and Nanofiltration*, vol M46 (American Water Works Association, AWWA, 1999), p. 173
11. S.J. Duranceau, Membrane processes for small systems compliance with the safe drinking water act, 3rd NSF Intl. Symp. on Small Drinking Water and Wastewater Systems, April 22–25, 2001, (Washington D.C., USA, 2001)
12. J. Krukowski, Opening the black box: Regulations and recycling drive use of membrane technologies. Pollut. Eng. **33**, 20–25 (2001)
13. J.E. Miller, Review of water resources and desalination technologies, Sandia National Laboratories, SAND 2003–0800, U.S. Department of Energy, U.S. Department of Commerce (2003) (https://prod-ng.sandia.gov/techlib-noauth/accesscontrol. cgi/2003/030800.pdf)
14. R.E. Brunner, *Electrodialysis in Saline Water Processing. Hans-Gunter Heitmann* (VCH Verlagsgesellschaft, Federal Republic of Germany, 1990), pp. 197–217
15. AMTA, *Electrodialysis reversal desalination*, (American Membrane Technology Association 2018) (https://www.amtaorg.com/electrodialysis-reversal-desalination)
16. NAP, *Desalination: A National Perspective, Committee on Advancing Desalination Technology Water Science and Technology Board, National Research Council* (The National Academies Press, Washington, D.C., 2008). http://nap.edu/12184

17. Lenntech, *Electrodialysis*, (Lenntech BV 2018) (https://www.lenntech.com/electrodialysis.htm)
18. J. Lee, T. Younos, Sustainability strategies at the water-energy nexus: Renewable energy and decentralized infrastructure. J. Am. Water Works Ass. **110**, 32–39 (2018)
19. T. Younos, K.E. Tulou, Desalination: Energy needs, consumption and sources. J. Contemp. Water Res. Educ. **132**, 27–38 (2005)
20. DESWARE, (Encyclopedia of Desalination and Water Resources, 2013) (http://www.desware.net/Energy-Requirements-Desalination-Processes.aspx)
21. H. Shih, T. Shih, Utilization of waste heat in the desalination process. Desalination **204**, 464–470 (2007)
22. T.M. Manth, E. Gabor, J. Oklejas, Minimizing RO energy consumption under variable conditions of operation. Desalination **157**, 9–21 (2003)
23. M.A. Darwish, N. Al-Najem, Cogeneration power desalting plants in Kuwait: A new trend with reverse osmosis desalters. Desalination **128**, 17–33 (2000)
24. T.C. Hung, M.S. Shai, B.S. Pei, Cogeneration approach for near shore internal combustion power plants applied to seawater desalination. Energy Convers. Manag. **44**, 1259–1273 (2003)
25. B. Alspach, I. Watson, Sea change. Civil Eng. **74**, 70–75 (2004)
26. E. Cardona, S. Culotta, A. Piacentino, Energy saving with MSF-RO series desalination plants. Desalination **153**, 167–171 (2002)
27. B. Van der Bruggen, C. Vandecasteele, Distillation vs. membrane filtration: Overview of process evolutions in seawater desalination. Desalination **143**, 207–218 (2002)
28. T. Younos, The feasibility of using desalination to supplement drinking water supplies in Eastern Virginia (U.S.), VWRRC Special Report SR25–2004, (Virginia Water Resources Research Center, Virginia Tech, Blacksburg, 2004)
29. S. Nisan, G. Caruso, J.R. Humphries, G. Mini, A.N. Naviglio, B. Bielak, O. Auuar Alonso, N. Martins, L. Volpi, Seawater desalination with nuclear and other energy sources: The EURODESAL project. Nuclear Eng. Design. **221**, 251–275 (2002)
30. IAEA, New technologies for seawater desalination using nuclear energy, (International Atomic Energy Agency, 2015) IAEA-TECDOC-1753
31. M.A. Darwish, N. Al-Najem, Energy consumption by multi-stage flash and reverse osmosis desalters. Appl. Therm. Energy **20**, 399–416 (2000)
32. ENGINEERS Australia, (World's first wave-powered desalination plant now operational in Perth 2018) (https://portal.engineersaustralia.org.au/news/worlds-first-wave-powered-desalination-plant-now-operational-perth)
33. M.M. Abou-Rayan, B. Dejbedjian, Advances in desalination technologies: Solar desalination, in *Potable water – Emerging global problems and solutions, the handbook of environmental chemistry*, ed. by T. Younos, C. A. Grady, vol. 30, (Springer, New York, 2014)
34. J.F. Manwell, J.G. McGowan, Recent renewable energy driven desalination system research and development in North America. Desalination **94**, 229–241 (1994)
35. D.G. Harrison, G.E. Ho, K. Mathew, Desalination using renewable energy in Australia. Renew. Energy **8**, 509–513 (1996)
36. S. Kalogirou, Parabolic-trough collectors. Appl. Energy **60**, 65–68 (1998)
37. L. García-Rodríguez, Seawater desalination driven by renewable energies: A review. Desalination **143**, 103–113 (2002)
38. B. Bouchekima, A solar desalination plant for domestic water needs in arid areas of South Algeria. Desalination **153**, 65–69 (2002)
39. M. Thomson, D. Infield, A photovoltaic-powered seawater reverse-osmosis system without batteries. Desalination **153**, 1–8 (2002)
40. T. Younos, Environmental issues of desalination. J. Contemp. Water Res. Educ. **132**, 11–18 (2005)
41. T. Younos, Desalination: Permits and regulatory requirements. J. Contemp. Water Res. Educ. **132**, 19–26 (2005)

42. B.K. Pramanik, S. Li, V. Jegatheesan, A review on the management and treatment of brine solutions. Environ. Sci.: Water Res. Technol. **3**, 625–658 (2017)
43. P. Mahi, Developing environmentally acceptable desalination projects. Desalination **138**, 167–172 (2001)
44. N.X. Tsiourtis, Desalination and the environment. Desalination **141**, 223–236 (2001)
45. T. Hoepner, S. Lattemann, Chemical impacts from seawater desalination plants-a case study of the northern Red Sea. Desalination **152**, 133–140 (2002)
46. M.C. Mickley, Membrane concentrate disposal: Practices and regulation, desalination and water purification research and development program, Report No. 123, (U.S. Department of Interior, Bureau of Reclamation, 2006) (https://www.usbr.gov/research/dwpr/reportpdfs/report123.pdf)
47. Charisiadis, C. (2018), Brine zero liquid discharge (ZLD) fundamentals and design,. LENNTECH. (https://www.lenntech.com/Data-sheets/ZLD-booklet-for-Lenntech-site-min-L.pdf)
48. T. Younos, The economics of desalination. J. Contemp. Water Res. Educ. **132**, 39–45 (2005)
49. IAEA, Desalination Economic Evaluation Program, (DEEP-3.0) User's manual, Computer Manual Series No. 19. (International Atomic Energy Agency, Vienna, 2006) (https://www-pub.iaea.org/MTCD/Publications/PDF/CMS-19_web.pdf)
50. U.S. Bureau of Land Reclamation, *WT Cost II Modeling* (Desalination and Water Purification Research and Development Program, Report No, 2008), p. 130. https://www.usbr.gov/research/dwpr/reportpdfs/report130.pdf
51. M. Elimelech, W.A. Phillip, The future of seawater desalination: Energy, technology, and the environment. Science **333**, 712–717 (2011)
52. U.S. Department of Energy, Bandwidth study on energy use and potential energy savings opportunities in U.S. seawater desalination systems, (U.S. Department of Energy, Advanced Manufacturing Office, 2017) Contract No. DE-AC02-05CH11231. (https://www.energy.gov/sites/prod/files/2015/08/f26/petroleum_refining_bandwidth_report.pdf)
53. California Desalination Planning Handbook, (California State University, Sacramento, 2008) (https://water.ca.gov/LegacyFiles/desalination/docs/Desal_Handbook.pdf)

Chapter 2
Thermal Desalination: Performance and Challenges

Osman Ahmed Hamed

2.1 Introduction

Desalination is defined as the process that removes salts from water. The two major types of desalination technologies currently in use are thermal and membrane-based. Both the technologies need energy to operate and produce fresh water.

Thermal desalination mimics the natural hydrologic water cycle where it is generally achieved through two successive steps. In the first step, saline water is heated generating a salt free vapor. In the second subsequent stage, the generated vapor condensates to form distilled water. The two main industrial thermal desalination processes are multi-stage flash (MSF) and multi-effect desalination (MED). Both processes are usually evaluated in terms of thermal energy performance through widely accepted energy factors such as performance ratio (PR) and gain output ratio (GOR) [1].

In the present chapter, the operational and design developments of thermal desalination processes will be covered. Emphasis is given to desalination process challenges such as scale control, material selection and methods to increase distiller production capacity. Operational experiences gained during the last four decades will be highlighted. Challenges that have to be addressed to enhance developments of thermal desalination processes such as introduction of innovative methods to reduce specific energy consumption, will be discussed.

O. A. Hamed (✉)
Desalination Technologies Research Institute, Saline Water Conversion Corporation,
Jubail, Saudi Arabia
e-mail: OHamed@swcc.gov.sa

© Springer Nature Switzerland AG 2020
V. S. Saji et al. (eds.), *Corrosion and Fouling Control in Desalination Industry*,
https://doi.org/10.1007/978-3-030-34284-5_2

2.2 Thermal Desalination Technologies

Two thermal desalination technologies are well-established globally; MSF and MED. The MSF process distils seawater by flashing a portion of the water into steam in multiple stages under vacuum. The produced vapour is condensed into fresh water on the tubular exchanger at the top of the stage. In a MED process, the feed water is heated by steam in tubes, usually by spraying saline water outside the tubes where a portion of water gets evaporates, and this steam flows into the tubes of the next effect heating, evaporating more water. Each effect essentially reuses the energy from the previous stage, with successively lower temperatures and pressures after each one. In the following sections, an overview of these two technologies will be presented with emphasis on their performances and major challenges.

2.2.1 Multi-Stage Flash (MSF) Desalination

2.2.1.1 Process Description

The most adopted MSF desalination process is the brine recycle configuration as illustrated in Fig. 2.1. The system includes three major sections: the brine heater, the heat recovery section, and the heat rejection section. The brine heater drives the flashing process through heating the recycle brine stream to the top brine temperature (TBT), which is a very important design parameter. Number of stages in the heat recovery section is higher than that in the heat rejection section.

Flashing occurs in each stage where a small amount of product water is generated and accumulated across the stages in the two sections (heat recovery and heat rejection). Vapor formation results because of the reduction of the brine saturation temperature; therefore, the stage temperature decreases from the hot to cold side of

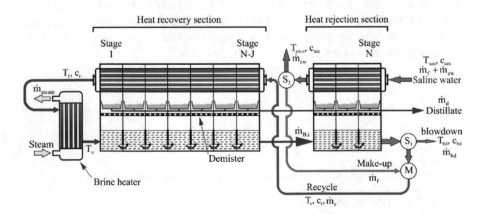

Fig. 2.1 Flow chart of brine recycle desalination process. Reproduced with permission from Ref. [2]; Copyright 2018 @ Elsevier

the plant. This allows for brine flow across the stages without the aid of pumping power. The flashed-off vapors condense on the tubes of the preheater/condenser units. The released latent heat by the condensing vapor is used to preheat the brine recycle stream. On the cold side of the plant, the feed and the cooling seawater are introduced into the condenser/preheater tubes of the last stage in the heat rejection section. As this stream leaves the heat rejection section, the cooling seawater is rejected back to the sea and the feed seawater is mixed in the brine pool of the last stage in the heat rejection section. Also, two streams are extracted from the brine pool in this stage, which include the brine blow down and the brine recycle. The rejection of brine is necessary to control the salt concentration in the plant. As is shown, the brine blowdown is withdrawn prior to mixing of the feed seawater and the recycled brine is withdrawn from a location beyond the mixing point. The brine blow down is rejected to the sea and the brine recycle is introduced to the last stage in the heat recovery section. Additional units in the desalination plant include pre-treatment of the feed and cooling seawater streams. Treatment of the intake seawater is limited to simple screening and filtration. On the other hand, treatment of the feed seawater is more extensive and it includes deaeration and addition of antiscalant and foaming inhibitors. Other basic units in the system include pumping units for the feed seawater and brine recycle. Also, gas-venting systems operate on flashing stages for removal of non-condensable gases.

2.2.1.2 Performance Indicators

In general, the performance of desalination plants reflects its efficiency to produce fresh water. Many performance criteria are adopted including amount of produced water, steam consumption, specific heat consumption, thermal efficiency, performance ratio, availability factor, conversion factor, and plant factor [3].

However, MSF desalination plant's performance is usually assessed through two major indicators: PR and GOR according to the following equations:

$$PR = \frac{m_d}{m_s} \times \frac{\Delta H_{ref}}{\Delta H_{BH}} \tag{2.1}$$

$$GOR = \frac{m_d}{m_s} \tag{2.2}$$

Where:
m_d = mass of distillate (kg/s)
m_s = mass of steam (kg/s)
ΔH_{ref} = distillate reference latent heat (kJ/kg)
ΔH_{BH} = steam specific enthalpy drop across brine heater (kJ/kg)

The PR is more accurate than GOR for thermal assessment because it takes into account the effect of temperature on the latent heat of steam [1]. Both the factors determine the water production cost which reflects desalination plant operation efficiency. The interaction between performance indicators, design parameters and process variables (seawater temperature, top brine temperature, number of stages/effects, concentration factor) was subject to multiple mathematical modelling efforts [1, 4, 5]. Despite the considered assumptions of these models, the obtained results are still acceptable to describe thermal desalination system's behavior.

A comprehensive study revealed that MSF distillers which are over 20 years old, instead of being derated due to ageing, actually maintained production and performance ratios that equaled or, in most cases, surpassed the original design specifications [6–10]. Thus, the service life of MSF distillers are expected to exceed 30 years which in turn, enhances the cost-effectiveness of MSF process. The reasons for such good thermal performance were attributed to several design and operating conditions, such as:

1. Conservative design fouling factors
2. Effective alkaline scale control
3. Proper material selection
4. Strict operation and maintenance practices

The maximum available PR of existing MSF desalination plant operating TBT temperature of 112 °C and 24 stages is limited to 8.6 kg/2326 kJ.

2.2.1.3 MSF Challenges

MSF desalination process is a robust and mature technology with more than 60 years of record. Performance sustainability depends on the control of two major challenges including scale and corrosion. In the following, brief discussion of these challenges are provided from practical experience perspective.

a. Scale control

Control of scale formation on heat transfer surfaces is one of the basic problems in the distillation processes. Formation of scale on heat transfer surfaces impedes the rate of heat conducted and increases energy consumption. The main scale forming constituents of seawater are calcium, bicarbonate, magnesium salts and calcium sulfate. On heating, bicarbonate yields carbonate which can precipitate with calcium if the saturation limit is exceeded and coexist with magnesium hydroxide [11]. Calcium carbonate ($CaCO_3$) and magnesium hydroxide ($Mg(OH)_2$) are generally referred to as 'alkaline scales'. This is for identification since they form under alkaline conditions with pH values greater than 8. On the other hand, calcium sulphate ($CaSO_4$) is distinguished as 'non-alkaline scale'. Formation of alkaline scale is controlled by either lowering the pH through bicarbonate depletion or by threshold additive chelation. Conversely, $CaSO_4$ formation is primarily controlled, so far in commercial plants, by operating at top brine temperatures that are lower than its appreciable precipitation limits, i.e., TBT < 120 °C in view of the raised salt concentration in MSF brine streams. Prevention of scale deposition can be controlled by depletion of carbonate through pH adjustment by acid addition or by controlling scale precipitation through the addition of much less than stoichiometric quantities of threshold (antiscalant) agents which are mainly inorganic or organic polymer compounds. As the problems associated with acid operation became dominant, the opportunity for high temperature scale control additives (antiscalants) to replace acid while giving commensurable performance was noted from the mid of 1970 onwards [12]. The commonly used antiscalants are derived from three families: condensed polyphosphates, organophosphonate and poly-electrolytes. Polyelectrolytes are mostly poly-carboxylates which include poly-acrylicacid, polymethaerylic acid and polymaleic acid. Mixtures of sodium tripolyphosphate dispersing agent have been used to inhibit alkaline scale deposition in seawater evaporators since the 1950s [12].

Many evaluation studies were conducted to assess the performance of commercial available antiscalants [13–16]. The criteria adopted to judge the antiscalant performance is mainly based on the heat transfer measurements [17]. On the other hand, the variety of antiscalant chemistry lead to a continuous debate on the mechanism of scale inhibition among scientists. Further details on this aspect can be found in Chaps. 14 and 15. The major problem with polyphosphate based inhibitors was found to be the thermal degradation of polyphosphate at temperatures above 90 °C and the subsequent loss of threshold effect of the product. This restriction in top brine temperature to 90 °C (by hydrolysis) limited the thermal efficiency of evaporator designed for threshold treatment. Acid dosing was introduced in the 1960s as a means of overcoming the temperature limitations and the poor performance of polyphosphate [12]. Acid dosing, by removing the bicarbonate from the feed water, allowed evaporators to operate at increased top temperatures, close to the calcium sulfate solubility limits. It was found to have certain drawbacks such as the mandatory careful control and monitoring of the dose level which was quite essential to minimize the risks of plants' corrosion or scale formation and to ensure a reasonable plant life. Low molecular weight polymeric/carboxylic acid and phosphorous base alkaline were developed as high temperature additives. Phosphonate based polymers do not hydrolyse as easily as polyphosphates due to the greater stability of the C-P bond in phosphonates as compared with the P-O bond in phosphates.

The dosing rate of scale control additives is one of the most important operating parameters. Optimum antiscalant dosing rate should be established where under-dosing leads to scale formation, while overdosing is believed to enhance sludge formation. A number of optimization tests have been carried out resulting in successful operation of MSF distillers at low antiscalant dosing rates [18–22]. This can be attributed to several factors such as plant operators' awareness to reduce chemical dosing while maintaining effective plant performance, adoption of on-line sponge ball cleaning and competition among various additive suppliers to provide the most cost-effective dose levels to meet the needs of desalination plant owners.

Although, a number of methods have been developed to minimize or prevent the formation of alkaline scales, no commonly accepted method for avoiding the formation of scales due to calcium sulfate salts at high temperatures and brine concentrations in large commercial MSF plants is available. The only commercialized

Table 2.1 On-load sponge ball cleaning

Plant	Chemical treatment	Ball/Tube ratio		Frequency of cleaning operation	No. of cycles per operation
		BH	HRC		
A	Antiscalant	0.450	0.427	3 times / day	8
B	Antiscalant	0.342	0.324	3 times / day	8
C		0.270	0.257		
D		0.300	0.302		
E	Acid	0.296	0.236	1 time / week	3
F	Antiscalant	0.29	0.665	3 times / day	4
G		0.251	0.370	2 times / week	10
H		0.453	0.458	3 times / day	9
I	Antiscalant	0.243	0.249	3 times / day	12
	Acid	0.243	0.249	1 time / week	12
J	Antiscalant	0.22	0.22	3 times / day	8 (16 for high TBT)
K		0.251	0.253	3 times / day	3
L		0.351	0.351	1 time / day	9

approach so far, i.e., currently used to prevent the formation of calcium sulfate is to operate the plant below the solubility limits of calcium sulfate.

Although, the formation of scale is combated and controlled by threshold treatment with the use of antiscalant, its complete prevention is impracticable. The combined use of chemical additives and on-line tube cleaning has been proved to be the most cost effective means to combat scale formation and to avoid acid cleaning [23–25]. Most desalination plants employ on-load sponge ball cleaning. The chemical treatment, ball to tube ratio and frequency of cleaning in different MSF plants are shown in Table 2.1.[1,2]

The ball to tube ratio for plants using chemical additive treatment varies from as low as 0.22 up to about 0.45 with average frequency of three operations of ball cleaning per day. The ball to tube ratio in MSF plants, thus in most cases, lie within the reported accepted range [26, 27]. Larger number of ball to tube ratio may cause problems by several balls passing one tube simultaneously and getting stuck while smaller ratio is not capable to reach all tubes [28]. The wide variation of ball to tube ratio reveals that ball cleaning operation is not yet well-established. This can be attributed to its dependence on many interacting operating and design factors such as brine chemistry, type of inhibitor and control regime, ball type and MSF design parameters such as temperatures, number of stages and tube length, flow pattern and arrangement of ball injection points.

b. Corrosion control

Corrosion in thermal desalination plant was subject to numerous studies based on practical experience. To mitigate the corrosion in the existing and old desalina-

[1] Brine Heater

[2] Heat recovery section

tion plant, material selection is perhaps the main solution. In this way, the first generation of desalination plants installed in the Gulf region used carbon steel as the main construction material for the evaporator shells and internals [29]. Some significant changes had then occurred in the material selection specified for the second generation of desalination plants designed and constructed in the last decades due to proper understanding of the operating conditions occurring inside the evaporator.

The most commonly used materials of construction in MSF plants are carbon steel, stainless steel, copper-nickel alloy and titanium [30, 31]. The shell of brine heaters of all plants is made of carbon steel and the tubes are either 70/30 or 90/10Cu-Ni or modified 70/30 copper nickel (with 2% Fe and 2% Mn). The material of construction of flash chambers is carbon steel with and without cladding. In some plants, the first high temperature stages are cladded with SS 316 L. In some plants flash chambers are completely cladded with 90/10Cu-Ni or with SS 316 L or 317 L. The material of construction of the heat rejection tubes are normally made of titanium.

The following measures are normally employed to control corrosion in MSF distillers [32]:

1. Install sacrificial anodes in the water box of the heat rejection, heat recovery and brine heater for controlling tube inlet, outlet and tube plate corrosion.
2. The liquid load of the external deaerator not to exceed 20 kg/sm^2 to limit dissolved oxygen concentration in the deaerator effluent less than 20 ppb.
3. To guarantee low dissolved oxygen in the brine recycle, addition sodium sulphite in the deaerator effluent is normally recommended.

The addition of sodium sulphite (Na_2SO_3) to deaerated seawater will be necessary when the performance of the deaerator will not yield a 20 ppb dissolved oxygen. It has been reported that addition of Na_2SO_3 to seawater containing up to 20 ppb dissolved oxygen has very little effect on the corrosion rates of evaporator and heat exchanger alloys [33]. Further discussion on the corrosion control in thermal desalination is provided in Chap. 5.

2.2.2 Multi-Effect Desalination

2.2.2.1 Process Description

As shown in Fig. 2.2, the MED evaporator consists of several consecutive cells maintained at a decreasing level of pressure (and temperature) from the first (hot) to the last (cold). Each cell (also called effect) contains a horizontal tube bundle. The top of the bundle is sprayed with seawater make-up that flows down from tube to tube by gravity.

Heating steam flowing inside the tubes are externally cooled by pretreated seawater make-up flow. The internal steam condenses into distillate (fresh water) inside the tubes and the released heat by condensation (latent heat) warms up the seawater outside the tubes and partly evaporates it. The vapour raised by seawater evaporation

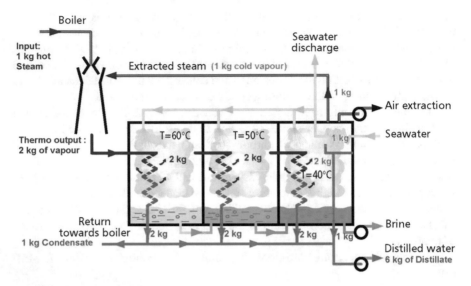

Fig. 2.2 Flowchart of MED desalination process [34]

is at a lower temperature than the heating steam. However, it can still be used as a heating medium for the next effect where the process repeats. Due to evaporation, seawater slightly concentrates when flowing down the bundle and gives brine at the bottom of the effect.

In the last cell, the produced steam condenses in a conventional shell-and-tubes-heat exchanger. This exchanger, called 'distillate condenser' or 'final condenser' is cooled by ambient seawater. At the outlet of the final condenser, part of the warmed seawater is used as make-up water of the unit, while the other part is rejected to the sea. Brine and distillate are collected from cell to cell till the last one, where from they are extracted by centrifugal pumps. The heating steam of the first effect is generally low pressure condensing steam (as low as 0.3 bar abs). Other heating media (such as hot water) may be used.

To take benefit of the available steam pressure (when this pressure is sufficient) (above 2 bar abs), MED desalination units are coupled with thermo vapor compressor (TVC) which enhance the unit thermal performance. The incoming (motive) steam is fed into the thermocompressor through a sonic nozzle. Its expansion will allow low pressure steam from a cell of the evaporator to be sucked out. Both steams will be mixed in the thermocompressor body. The mixture is then compressed to the pressure of the first effect's bundle through a shock wave. The latent heat of the sucked vapour is thus recycled in the evaporator and is again available for desalination, leading to energy savings.

The performance of a thermocompressor is expressed by the ratio (w) of the sucked steam mass (in kg) per kg of motive steam. A higher motive steam pressure corresponds to a higher value of w (for a given temperature difference between the extraction of the sucked steam and the condensation temperature of first cell). On the other hand, for a given motive steam pressure, the higher the temperature difference, the lower the value of w.

2.2.2.2 Performance Indicators

Very high GOR can be obtained with MED-TVC units. For example, Sidem has designed units with twelve cells and a motive steam pressure of 30 bar capable to reach a GOR of 17 (i.e. 17 kg of distillate water produced per kg of steam fed into the thermocompressor). They have been in operation since 1989 in the French West Indies. MED-TVC is the only process that can produce such a high performance.

Historically, MED process is an old process which suffered from scaling problems associated with the old design of these early units. Recently considerable improvements in MED desalination systems have been introduced to reduce the undesirable characteristics of the old MED submerged tube evaporators such as low heat transfer rate and high scale rate formation. Falling film evaporators such as vertical tube evaporator (VTE) and horizontal tube evaporator (HTE) found in new MED plants have a number of distinct advantages [1, 35–38]. They provide higher overall heat transfer coefficients and low specific heat transfer surface area when compared to MSF desalination systems. They do not employ recycling and are thus based on the once through principle and have low requirements for pumping energy. Power consumption of MED-TVC plants is only around 2 kWh/m³ as there are no requirements to recirculate large quantities of brine. The combination of high performance ratio and low power consumption results in lower overall energy costs. MED also offers the possibility of reducing plant size and footprint. However, there are some problems which are associated with MED systems such as the complexity of morphology and the limitation of production capacities.

MED process has recently made substantial progress and problems which are normally associated such as limitation of production capacities have been addressed. Historical evolution of the increase of the capacity of MED desalination plants in the GCC states is shown in Fig. 2.3. MED production capacity increased during the period 2000–2010 exponentially from 0.24 million m³/day to 2.64 million m³/day and increased to 3.0 million m³/day in 2018.

To reduce the specific installation and operational costs, MED-TVC designers started to adopt large production capacities units. The large size of these units implies high efficiency at steady state and relatively small flexibility during load variations. With fewer units of higher production capacity, the need for interconnection and control piping will be much reduced. The single unit is also simple to operate, and the number of operators required is smaller.

The new trend of combining TVC with conventional MED has allowed this unit capacity to increase from 2 to 20 MIGD[3] in the last decade as shown in Fig. 2.4.

---

[3] Millions of gallon per day

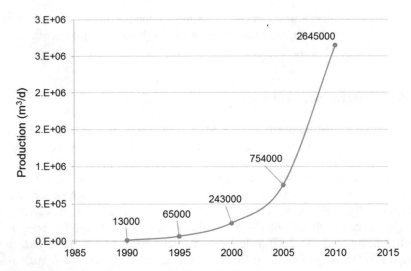

Fig. 2.3 Historical evolution of MED-TVC desalination water production in the Gulf council corporation (GCC) countries [34]

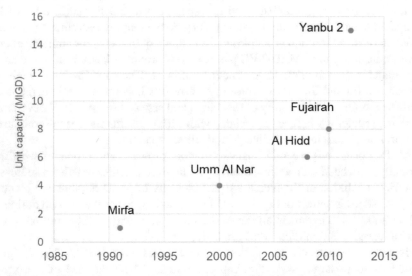

Fig. 2.4 Historical evolution of MED-TVC unit size. Modified from Ref. [39]

This considerable increase in capacity, poses a real competition to the MSF as a large-scale production plant with lower operation temperatures.

The applied MED technology consists in horizontal tubes with falling film evaporators and thermal vapor compression [40]. As shown in Fig. 2.5, each MED units consists of three modules. In order to prevent up-scaling difficulties and vapor flow

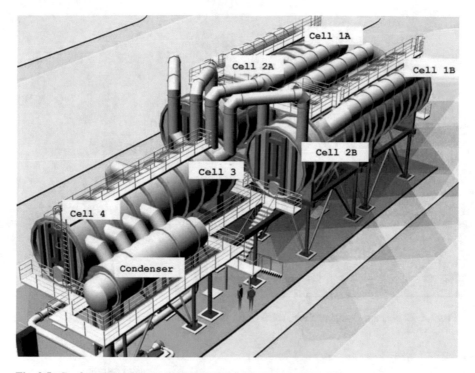

Fig. 2.5 Configuration of Sharjah MED-TVC plant. (Source: MIT) [41]

rate limits, two modules each containing 2 cells and each integrated with TVC are arranged in two identical rows. The top brine temperature is kept at only 63 °C, in order to minimize scale forming on the tubes. A third module contains three cells and without TVC, is installed downstream.

During the period 2012 and 2018, the modular design of MED-TVC was implemented by Saline Water Conversion Corporation (SWCC). Figure 2.6 shows Yanbu MED-TVC modular design concept. The distiller consists of four modules where each module consists of effects arranged in series and produces 3.75 MIGD. The first high temperature effects are combined with a TVC. Each module is divided into a number of separate tube bundles. The four modules are combined to produce a total of 15 MIGD.

In the year 2018, SWCC built and commissioned the World's largest MED-TVC distiller to produce 20 MIGD at Shuaiba. The distiller consists of four modules. Each module is divided into four tube bundles and produces 5 MIGD. The distiller is designed to operate at top brine temperature up to 70 °C and with an unprecedented PR of 14.6 kg distillate/2326 heat input as noticed in Fig. 2.7.

3.75 MIGD Modular Unit

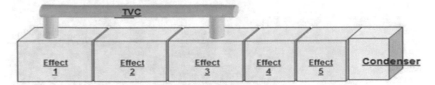

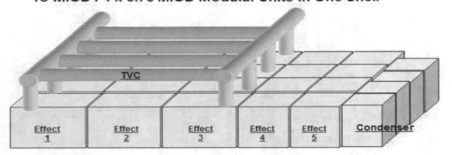

15 MIGD : 4 x 3.75 MIGD Modular Units in One Shell

Fig. 2.6 Conceptual design of Yanbu MED-TVC plant. (Source: Doosan) [42]

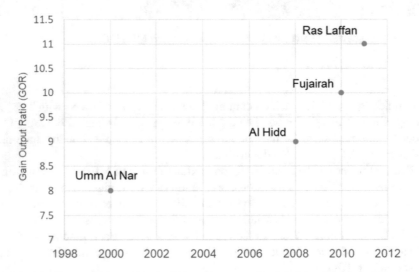

Fig. 2.7 Historical evolution of MED-TVC performance ratio. Modified from Ref. [39]

2.2.2.3 MED Challenges

In order to mitigate scaling difficulties and vapor flow rate limits and to limit the maximum TBT to 65 °C, currently available commercial MED-TVC units consist of maximum of 12 effects. The low value for the available overall temperature difference puts a limitation to the possible number of effects. Increasing the number of

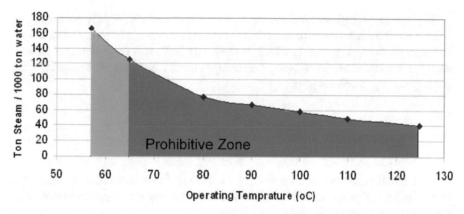

Fig. 2.8 Impact of variation of operating temperature on the energy consumption of the MED Process [34]

effects more than 12 will significantly reduce temperature difference (thermal driving force) in each effect and hence overall heat transfer area will be increased exponentially. Once TBT is increased more than 75 °C, more number of effects can be installed which can increase the PR.

Increasing the TBT above 65 °C provides the opportunity to increase the overall temperature difference and consequently provides the opportunity to increase the number of effects and this in turn will be reflected in the significant increase of the GOR and reduces the steam and power consumption. Unlike the MSF distiller, increase of TBT of the MED distiller results in a significant increase of the GOR ratio. As shown in Fig. 2.8, increase of TBT from 65 °C to 125 °C shall increase the GOR by more than two-folds.

2.3 Thermal Desalination Development Efforts

2.3.1 Desalination Units with Enhanced Configuration

A promising approach to pretreat seawater using nanofiltration (NF) membranes was developed to overcome the main challenge of thermal desalination process [43, 44]. The application of NF technique for pretreating seawater resulted in the reduction of salt concentration and removal of most of the hardness ions (Ca^{2+}, Mg^{2+}) and co-ions (SO_4^{2-}, HCO_3^-). Pretreating raw seawater by NF allow to cross the prohibitive operating zone (Fig. 2.8) and provides an opportunity to operate thermal seawater desalination plants above their present TBT limit. Increasing the TBT induces a large flash range, larger heat transfer area. Then, highly cost-effective water production with improved PR would be achieved.

Exploratory experimental work was carried out on an MED-TVC pilot plant with a capacity of 24 m³/d operating within the context tri-hybrid NF/RO/MED configuration as shown in Fig. 2.9. The TBT of the MED-TVC was successfully increased steeply from 65 °C to 125 °C using a make-up formed from softened seawater, without any scaling problems [44].

2.3.2 Hybridization

Hybrid desalination systems integrating thermal and membrane desalination process with power generation in the same site are currently considered a viable alternative to dual evaporation plants (Fig. 2.10).

The advantages of triple hybrid power/MSF/SWRO over the dual power-MSF and single purpose MSF or RO plants were reported [46–51]. Integrating a seawater reverse osmosis (SWRO) unit with a multistage flash distiller provides the opportunity to blend the products of the two processes. Such arrangement allows to operate the RO unit with relatively high TDS and consequently allows to lower the replacement rate of the membranes. If the useful life of the RO membrane can be extended from 3 to 5 years the annual membrane replacement cost can be reduced by nearly 40% [49]. Blending the products of the thermal and SWRO allows for the use of a

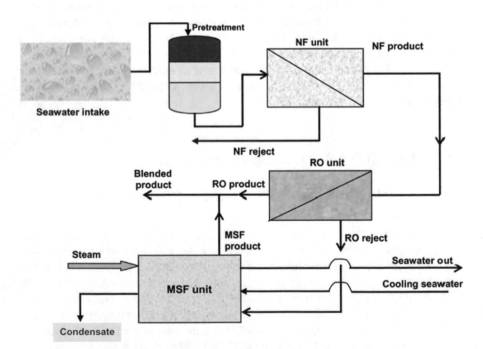

Fig. 2.9 Trihybrid NF/RO/MED desalination configuration. Reproduced with permission from Ref. [45]; Copyright 2006 @ Elsevier

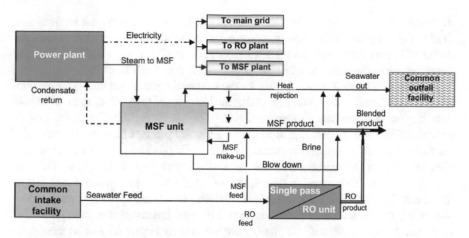

Fig. 2.10 Schematic diagram of commercially available simple hybrid desalination plants. Reproduced with permission from Ref. [45]; Copyright 2006 @ Elsevier

single stage SWRO instead of the two stage SWRO plant normally employed in standalone SWRO plants. Combining thermal and membranes desalination plant in the same site will allow using common intake and outfall facilities with less capital cost. An integrated pretreatment and post-treatment operation can reduce cost and chemicals.

During cold seasons, the preheated seawater leaving the heat rejection of the MSF distiller or the last effect of the MED plant can be used as feed water for RO plant. Increase of seawater feed temperature by 1 °C will increase the water production of SWRO by 3%. Experiments were carried out in which RO seawater feed was withdrawn from the MSF reject stream [52]. It has been reported that a 42–48% gain in RO product water recovery was obtained for seawater feed temperature of 33 °C compared to an isolated RO plant using surface seawater temperature at a temperature of 15 °C.

An optimization methodology was proposed for the design of fully integrated tri-hybrid power-MSF-RO plants [53]. The optimal design is based on exergo-economics and on profit optimization. The optimization parameters are extraction pressure (or temperature) of steam from turbine and the capacity ratio between the MSF and RO sections. The proposed model is flexible and suitable for comparative applications. A hybrid system integrating MSF and RO desalination processes where MSF is fed by brine reject of the RO is proposed [54, 55]. The MSF represents a second stage in series to increase the amount of water produced. Such configuration provides the opportunity to reduce the size of the pretreatment unit with consequent decrease of cost. It will also make it possible to use less expensive membranes and improve the thermal efficiency of the cogeneration plant. Permeate from RO and distillate from MSF can be mixed which will consequently allow the use of RO membranes with low salt rejection. It has been reported that the blow downstream leaving the MSF plant which is sterile and deaerated can be used as a feed to the RO plant [44]. It has been argued that although a higher TDS at the inlet of the RO plant decreases the membrane flux, the higher temperature of the blow down increases the flux, thereby compensating the negative effect.

A study was carried out to determine the economic impact on the water production of an RO unit when integrated with an MSF plant [56]. A number of hybrid desalination configurations were analysed. The study revealed that the simple hybrid MSF/RO desalination plant in which the RO product is blended with MSF product results in significant reduction of RO water production cost and is around 13% lower than the water production cost of non-hybridized RO plant. The cost savings resulted from smaller intake, use of single-stage RO process and longer membrane life. When a fully integrated MSF/RO desalination plant is used, selection of RO membranes of high flux and low salt rejection coupled with the blending of the products of MSF and RO plants have a relatively higher impact on the reduction of water production cost of the non-hybrid RO plant. An optimization study for the prediction of the minimum water cost of seven different designs of RO/MSF hybrid desalination plants was reported [53, 57]. The MSF plants are either of brine recycle or once-through flow configurations and are coupled to single stage SWRO with energy recovery system. For comparison the water cost of a standalone two-stage SWRO unit and a stand-alone brine recycle MSF plant, was also determined. The study revealed that the two-stage RO plant yields the lowest water production cost. Hybridization of RO and MSF processes hence would result in better economics and operation characteristics than those corresponding to the stand alone MSF process.

2.4   Conclusions and Outlook

Although MSF desalination plants which have been built more than 30 years back are still operating with high load and availability factors and with remarkable reliability and robustness; they are suffered by relatively high specific energy consumption.

MED-TVC process has recently made substantial progress and problems which are normally associated with this such as limitation of production capacities have been successfully addressed. The MED-TVC unit capacity has been increased up to 20 MIGD getting use of economy of scale. The GOR which is an indicator of specific energy consumption has been increased up to 16. There are further opportunities to reduce the specific energy consumption through proper pretreatment operations. Due to the synergy between power generation cycles and MED-TVC, this desalination process can perform better than any other desalination processes if employed within the context of hybrid dual-purpose configuration integrating high efficient power cycle with either MED-TVC or with hybrid MED-TVC/SWRO.

References

1. O.A. Hamed, Thermal assessment of a multiple effect boiling (MEB) desalination system. Desalination **86**, 325–339 (1992)
2. S. Shaaban, Performance optimization of an integrated solar combined cycle power plant equipped with a brine circulation MSF desalination unit. Energy Convers. Manag **198**, 111794 (2019)
3. S. Al-Hengari, M. El-Bousiffi, W. El-Mudir, Performance analysis of MSF desalination unit. Desalination **182**, 73–85 (2005)
4. O.A. Hamed, Thermal performance of multistage flash distillation plants in Saudi Arabia. Desalination **128**, 281–292 (2000)
5. N.H. Aly, A.K. El-Fiqi, Thermal performance of seawater desalination systems. Desalination **158**, 127–142 (2003)
6. M. Al-Sulaiman, A. Al-Mubayed, A. Ehsan Modification done/proposed for improvement of productivity of quality at SWCC, Al-Jubail MSF Plant, Proceedings of IDA Conference, Madrid, (1997), p. 179–189
7. T. Bahamdan, A. Ahsan, A. Al-Sabilah, A. Al-Mobayed, Titanium tube MSF evaporator two decades of operating experience, IDA, Bahrain, (2002), https://ida.memberclicks.net/assets/docs/1113.1.pdf
8. A.A. Al-Sabilah, M.A. Ahsan, Effective utilization of available as built capacity for production and performance enhancement in Multi-Stage Flash (MSF) desalination plants, Arwadex Conference, Riyadh, 2007
9. O.A. Hamed, M.A.K. Al-Sofi, G.M. Mustafa, K. Bamardouf, H. Al-Washmi, Overview of design features and performance characteristics of major Saline Water Conversion Corporation (SWCC) MSF plants, WSTA, 5th gulf water Conf., Doha, 24–28 March 2001, http://citeseerx.ist.psu.edu/viewdoc/download?doi=10.1.1.527.9429&rep=rep1&type=pdf
10. A. Al-Mansour, S. Al-Zahrani, E. Taha, Experience in rehabilitation plan of six major plants of SWCC (IDA Singapore, 2005), p. 219
11. A.M. Shams El Din, R.A. Mohammed, Brine and scale chemistry in MSF distillers. Desalination **99**, 73–111 (1994)
12. M.A. Finan, S. Smith, C.K. Evans, J.W.H. Muir, Belgard EV-15 years of experience in scale control. Proceedings of the 4th world congress on desalination and water reuse, 4–8 November 1989, p. 341–357
13. A.M.A. Al-Mudaiheem, R.M. Szostack, (1986) Evaluation of chemical additives for scale control in MSF evaporators. Topics in Desalination (No. 8), SWCC
14. S. Al-Zahrani, A.M. Al-Ajlan, A.M. Al-Jardan, Using different types of antiscalants at the Al-Jubail power and desalination plant in Saudi Arabia, Proceedings of IDA and WRPC world Conference on desalination and water treatment, Yokohama, Vol. 1, 1993, p. 421–431
15. N. Nada, Evaluation of various additives at Al-Jubail Phase I during reliability trials, Topics in Desalination (No. 13) SWCC, 1986
16. O.A. Hamed, M.A.K. Al-Sofi, G.M. Mustafa, A.G.I. Dalvi, The performance of different antiscalants in multistage flash distillers. Desalination **123**, 185–194 (1999)
17. M.A.K. Al-Sofi, E.F. El-Sayed, M. Imam, G.M. Mustafa, T. Hamada, Y. Tanaka, S. Haseba, T. Goto, Heat transfer measurements as a criterion for performance evaluation of scale inhibitor in MSF plants, 7th IDA International Conference, 1995
18. O.A. Hamed, M.A.K. Al-Sofi, M. Imam, K. Bamardouf, A. Al-Mobayed, A. Ehsan, Evaluation of polyphosphonate antiscalant at a low dose rate in Al-Jubail Ph-II MSF plant, Saudi Arabia. Desalination **128**, 275–280 (2000)
19. M.A.K. Al-Sofi, M.A. Al-Hussain, S. Al-Zahrani, Additive scale control optimization and operation modes. Desalination **66**, 11–32 (1987)
20. M.A.K. Al-Sofi, S. Khalaf, A. Al-Omran, Practical experience in scale control. Desalination **73**, 313–325 (1989)

21. O.A. Hamed, H.A. Al-Otaibi, Prospects of operation of MSF desalination plants at high TBT with low antiscalant dose rate. Desalination **256**, 181–189 (2010)
22. O.A. Hamed, Scale control in multistage flash (MSF) desalination plants- lessons learned. Desalin. Water Treat **81**, 19–25 (2017)
23. M.A.K. Al-Sofi, M.A. Al-Hussain, A.A. Al-Omran, K.M. Farran, A full decade of operating experience on Al-Khobar II multistage flash (MSF) evaporators (1982-1992). Desalination **96**, 313–323 (1994)
24. O.A. Hamed, Lessons learnt from the operational performance of SWCC MSF desalination plants. Desalin. Water Treat **18**, 321–326 (2010)
25. A. Malik, F.A. Aleem, Scale formation and fouling problems and their predicted reflection on the performance of desalination plants in Saudi Arabia. Desalination **96**, 409–419 (1994)
26. H.Böhmer, On-load sponge ball cleaning system. Encyclopedia of Desalination and Water Resource (DESWARE). 1998, http://www.desware.net/Sample-Chapters/D09/D11-023.pdf
27. H. Böhmer, On-load tube cleaning systems and debris filters for avoidance of microfouling in MSF desalination systems. Desalination **93**, 171–179 (1993)
28. M.A.K. Al-Sofi, Fouling phenomena in Multistage Flash (MSF) distillers. Desalination **126**, 61–76 (1999)
29. C. Sommariva, H. Hogg, K. Calister, Forty-year design life: The next target material selection –conditions in thermal desalination plants. Desalination **136**, 169–176 (2001)
30. A.U. Malik, P.C. Mayan Kutty, Corrosion and material selection in desalination plants. Proceedings of seminar on operation and maintenance of desalination plants, SWCC, 27–29 April 1992, p. 274–307
31. T. Hodgkiess, Current status of materials selection for MSF distillation plants. Desalination **93**, 445–460 (1993)
32. N. Nada, O. Mughram, Design features of Shoaiba phase II 500 MW -100 MIGD power and desalination plant. Proceedings of the 2nd Gulf Water Conference, 5–9 November, 1994
33. A.U. Malik, N.A. Siddiqi, I.N. Andijani, P.C. Mayan Kutty, T.S. Thankachan, Effect of deaeration and sodium sulfite addition to MSF make-up water on corrosion of evaporator and heat exchanger materials. J. King Saud Univ. Eng. Sci **8**, 21–35 (1996)
34. O.A. Hamed, Evolution of thermal desalination processes. Water Arabia February 4–6, Khobar, (2013), http://www.sawea.org/pdf/waterarabia2013/Session_A/Evolution_of_Thermal_Desalination_Processes.pdf
35. T. Michels, Recent achievements of low temperature multiple effect desalination in the Western Areas of Abu Dhabi. Proceedings of the Arabian Gulf regional water desalination symposium, Al-Ain, 1992
36. C. Sommariva, V.S.N. Syambabu, Increase in water production in UAE. Desalination **138**, 173–179 (2001)
37. J.W. Vermey, D. Berand-Sudreau Breakthrough of MED technology in very large scale applications, the Taweelah Al-Case, Proceedings of the IDA Conference, 2001, https://ida.memberclicks.net/assets/docs/1106.1.pdf
38. V. Banjat, T. Bukato, Research and development towards the increase of MED unit capacity, Proceedings of the IDA Conference, Bahamas, 28 September – 03 October 2003
39. I.S. Al-Mutaz, I. Wazeer, Current status and future directions of MED-TVC desalination technology. Desalin. Water Treat **55**, 1–9 (2014)
40. O.A. Hamed, Development of 7 MIGD MED-TVC distiller within the context of trihybrid NF/RO/MED configuration, IDA World Congress, Perth, 4–9 September 2011
41. C. Somariva, Desalination and water purification course. MIT opencourseware. (2009), https://ocw.mit.edu/courses/mechanical-engineering/2-500-desalination-and-water-purification-spring-2009/readings/MIT2_500s09_lec18.pdf
42. K.S. Park, I.S. Song, Y.W. Yoo, Y.W. Park, A.A. Alyoubi, A. Almutairi, A. Bajbaa, Design and operation of the largest MED desalination plant, Doosan (2012), https://docplayer.net/61883706-Design-and-operation-of-the-largest-med-desalination-plant.html

43. A.M. Hassan, M.A.K. Al-Sofi, A.S. Al-Amoudi, A.T.M. Jamaluddin, A.M. Farooque, A. Rowaili, A.G.I. Dalvi, N.M. Kither, G.M. Mustafa, I. Al-Tisan, A new approach to membrane and thermal seawater desalination process using nanofiltration membranes. Desalination **118**, 35–51 (1998)
44. O.A. Hamed, M.N. Al-Ghannam, K.A. Al-Shail, A.M. Farooque, R. Al-Rasheed, S. Al-Fozan, A. Al-Arifi, M. Hirai, Y. Taniguchi, S. Araki, K. Harada, K. Maekawa, Successful operation of MED/TVC desalination process at TBT of 125 °C without scaling. IDA World congress, Dubai, 7–12 November 2009
45. O.A. Hamed, Overview of hybrid desalination systems-current status and future prospects. Desalination **186**, 207–214 (2005)
46. B.M. Misra, P.K. Tewari, B. Bhattacharjee, Futuristic trends in hybrid system for desalination. IDA Conference, San Diego, pp. 311–320, 28 August – 03 September 1999
47. L. Awerbuch, Power-desalination and the importance of hybrid idea's. IDA world congress, Madrid, 06–09 October 1997
48. M.A.K. Al-Sofi, A.M. Hassan, E.F. El-Sayed, Integrated and non-integrated power/MSF/SWRO plants. Desalin. Water Reuse. Part 2, **4**, 42–46 (1992)
49. L. Awerbuch, S. May, R. Soo-Hoo, V. Van Der Mast, Hybrid desaling systems, Proceedings of the 4th world Congress desalination and water reuse, IDA, 04–08 November 1989
50. L. Awerbuch, V. Van Der Mast, R. Soo-Hoo, Hybrid desalting systems – A new alternative. Desalination **64**, 51–63 (1987)
51. I.S. Al-Mutaz, M.A. Soliman, A.M. Daghthem, Optimum design for a hybrid desalting plant, Proceedings of the 4th world Congress desalination and water reuse, IDA, 04–08 November 1989
52. E. El-Sayed, M. Abdeljawad, S. Ebrahim, A. Al Saffar, Performance evaluation of two membrane configurations in a MSF/RO hybrid system. Desalination **128**, 231–245 (2000)
53. A.M. Helal, A.M. El-Nashar, E. Al-Katheeri, S. Al-Malik, Optimal design of hybrid RO/MSF desalination plants: Part-1 modeling and algorithms. Desalination **154**, 43–66 (2003)
54. E. Cardona, A. Piacentino, Optimal design of cogeneration plants for seawater desalination. Desalination **166**, 411–426 (2004)
55. E. Cardona, S. Culotta, A. Piacentino, Energy saving with MSF-RO series desalination plants. Desalination **153**, 167–171 (2002)
56. M. Sherman, Hybrid desalination system, MEDRC series of R&D reports, Project 97-AS-008b (2000)
57. A.M. Helal, A.M. El-Nashar, E. Al-Katheeri, S. Al-Malik, Optimal design of hybrid RO/MSF desalination plants: Part-1I results and discussion. Desalination **160**, 13–27 (2004)

Chapter 3
Reverse Osmosis Desalination: Performance And Challenges

A. Mohammed Farooque

3.1 Introduction

Producing potable water of acceptable quality with minimum cost is the major goal for water industry professionals. Fresh water production by seawater desalination is an expensive affair. For several years, thermal processes have been dominating the seawater desalination over the membrane process and during the last few years, seawater reverse osmosis (SWRO) desalination technology has gone through a remarkable transformation and gained widespread acceptance, which is evident from the increased share of SWRO. The major reasons for the increase in popularity of SWRO desalination process are its simplicity and significant reduction in capital and operational costs, which was possible due to various advancements made in membranes, energy recovery devices (ERDs), materials, etc. [1, 2].

Loeb and Sourirajan were the first to report the discovery of a practical RO membrane, comprised of an asymmetrically structured cellulose acetate (CA) film during the year 1959–1960 [3] and CA based RO membranes, which were then commercialized at Gulf General Atomic, San Diego, California, USA in the early 1960s [4]. During the late 1960s, brackish water RO plants were successfully deployed [5]. In the following decade, as a result of membrane material improvements, RO membranes were applied to seawater desalination [6] and over the past 50 years, dramatic improvements in RO membrane technology elevated RO to be the primary choice for new desalination facilities throughout the world.

In RO process, a semi-permeable membrane is used to remove salts from saline water by applying sufficient pressure to overcome the osmotic pressure of the same. RO membranes allow only the passage of water through it while rejecting the dissolved salts thus enabling to obtain product water with low salinity.

A. M. Farooque (✉)
Desalination Technologies Research Institute, Saline Water Conversion
Corporation, Jubail, Saudi Arabia
e-mail: amfarooque@swcc.gov.sa

© Springer Nature Switzerland AG 2020
V. S. Saji et al. (eds.), *Corrosion and Fouling Control in Desalination Industry*,
https://doi.org/10.1007/978-3-030-34284-5_3

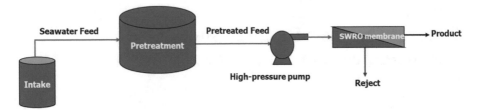

Fig. 3.1 Schematic diagram of a typical SWRO plant

Typical SWRO desalination plant consists of four major systems, a seawater intake system, a pre-treatment system, a high-pressure pumping system, and a membrane system (desalting module) (Fig. 3.1). An intake system is the one, which provides consistent supply of seawater required for the desalination. Whereas the pre-treatment system adjusts the source feed seawater quality to make it compatible with the SWRO membrane. High-pressure pumping system raises the feed pressure to the SWRO membranes to overcome the osmotic pressure of the pre-treated seawater fed to the membrane and produce sufficient quantity of potable water. The membrane system consists of semi-permeable RO membranes arranged in multiple pressure vessels that are capable of rejecting dissolved solids in the seawater and allowing water to pass through the same thus producing potable water.

The performance of the entire system is dependent upon the proper design and operation of each component. It is desirable that the RO plant is capable of sustainably producing the required amount of potable water with acceptable quality at minimum cost.

Three major challenges are observed for SWRO: high water production cost, sustainable plant performance and the impact on the environment. The former challenge is due to the high energy consumption and the use of expensive corrosion resistant alloys (CRAs). The sustainability of the plant performance depends on the fouling phenomenon. Finally, the use of chemicals and the brine discharge still debated in terms of negative impact on the environment.

The chapter discusses various SWRO performance assessment parameters. This includes the product flow rate, salt rejection, recovery rate and differential pressure drop across the membrane. It also addresses various challenges faced by SWRO plants and measures adopted to overcome these challenges to maintain the best plant performance. Recent advancements with different system components are also discussed.

3.2 Process Components Performance and Challenges

3.2.1 Intake System

The intake is one of the key systems for SWRO desalination plant. Its purpose is to ensure adequate and consistent flow of source water over the entire useful life of the desalination plant. Moreover, it should be capable of providing a reliable quantity of clean seawater with a minimal ecological impact. Seawater intakes are broadly categorized as surface and subsurface intakes. In surface intakes, the feed seawater is collected from an open sea above the seabed via on-shore or off-shore inlet structure and pipeline interconnecting this structure to the desalination plant. Whereas in subsurface intakes the feed seawater is collected via beach wells, infiltration galleries, or other locations which lie beneath the seabed.

A good intake design will not only protect downstream equipment and reduce environmental impact on marine life, but it will improve the performance and reduce the operating cost of the pretreatment system. To meet these objectives, it is essential that a thorough assessment of site conditions and careful environmental impact study be conducted prior to the design of the rest of desalination plant. Physical characteristics, meteorological and oceanographic data, marine biology, and the potential effects of fouling, pollution, and navigation must be evaluated, and an appropriate intake design should be employed.

Performance of intake systems is usually evaluated by the assessment of the source water quality. This include but not limited to low values of Total Dissolved Solids (TDS), Total Suspended Solids (TSS), turbidity and Total Organic Carbon (TOC).

Large seawater desalination plants have traditionally adopted open sea, surface water intake configuration. The reason behind this choice is that they are suitable for all sizes of desalination plant and more predictable and reliable in terms of productivity and performance. Moreover, they are easy and economical to manage in terms of operation and maintenance. Open sea intakes offer also a better economy of scale for desalination systems of capacity greater than five million gallons per day (MGD) [7].

Although open intake usually provides a sufficient quantity of feed seawater, in many instances it fails to provide quality feed water and may lead to more burden on downstream pretreatment system. Moreover, seasonal and daily variations in seawater quality is challenging to seawater desalination plants. To overcome this challenge, SWRO plants collect seawater at sufficient depth (typically ≥35 m), where such an impact is minor. However, at many locations such a desirable depth is reached only at a far distance of few kilometers away from the shore, which increases the capital expenditure (CAPEX) of the SWRO plant.

A subsurface intake such as beach wells can be used, where geologic conditions support water extraction and are known to have the advantage of delivering "prefiltered" water that may greatly reduce additional pretreatment requirements. Beach wells are typically located on the seashore, in close proximity to the ocean (Fig. 3.2).

They are relatively simple to build and the seawater collected is pretreated via slow filtration through the subsurface seabed formations in the area of source water extraction. Therefore, source seawater collected using beach well intakes is usually of better quality in terms of solids, silt, oil and grease, natural organic contamination and aquatic microorganisms as compared to open seawater intakes. Although beach wells appear to be desirable for developing feed water intake for SWRO desalination plants, they do have their limitations. There are a number of key factors that have to be taken into consideration, when assessing the viability of using a beach well intake for a large desalination plant, and these are site conditions, seashore

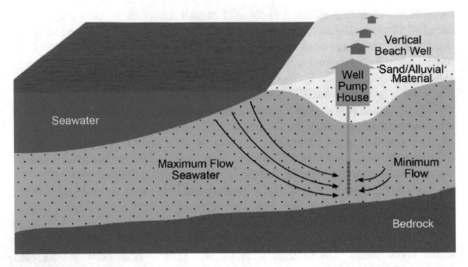

Fig. 3.2 Schematic diagram of a typical beach well. Reproduced with permission from Ref. [8]; Copyright 2013 @ Elsevier B.V

impacts, visual impact, beach erosion, lifecycle costs, need for source water pretreatment, source water quality variation and dissolved oxygen concentration.

To determine the efficacy of beach wells for a particular site, exploratory drilling to determine the aquifer characteristics (depth, strata) and pump tests on boreholes should be conducted. A preliminary idea of required depth, potential discharge and number of required beach wells in the site area can be obtained to match the demands of the proposed plant. Spacing between wells at proposed withdrawal rates is another important factor in design. The use of beach wells as a pretreatment strategy for the treatment of seawater is attractive because of the potentially lower operations and maintenance cost as compared to other pretreatment options, including media or cartridge filtration. The use of traditional vertical beach wells, however, is limited to smaller systems due to the large number of wells that would need to be drilled in order to fulfill the pretreatment needs and are an economical alternative to open sea intakes for desalination plants with capacities less than 20,000 m³/d only [9].

Open ocean intakes are currently the method of choice to provide high-capacity SWRO systems with feed water because of their general reliability. However, the high levels of pretreatment required make the operation of SWRO facilities costly and the heavy use of chemicals in these processes make the discharge impacts more environmentally unfriendly. Also, the increase in frequency of harmful algal blooms makes a subsurface intake system more attractive because the SWRO pretreatment processes can be overwhelmed by the high organic load during these events, causing plant shutdown and/or damage [10].

A considerable amount of research is ongoing to make seabed gallery systems less expensive to construct and to achieve economic benefits that can make them competitive with conventional open ocean intake systems [11]. Over the past years, many improvements have been made in the design and operation of seawater intake systems, which are aimed at reducing the impact on the surrounding environment. This includes the invention of the velocity-cap offshore intake system to reduce the entrainment of fish, the use of passive screen intakes to further reduce the environmental impacts of open-ocean surface intake systems by near elimination of impingement, and a reduction in entrainment (due to low inflow velocity and the small slot size of the screens).

Other than conventional vertical beach-wells, relatively new well types, such as slant wells, horizontal wells, and radial collectors, have been introduced to provide feed water for SWRO facilities. Gallery intake systems were developed for use in a wider range of required capacities from medium to large. The concept of the self-cleaning beach gallery intake was developed and has not yet been installed for use in a medium or large capacity SWRO facility, but this intake design is claimed to have a high potential for success [12].

3.2.2   Pretreatment System

In order to obtain the best performance from the SWRO membrane and also to operate the membrane in a sustainable way, it is very essential to have extremely good quality pretreated feed, free from suspended solids and microbes, which would otherwise cause membrane fouling. Pretreatment using conventional coagulation-filtration is very popular. It is widely used for the seawater feed pretreatment because of its simplicity, low cost and in many cases the pretreated feed quality is adequate for the satisfactory performance of the downstream SWRO membranes. A typical conventional pretreatment includes chlorination, coagulation, acid addition, multi-media filtration, micron cartridge filtration and de-chlorination. The type of pretreatment to be used largely depends on the feed water characteristics, membrane type and configuration, recovery ratio and desired product water quality [13].

The performance of the pretreatment is determined through the feed seawater quality. Pretreated quality of water is usually measured in terms of fouling tendency through the Silt Density Index (SDI) where a value lower than 3 is considered as threshold for a very good performance of membrane [14, 15]. However, when the feed water quality varies significantly the conventional pretreatment may not perform as expected, i.e., may not yield a steady good quality (SDI < 3) feed water. This is mainly due to the fact the coagulation-filtration process does not act as an absolute physical barrier, which is capable of preventing, especially the fine and very fine suspended particles, from passing through the filters. In addition to SDI value, the turbidity and TOC should be less than 0.5 NTU (nephelometric turbidity units), and 2 ppm, respectively [16].

An index similar to the SDI, the modified fouling index (MFI), has been developed to better correlate membrane fouling, flux decline, and particle concentration. The original MFI method used a 0.45 µm microfiltration membrane in dead-end filtration and provided a linear correlation between the index and the particle concentration [17]. However, the MFI do not predict always accurately the fouling observed in membrane systems. This is due to the number of small particles that pass through the 0.45 μm membrane. Later on, a modified MFI, the MFI-UF, was developed which uses ultrafiltration membranes to retain a larger portion of the small particles that can pass through microfiltration membranes but will foul a RO membrane. The MFI-UF has subsequently been used to analyze pretreatment performance and RO membrane fouling potential during plant operation [18]. However, majority of commercial SWRO plants tend to use SDI due to its simplicity.

Better pretreated water quality means that the RO membrane can be operated at higher flux as well as at higher water recovery without any adverse effect on membrane performance. The practical average flux rate of SWRO membrane is between 12–18 $L/m^2/h$. The lower limit of flux is for feed water with poor quality and the upper value is for feed with very good quality, e.g. beach well, where SDI value of about 1 can be achieved. Operation of SWRO membranes at higher flux rate is desirable, as it requires only less number of membranes, thus reducing the size of the RO unit, which results in cost savings from a CAPEX perspective. However, the conventional coagulation-filtration process produces mostly a pretreated feed with SDI between 2 to 3.

Another major problem with conventional pretreatment process is the use of many chemicals, especially in the form of coagulants, e.g., $FeCl_3$ and coagulant aids (organic polyelectrolyte), in addition to the disinfectants such as Cl_2 as well as sodium metabisulfite used for de-chlorination. Also, to prevent scale deposition on membrane surface, proprietary antiscalants or acids such as sulfuric acid are used in the pretreatment. Chemicals are not only increasing the total cost of water production but also are detrimental to the environment [19, 20]. Thus reducing the chemicals consumption or even completely eliminating them by an alternate pretreatment process will lead to great advantages in process economics as well as in minimizing damage to the precious environment. Recently, efforts are taken to convert SWRO plants to more environmental friendly operation with only a biocide (copper sulfate) without the use of antiscalant and sulfuric acid, since early 2017 [21].

Thus, desalination industries always seek alternate pretreatment processes, which are capable of eliminating all the above-mentioned problems associated with conventional pretreatment. In this context, the application of microfiltration (MF) as well as ultrafiltration (UF), which emerged as an efficient method in treating surface water becomes quite important [22]. Both UF as well as MF membranes offer good physical barrier to colloids, suspended particles as well as microbes and they are differentiated based on the ability to filter out particles of various sizes. UF membrane is capable of filtering out particles in the ranges between 0.01 μm to 0.1 μm, whereas MF with slightly higher ranges of 0.1 μm to 1 μm.

MF/UF membranes available in the market are made of polymers with different chemical composition, configuration and operating procedure. As there are many different MF/UF membranes available in the market with various claims, it is essential for the end-user to evaluate them to choose the best performing one.

An important improvement made in the operation of MF/UF to reduce the cost is to operate them in "direct filtration" mode rather than the conventional "cross-flow filtration" mode by which the water recovery can be maximized [23]. But this leads to the settling of suspended solids and colloidal particle matters on the membrane surface, which impede the flow through the membrane. In conventional cross-flow mode filtration process, these suspended matters are rejected in the reject stream whereas in direct filtration mode they accumulate on the membrane surface. Thus it requires frequent removal of these settled particles or matter from the membrane surface during the direct filtration mode. In order to remove these settled matters backwash treatment is applied to the membrane. Thus, during the backwash, the membrane fails to produce any pretreated water and part of the pretreated feed water is used for the backwash operation. Depending on the amount of foulants on the surface as well as the flux rate of the pretreatment, this automated backwash has to be repeated every 15–60 min of operation for duration of 20–40 s. Hence, all the recent MF/UF systems do have automatic PLC controlled backwashing and operating system which slightly adds to the system cost and by which the backwash frequency and duration can be controlled depending either on time or on the value of Trans Membrane Pressure[1] (TMP) across the membrane.

MF/UF membranes typically operate in the range of feed pressure 2–4 bars and as the foulants build up on the membrane surface, the TMP increase if unit is operated at constant flow rate mode. Once a predefined TMP is reached, a backwash is performed or it can be performed based on fixed duration of time regardless of the TMP value. Usually with backwash, the TMP value does not come back to the original value because the foulants cannot be completely removed and moreover, with time the TMP cannot be reduced below certain value, which necessitates an offline chemical cleaning of the membrane using chemical agents such as detergent, NaOCl, citric acid, etc. These cleaning can take typically 1–5 h and may be performed every 1–6 months depending on fouling tendency, water quality and membrane. Both the backwashing as well as chemical cleaning contribute to the final water production cost.

During the regular operation, some of the membranes use chemicals such as coagulant for improving the quality of water, flux rate as well as backwashing efficiency. Moreover, chemicals such as NaOCl, as well as H_2O_2 are also used during backwashing in addition to air scouring. Thus the ultimate cost of MF/UF pretreatment depends on all the factors, viz., its capital cost, operating cost, chemicals cost, etc. Most of the MF/UF membranes operate at a flux rate below 100 L/m^2/h and some even at very low flux rate, e.g., 30 L/m^2/h, but they have the advantage that they use very little chemicals. Thus it is essential to look into balance between the capital and operating cost and of course the quality of final pretreated water to be able to decide on which type of membrane is most suitable for a particular seawater pretreatment. In this context, a field test of MF/UF membrane may become very important.

One of the major hindrances for the popularization of the MF/UF in seawater application is its high capital investment cost compared to the conventional pretreatment process. However, recent developments in membrane manufacturing, as well

---

[1] The difference in pressure between the feed side and the permeate side of the membrane

as competition among various manufactures, made the cost of MF/UF units to be comparable in cost with the conventional pretreatment process. Moreover, it becomes essential to use these membranes at places where the conventional pretreatment fail due to large seasonal variation in the feed water quality. In order to operate the MF/UF units economically the size of the unit need to be reduced, which occupies only small footprints and the recent trend is to have them in capillary configuration and most of the MF/UF membranes available in the market are capillary type. In this way, large surface area of membrane can be accommodated in small volume of space.

Although many of the studies on both pilot plant scale as well as large commercial plant scale show good quality filtrate from MF/UF treatment in terms of SDI and turbidity, the effect of the filtrate quality on the performance of SWRO membrane and its advantages are still debatable. This is especially due to the fact that although MF/UF membranes offer good barrier to microorganism and suspended solids, it fails to remove small natural organic molecules (NOM). These molecules are one of the major culprits for the fouling, especially biofouling of SWRO membranes [24]. Moreover, the claim of steady filtrate quality from membrane filtration unit has been questioned by some of the field study results and also failure of UF system at places, where algal bloom occurred. One has to carefully study the situation by pilot testing before venturing into membrane pretreatment for SWRO membrane. The claim of increased SWRO flux is to be addressed carefully in view of the failure of MF/UF in removing organics from the influent. Also at some locations, based on evaluation of both UF and conventional pretreatment, it was found that membrane pretreatment did not offer many advantages, and hence conventional pretreatment was preferred. A detailed review on selection of the right pretreatment system for SWRO plants can be found elsewhere [25].

3.2.3   High-Pressure Pumping (HPP) System

The high salt concentrations found in seawater require elevated hydrostatic pressures (up to 70 bar) and the higher the salt concentration, the greater the pressure (pumping power) needed to produce a desired permeate flux [16]. The required hydrostatic pressure must be greater than the osmotic pressure on the feed (concentrate) side of the membrane. As the recovery of a RO unit increases, the osmotic pressure increases on the feed side of the membrane, thus increasing the feed pressure required.

ERDs have been developed to recover some of the energy typically lost from the pumps and membrane system. The primary objective is to recover much of the energy held in the pressurized RO concentrate stream before its disposal or treatment. Part of the recovered energy is used to power the process pumps.

Recent advances in SWRO that has allowed a drastic reduction in the cost of desalinated water include the application of highly efficient ERDs and the utiliza-

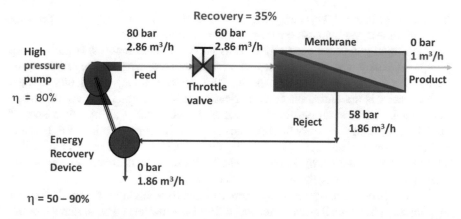

Fig. 3.3 Typical SWRO high-pressure pumping system with ERD arrangement

tion of low-pressure RO membranes. It is well established that energy is a major contributor to water production cost. It represents about 30–50% of the total production cost of water and can be as much as 75% of the operating cost, depending on the cost of electricity. It is also reported that 75–85% of total cost of water production is energy use and capital amortization [26]. Hence reducing the energy cost, which is mainly due to wastage of energy in high-pressure brine, is one of the major challenges of desalination industries.

By means of ERDs, it is possible to reduce the Specific Energy Consumption[2] (SEC) of HPP which constitute more than 80% of the energy required to produce a unit volume of permeate. The amount of energy saving mainly depends on the efficiency (η) of the ERD unit. It is a fact that due to the low recovery (about 35%) of SWRO process, a lot of water had to be pretreated, and then pumped to high pressure before dumping about 65% of this pressurized water to the sea as reject (Fig. 3.3).

SEC is largely dominated by two factors; the amount of trans-membrane pressure difference as well as the design and efficiency of the HPP in combination with the respective ERD system. The trans-membrane pressure difference is required in order to achieve the necessary permeate flow rate at various mass transfer conditions. Hence, it is essential to utilize both HPP and ERD of highest η to have minimum possible SEC.

However, hydraulic devices i.e. feed pump (typically centrifugal) and ERD (typically turbine), do have a very specific duty point in terms of flow rate and pressure, at which they will perform with optimum energetic efficiency. It is reported that the performance of ERDs such as Pelton wheel, Francis turbine as well as Turbo charger are affected by variation in flow and pressure [27]. Since non-ideal operating parameters cause deviation from optimum component performance, the combined

---

[2] The total energy consumed by the high pressure system

interaction effects of two energy conversion devices (pump and ERD) operating under non-ideal, yet realistic, conditions are to be considered.

Centrifugal pumps are generally used as HPP in SWRO process. High-pressure pumps efficiency typically increases with increase in capacity and an efficiency of 88.7% was reported for HPP of capacity 2270 m³/h. The major difficulty to have high capacity is the limitation in its fabrication as well as train size. As the membrane banks do have an optimum size, it may not match with optimum size of HPP. Indeed, energy recovery has become one of the hottest topics in SWRO desalination and several new devices are being developed in the market, which claims superiority over others [28]. Many of these ERDs work on different principles and the major players among them are described here.

Pelton wheel is one of the old systems claimed to be highly efficient depending on plant capacity as well as water recovery. The Pelton wheel takes advantage of the high-pressure energy, which remains in the reject (brine) of the RO process. The high-pressure concentrate is fed into the Pelton wheel hydraulic impulse turbine, which then produces rotating power output, which is used to assist the main electric motor in driving the high-pressure pump (Fig. 3.4). This concept allows a smaller, less costly motor to be utilized for HPP and saves a very considerable proportion of the power and, therefore, cost necessary to drive the pump.

The system is very easy to operate with only one control and consists of an adjustable input nozzle to convert water pressure into kinetic energy contained in a high-velocity jet. This is directed to a series of buckets- or metal vanes in the modern sense- around a rotating shaft that intercepted the jet stream and converted the kinetic energy into rotational energy to turn a shaft and then finally discharges the water at atmospheric pressure.

Fig. 3.4 Pelton wheel energy recovery device. Reproduced with permission from Ref. [29]; Copyright 2020 @ Rickly Hydro

Fig. 3.5 Turbocharger energy recovery device. Reproduced with permission from Ref. [30]; Copyright 2020 @ Energy Recovery, Inc

It is also claimed that its efficiency stays relatively high over the full operating range and the changes in flow and pressure, basically, do have only a small effect on operation of the turbine. However, in reality this unit suffers from loss of efficiency, especially when operating in the off duty range. Moreover, if the unit is not properly positioned and designed in an RO plant, which is the normal practice, the system could suffer from loss of efficiency. Here, energy recovery starts at about 40% of system pressure and the inlet nozzle acts as a brine control valve and no further pressure control is used on the RO system.

The turbo charger has been specifically designed for RO systems. This device transfers hydraulic energy from one liquid stream, the RO brine, to a second fluid stream, the feed (Fig. 3.5). The two flows may be at different pressures and rates. The system is entirely powered by the brine; it has no electrical cooling or pneumatic requirements.

Turbo charger consists of a hydraulic turbine and a pump, thus it is an integral turbine driven centrifugal pump. The turbine section is a single-stage radial inflow type (similar to a reverse running pump). The pump portion is a single-stage centrifugal with its impeller mounted on the turbine shaft. The energy transfer results in a feed pressure increase. The entire rotating element is dynamically balanced as a unit. The device has a by-pass around it that enables the operator to control and balance the flow. This by-pass is needed when second stage brine flow is more than that is required for the boost pressure, especially when the feed is subjected to large temperature variations as are usually seen in surface intake plants and/or for membrane ageing, this arrangement becomes important. Here the energy saving is achieved because the main high-pressure pump's required discharge pressure is reduced.

Devices using the principle of positive displacement are commonly referred to as pressure exchangers. The two basic designs dominating the market are; one which use valves and pistons and another which only uses a single spinning cylindrical

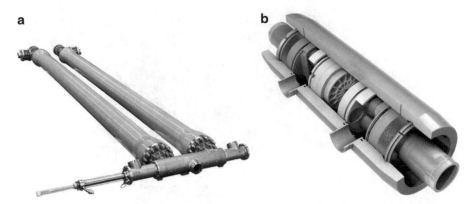

Fig. 3.6 Isobaric energy recovery devices, (**a**): DWEER, Reproduced with permission from Ref. [31]; Copyright 2020 @ Flowserve US, Inc., all rights reserved [31], (**b**): PX Pressure Exchanger, Reproduced with permission from Ref. [30]; Copyright 2020 @ Energy Recovery, Inc

rotor (Fig. 3.6). Flowserve's Work Exchanger Energy Recovery (DWEER) represents former, and Energy Recovery Inc.'s Pressure exchanger (PX), represents the latter (Fig. 3.6). They claim to have a flat performance curve but tend to be less effective at higher water recovery. Here energy saving is achieved by reducing the volumetric output required by the main high-pressure pump and claim to have efficiency up to 98%.

The PX claims to be of 98% efficiency in recovering the energy with small leakage (mixing of feed and reject) occurring. The PX technology is different from conventional ERD design, where the brine is passed through the PX unit and its pressure energy is transferred directly to a portion of the incoming seawater feed. This seawater stream, nearly equal in volume to the reject stream, then passes through a small booster pump, which makes up for the hydraulic losses through the SWRO system. This seawater stream then joins the seawater stream from the HPP without passing through it. Thus, the HPP is sized to match the permeate flow, not the full flow. The HPP also makes up the small volume of brine lost through the PX hydrostatic bearing. The PX's one moving part, a shaftless ceramic rotor with multiple ducts, is hydrostatically suspended within a ceramic sleeve. The rotor effects an exchange of pressure from brine to feed through direct contact displacement. High flow capacities are obtained by arranging multiple units in parallel. Since the main high-pressure pump flow equals the product water flow, the energy savings are actually achieved at lower conversion rates. Overall energy consumption of an SWRO plant using PX device has a low point at conversion rates of typically between 30–40%. Outside these conversion points the plant will start to consume slightly higher amount of power. The system claims to have easy start and stop procedure and requires high-pressure by-pass valve to control the pressure initially. Moreover, it has been applied in several SWRO plants worldwide since 2003, with more than 20,000 devices in operation.

Table 3.1 Ranking of ERDs

ERD	Simplicity	Capital cost	Efficiency
Pelton wheel	2	2	2
Turbo charger	1	1	3
PX/DWEER	3	3	1

DWEER works similar to PX, however, here instead of a rotor, positive displacement pistons are used. A booster pump to boost the required feed pressure equal to the feed pump pressure is also needed. DWEER transfers the fluid pressure in the brine stream to fluid pressure in the feed across a piston, where it reduces mixing of the brine and feed. For a piston designed for minimum drag, the transfer of energy in this scheme is essentially 100%. For this reason, the fundamental exchange of energy between the brine and seawater feed is more efficient than centrifugal devices relying on shaft conversion of power. However, in an actual RO system, there is a pressure drop between the feed entering the RO module and the brine exiting from it and entering the DWEER. Because of this loss, it is not possible for the effluent from the DWEER to flow back into the feed. Also, because the piston is at the pressure of the membrane array, it must be housed within a pressure vessel. The pressure vessel has a limited volume, so a valve causes the two vessels to exchange functions before the piston in that volume completes its stroke. By installing a booster pump, the flow exiting the DWEER is now able to match the discharge pressure of the HPP, allowing the system to operate in a loop. The flow rate of this booster pump is equal to the membrane brine flow rate less some small leakage. The HPP, therefore, pumps only the permeate flow.

If one considers ease of operation and simplicity among the modern ERDs, the Turbo charger ranks first followed by Pelton wheel and the pressure exchangers are slightly complicated. Capital cost-wise Turbo charger stands lowest followed by Pelton wheel and pressure exchangers are considered expensive (Table 3.1). However, while deciding the capital cost, the cost of HPP also should be considered as the pressure exchangers reduce the size of HPP, which shall also reflect on the total capital cost. But, there are concerns such as over flushing and mixing of salt resulting increase in feed salinity by about 3% for PX and other positive displacement ERD systems. Hence, before selecting a ERD for specific application, one has to look into all these aspects such as capital cost, installation cost, maintenance cost, ease of operation, reliability, availability and long term source of supply of spare parts in addition to SEC.

3.2.4 RO Membrane System

RO membranes are considered to be the heart of SWRO desalination plant as its performance decides ultimately the desalination plant performance. Usually, the performance of RO membrane is assessed by the quantity and quality of the final product water produced at lowest possible SEC. The quantity of water produced is related to membrane water permeability, which is an inherent property of SWRO membrane and it is desirable to produce more water at lowest possible feed pressure resulting in low SEC. On the other hand, the final product quality depends on the ability of membrane to reject the dissolved salt (membrane salt permeability). Membrane developers consider the inverse proportionality existing between water and salt permeabilities. Thus achieving very high water permeability with very high salt rejection is practically impossible. Hence, membrane manufacturers do have different models of SWRO membranes that meet various requirements of the customer.

3.2.4.1 Membrane Configuration and Chemistry

Desalination industry uses mainly two different types of SWRO; one is cellulose tri-acetate (CTA) based hollow fine fiber (HFF) membranes. The second one, which is more popular, is thin film composite (TFC) based spiral wound membranes. The advantage with CTA-HFF SWRO membranes is its ability to pack large surface area of membrane in a single module thus requiring only a small footprint compared to the plant using spiral-wound SWRO elements. These hollow fibers are of very small diameter of about 160 μm and hence a single module of same size could accommodate about 10 times the surface area of spiral wound membrane due to its compact nature. This also enables HFF SWRO membrane to operate at relatively lower flux rate compared to spiral wound membrane. Lower flux rate operation means lower pressure requirement and better salt rejection. However, since HFF membranes are made of CTA, it has inherently lower water permeability and higher salt permeability compared to synthetic TFC membranes. Thus, a comparable performance is achieved by low flux operation. Moreover, CTA membrane has better resistance to free chlorine, which is commonly used as disinfectant in SWRO plants that allows exposing the CTA membrane to free chlorine for short period. This approach helps in preventing membrane biofouling in CTA-HFF membrane compared to TFC based spiral wound membrane. However, CTA membranes are very sensitive to operating pH that requires large quantities of acid to be added to the feed seawater to lower pH from about 8 to about 6.4 at which the membrane is more chemically stable. Table 3.2 shows a comparison between the two types of SWRO membranes.

The advantage with polyamide-based TFC membrane is that they do have inherently very high water permeability along with high salt rejection and further R & D work is ongoing to improve the same. Also, it does have a better rejection of boron compared to CTA based membrane. They are made of three layers of synthetic polymers with polyamide-based separation layer (< 200 nm) deposited on the top of porous polyethersulfone or polysulfone layer (about 50 μm) over a non-woven polyester-based fabric support sheet.

Table 3.2 A comparison between spiral wound and hollow fine fiber SWRO membranes

Properties/Parameter	Hollow fine fiber		Spiral wound	
	Advantage	Disadvantage	Advantage	Disadvantage
Energy consumption	–	High	Low	–
Chlorine tolerance – Prevents biofouling	Tolerant to chlorine	–	–	Not tolerant to chlorine
Cost	–	High	Low	–
Productivity	–	Low	High	–
Feed water quality in terms of SDI	–	High <3	Low <4	–
Chemical stability	–	Less stable	More stable	–
Membrane preservation	–	Tend to degrade with time	Relatively stable	–
Rejection of boron and trihalomethanes (THMs)	–	Poor	Better	–
Feasibility for replacement with other manufactures membranes	–	Cannot be replaced as no other manufacturer available	Yes – several manufactures are available	–
Feed pH requirement – Acid consumption	–	Higher acid consumption	Lower acid consumption	–
Feed pH tolerance	–	Limited range pH = 4–8	Wide range pH = 3–10	–
Chemical cleaning options	–	Limited options due to pH limitations and configuration	Several options available	–
Handling	–	Slightly difficult	Easy handling	–
Footprint	Smaller	–	–	Larger

Standard spiral wound membranes are of size 8 inch in diameter and 40 inch in length (8 × 40). Smaller size membranes (4 × 40, 2.5 × 40) are also available for small scale units. They are spirally wound and packed in to create a single SWRO element, where a feed spacer is used as separation between different sheets of membrane. The spacer could be of different thickness, which shall decide the total membrane area that can be accommodated in a single SWRO element.

Regarding membrane surface area, the early industry standard used to be 33.5 m^2; and presently SWRO membranes with surface area of 37.2 m^2 and 41 m^2 are available by employing advanced manufacturing technology as well as by utilizing thinner spacer, which is usually used for low fouling feed water. Up to 6–8 membrane elements are arranged in series in a single pressure vessel to achieve water recovery in the range of 35–45%, depending on feed water quality. Recently, standard 8 inch spiral wound SWRO membranes with membrane area of 41 m^2 are available in the market with 99.8% salt rejection and 37 m^3/d flow, which is expressed at standard operation conditions of feed, TDS = 32,000 ppm, feed temperature = 25 °C and feed pressure of 55 bar. Membrane manufacturers claim to have produced high flow membrane with 64 m³/d but with lower salt rejection of 99.7% and also produced high salt rejection SWRO membranes with 99.89% but with lower flow of 31 m³/d.

3.2.4.2 Membrane Arrangements

A typical concept of equipment arrangement in desalination plants is based on several identical RO trains; each train includes a high-pressure pump, an energy recovery device and a bank of RO membranes, which are installed in multiple pressure vessels and are arranged in parallel. A desalination plant may consist of single or multiple trains depending on the requirement. These plants usually operate at a recovery of 35–45% depending on feed seawater conditions and are designed to produce specific amount of product water with acceptable product quality that works at fixed water recovery.

3.2.4.3 Membrane Performance

The membrane product flow tends to vary with feed temperature, feed TDS and in addition to membrane behavior, which includes fouling, membrane degradation and compaction. Similarly, salt passage could be affected by feed temperature, feed TDS, and membrane fouling condition. Thus target product flow usually is achieved and maintained by varying the feed pressure, which need to be controlled according to seasonal variation in feed temperature and to some extent to feed water TDS, where the variation is not very significant. Also, the membrane performance could be affected by membrane fouling or degradation or compaction, which affects the feed pressure requirements. Fouling could lead to increased feed pressure requirements, and membrane degradation could lead to increased salt passage in addition to reduced feed pressure. Hence, to differentiate between the variation in feed pressure due to seasonal variation in feed temperature and TDS from that of the membrane performance deterioration due to fouling or degradation, SWRO plants usually perform normalization of the data, which shall help to identify whether the performance variation of membrane is due to fouling or not. Here, the membrane performance, especially the product flow and salt passage, is usually monitored at standard conditions, which is called normalization, where the relevant software is provided by all membrane manufacturers. The normalization is based on standard practice developed by ASTM [32]. Standard conditions include temperature, pressure, feed TDS and recovery.

3.2.4.4 Membrane Fouling

Membrane fouling is one of the major challenges faced by SWRO plants, which prevent the plant to operate at the desired performance. That makes it difficult to produce sufficient quantity of water with acceptable quality at target specific energy

Fig. 3.7 Digital photographs of fouled HFF SWRO membrane

Fig. 3.8 Digital photographs of fouled spiral wound SWRO membrane

consumption. Fouling is defined as an undesirable deposition of external matter on the membrane surface, which shall result in increase of feed pressure and correspondingly the specific energy consumption and shall lower salt rejection resulting in inferior water quality. Photographs of fouled SWRO membranes of both HFF and spiral wound configurations, which are cut open, are shown in Figs. 3.7 and 3.8, respectively.

Fouling could be of biological in nature or organic or inorganic or a combination of thereof. Biological fouling is due to the attachment of living microorganisms on the membrane surface, which shall ultimately lead to increased feed pressure as well as differential pressure across the membrane. This can be tackled by adopting proper pretreatment and different approaches are adopted by the industry to tackle the same that include use of different types of disinfectants. Similarly, organic and inorganic fouling could be prevented by proper design of pretreatment and intake system. Even after having adopted excellent pretreatment scheme, with long term operation, there could some built up of foulants on the membrane surface, which is inevitable. This leads to increased feed pressure requirement, or reduce salt passage or increased differential pressure across the membrane. Usually a 15% variation in any of these parameters warrants chemical cleaning of membrane, which shall in many cases restore the plant performance. If chemical cleaning fails to restore the plant performance, partial replacement shall be carried to achieve the same. Usually, an annual replacement of 10–15% of membrane is carried out to maintain the membrane performance, which shall mainly depend on the pretreated water quality. Moreover, membranes, which are in operation for more than 5–7 years need to be replaced as its performance could deteriorate with time. In addition, utilizing low fouling RO membranes could also lead to sustainable performance with low frequency of cleaning and membrane replacement. That means production of required quantity of water of acceptable quality at the lowest specific energy consumption. Hence, for the best performance, one has to select membranes that are low energy, low fouling with high salt rejection and membrane manufacturers are working towards achieving the same.

3.3   Recent Developments

During recent years, a lot of improvements have been made in the RO process, which are reflected in the dramatic reduction of both capital and operation costs. Most of the progress has been made through improvements in membranes themselves. There has been a gradual increase in the RO train size reaching 25,100 m^3/d, and the world largest seawater RO plant has a production capacity of 624,000 m^3/d. The designers of desalination systems have tried to reduce the cost of the water by increasing the size of high-pressure pumps, since larger pumps are more efficient and at the same time, less expensive. However, enlargement, which is beneficial for the pump is detrimental to the membrane bank. Although large banks require less equipment and instruments, it has low availability. Each RO bank includes thousands of O-ring seals and plastic elements working under high pressure, and each O-ring seal malfunction could lead to a stoppage of the RO bank. Smaller banks increase plant availability by eliminating the need to stop all the membranes in the plant to replace one O-ring. Hence the concept of several identical RO trains was changed to a Three-Center Design, where it is claimed that HPPs, ERDs and membrane banks operate independently, flexibly and efficiently. The three center approach which is applied at Ashkelon plant [33] by utilizing what they claimed to be the best (optimum) capacity for all the three components, where a group of optimum sized HPPs, a group of optimum sized RO banks and a group of optimum sized ERDs are connected to each other through a common header and by this approach it is claimed that an optimum system can be created. In this design, the performance of all the components could lead to a reduced water cost.

The main array configuration in case of spiral SWRO membranes is the traditional six membrane elements in a single pressure vessel. However, recently there are many SWRO plants with seven elements and a few with eight elements per pressure system are being used. It is claimed that this approach could lead to a reduction in capital cost by 25% with same flux and recovery as that of six element pressure vessels. However, a slight increase in energy is expected due to pressure loss in the system resulting from longer array.

Another issue with SWRO plants is that many plants utilize a second pass brackish water reverse osmosis (BWRO) system to meet the demand of the TDS as well as the boron content in the final product. For this reason, instead of sending entire product from first-pass SWRO, it has been in practice to split the permeate into two portions, one exiting from feed side and other exiting from brine side, and to pass only the permeate exiting from brine side to second pass BWRO. In this way, the good quality permeate exiting from the feed side with low salinity is directly utilized and the only portion exiting from the brine side send as feed to second pass BWRO which was found to be economical.

3.4 Conclusions and Outlook

SWRO desalination plants performance has been discussed with emphasis on four major components, namely seawater intake, pretreatment, high-pressure pumping and SWRO membrane system. Overall performance of entire SWRO plant depends on each of these four components. Different types of approaches presently adopted by the plants to enhance the plant performance as well as various performance monitoring parameters employed are described in this chapter. Moreover, various challenges faced by each of the system components and the measures taken to overcome the same are highlighted.

Over the past few years, there has been a continuous effort to improve the SWRO desalination plant performance, which was mainly directed to reduce water production cost by reducing energy consumption utilizing advanced ERDs and pumps, advanced low energy RO membranes and system configuration which enable to operate the RO plant at higher recovery. In addition, efforts are directed to tackle membrane fouling by adopting fouling resistant membranes as well as smarter plant operation scheme. Moreover, in order to make SWRO desalination sustainable, plants are being operated with minimum amount of additive chemicals and also working towards zero liquid discharge (ZLD). This trend is expected to continue in the future to establish SWRO desalination as a reliable alternative for many coastal communities throughout the world.

References

1. N. Voutchkov, Energy use for membrane seawater desalination: Current status and trends. Desalination **431**, 2–14 (2018)
2. G. Amy, N. Ghaffour, L. Zhenyu, L. Francis, R. Linares, T. Missimer, S. Lattemann, Membrane-based seawater desalination: Present and future prospects. Desalination **401**, 16–21 (2017)
3. S. Loeb, The Loeb-Sourirajan membrane: How it came about. *Synthetic membranes,* Chapter 1, ACS Symposium Series, 153, (American Chemical Society, Washington, 1981). pp. 1–9, https://doi.org/10.1021/bk-1981-0153.ch001
4. J. Glater, The early history of reverse osmosis membrane development. Desalination **117**, 297–309 (1998)
5. Z. Amjad, *Reverse osmosis: Membrane technology, water chemistry, and industrial applications* (Chapman & Hall, International Thomson Publishing, New York, 1993). ISBN: 978-0442239640
6. B. Van der Bruggen, C. Vandecasteele, Distillation vs. membrane filtration: Overview of process evolutions in seawater desalination. Desalination **143**, 207–218 (2002)
7. Watereuse Association, *Overview of desalination plant intakes alternatives.* Watereuse Association White Paper (2011), https://watereuse.org/wp-content/uploads/2015/10/Intake_White_Paper.pdf
8. T.M. Missimer, N. Ghaffour, A.H.A. Dehwah, R. Rachman, R.G. Maliva, G. Amy, Subsurface intakes for seawater reverse osmosis facilities: Capacity limitation, water quality improvement, and economics. Desalination **322**, 37–51 (2013)
9. D. Gille, Seawater intakes for desalination plants. Desalination **156**, 249–256 (2003)
10. T.M. Missimer, R.G. Maliva, Environmental issues in seawater reverse osmosis desalination: Intakes and outfalls. Desalination **434**, 198–215 (2018)
11. A.H.A. Dehwah, T.M. Missimer, Seabed gallery intakes: Investigation of water pretreatment effectiveness of the active layer using a long-term column experiment. Water Res. **121**, 95–108 (2017)
12. T.M. Missimer, B. Jones, R.G. Maliva, *Intakes and outfalls for seawater reverse osmosis desalination facilities: Innovations and environmental impacts* (Springer, New York, 2015). ISBN: 978-3-319-38429-0
13. A.D. Khawaji, I.K. Kutubkhanah, J.M. Wie, Advances in seawater desalination technologies. Desalination **221**, 47–69 (2008)
14. V. Bonnelye, M.A. Sanz, J.P. Durand, L. Plasse, F. Gueguen, P. Mazounie, Reverse osmosis on open intake seawater: Pre-treatment strategy. Desalination **167**, 191–200 (2004)
15. C. Fritzmann, J. Lowenberg, T. Wintgens, T. Melin, State of-the-art of reverse osmosis desalination. Desalination **216**, 1–76 (2007)
16. L.F. Greenlee, D.F. Lawler, B.D. Freeman, B. Marrot, P. Moulin, Reverse osmosis desalination: Water sources, technology, and today's challenges. Water Res. **43**, 2317–2348 (2009)
17. J.C. Schippers, J. Verdouw, Modified fouling index, a method of determining the fouling characteristics of water. Desalination **32**, 137–148 (1980)
18. S.F.E. Boerlage, M. Kennedy, M.P. Aniye, J.C. Schippers, Applications of the MFI-UF to measure and predict particulate fouling in RO systems. J. Membr. Sci **220**, 97–116 (2003)
19. S. Lattemann, T. Höpner, Environmental impact and impact assessment of seawater desalination. Desalination **220**, 1–15 (2008)
20. S. Miller, H. Shemer, R. Semiat, Energy and environmental issues in desalination. Desalination **366**, 2–8 (2015)
21. A.M. Farooque, A. A. Alhajouri, E.M. Idris, A.S. Al Amoudi, Efforts towards chemical free operation: Two years experience at Umlujj SWRO desalination plant. Proceedings of the IDA world congress desalination water reuse, Dubai, 20–24 October 2019

22. L.R. Henthrone, E.R. Jankel, Analysis of MF and UF applications for RO pretreatment and applicability to the Arabian Gulf. Proceedings of the IDA world congress desalination water reuse, March 2002
23. J. Mallevialle, P.E. Odendaal, M.R. Wiesner, *Water treatment membrane processes* (McGraw-Hill, New York, 1996). ISBN: 9780070015593
24. V. Bonnélye, L. Guey, J. Del Castillo, UF/MF as RO pre-treatment: The real benefit. Desalination **222**, 59–65 (2008)
25. M. Badruzzaman, N. Voutchkov, L. Weinrich, J.G. Jacangelo, Selection of pretreatment technologies for seawater reverse osmosis plants: A review. Desalination **449**, 78–91 (2019)
26. A.M. Farooque, A.T.M. Jamaluddin, A.R. Al-Reweli, P.A.M. Jalaluddin, S.M. Al-Marwani, A.A. Al-Mobayed, A.H. Qasim, Parametric analyses of energy consumption and losses in SWCC SWRO plants utilizing energy recovery devices. Desalination **219**, 137–159 (2008)
27. E. Kundig, J.M. Linerio, Advancements and improvements on power recovery turbines, Proceedings of the IDA world congress desalination water reuse, Madrid, 6–9 October 1997, pp. 469–478
28. A. Subramani, M. Badruzzaman, J. Oppenheimer, J.G. Jacangelo, Energy minimization strategies and renewable energy utilization for desalination: A review. Water Res. **45**, 1907–1920 (2011)
29. https://ricklyhydrosystems.com/micro-mini-hydro-systems/turbines/pelton-wheel-turbine/
30. http://www.energyrecovery.com/water/energy-recovery-desalination/
31. https://www.flowserve.com/en/products/pumps/specialty-products/energy-recovery-device/energy-recovery-device-dweer
32. ASTM, *Standard practice for standardizing reverse osmosis performance data, D 4516–00 American Society for Testing and Materials* (ASTM International, West Conshohocken, 2010)
33. M. Taub, The world's largest SWRO desalination plant, 15 months of operational experience. Proceedings of the IDA world congress desalination water reuse, Maspalomas, Gran Canaria, 21–26 October 2007

Chapter 4
Advancements in Unconventional Seawater Desalination Technologies

Hasan Al Abdulgader and Sayeed Rushd

4.1 Introduction

According to the United Nation (UN) and the United States Census Bureau (USCB), the world population has exceeded 7 billion and is expected to reach 10 billions by 2050. Unfortunately, the dramatic increase in the population is met with diminishing resources of clean potable water; and high water consumption, mostly from developing countries [1, 2]. Additionally, the high oil prices over the past 15 years instigated increased attention for the advancement of more efficient desalination technologies [3–11].

The current conventional desalination processes (e.g., thermal distillation and RO) require a considerable amount of energy. RO, for example, demands a high hydraulic pressure to desalinate water that has high salinity. Additionally, commercial RO membranes are limited to seawater salinity. Brine and produced water with salinity higher than 70,000 ppm is currently being treated via thermal based and energy intensive technologies. As a result, extensive research and development during the last two decades have been directed for innovative desalination technologies and processes. Many novel hybrid desalination processes have been proposed while few showed promising results [12, 13]. These hybrid processes entail the integration of well-developed technologies to provide a more effective way for desalinating saline water. More details on these hybrid processes can be found elsewhere [14–18].

H. Al Abdulgader (✉)
Research & Development Center, Saudi Aramco, Dhahran, Saudi Arabia
e-mail: hasan.alabdulgader@aramco.com

S. Rushd
Department of Chemical Engineering, College of Engineering, King Faisal University, Al Ahsa, Saudi Arabia

© Springer Nature Switzerland AG 2020
V. S. Saji et al. (eds.), *Corrosion and Fouling Control in Desalination Industry*,
https://doi.org/10.1007/978-3-030-34284-5_4

This chapter focuses on promising and emerging desalination technologies, namely membrane distillation (MD), forward osmosis (FO), adsorption desalination (AD) and freeze desalination (FD). Principles, system components, challenges and opportunities of these desalination systems are explained.

4.2 Membrane Distillation (MD)

4.2.1 Principles

Membrane distillation (MD) technology was first introduced in late 1960 [19, 20]. However, the technology was not realized commercially until a few decades later due to the lack of suitable membrane and the economic limitations compared to conventional desalination technologies (e.g. MSF, MED, RO) [21]. Nevertheless, MD has numerous advantages over conventional desalination technologies. For example, unlike RO, one of the key features of MD technology is its ability to desalinate independent of feed TDS. RO, on the other hand, is limited to operation pressure of around 70 bar, which means it is unable to desalinate high salinity feed water (i.e. seawater brine, produced water, etc.). MD can handle elevated levels of feed salinity and often able to remove almost all dissolved salt in water in one-step. Furthermore, MD is less prone to organic and colloidal fouling compared with RO process.

MD is a separation technology driven by vapor pressure difference across the membrane (see Fig. 4.1). The driving force is created through a heating source that can maintain a temperature difference between the two sides of the membrane. The microporous and hydrophobic membrane is intended to allow only water vapor to pass through the membrane. MD can operate at atmospheric pressure and temperatures much lower than the boiling point of water and, therefore, provides the opportunity to use waste heat or low-grade energy. The main desired characteristics of MD membranes include low thermal conductivity, high porosity (i.e. large pore density), high hydrophobicity and low membrane thickness.

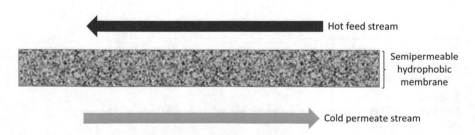

Fig. 4.1 Schematic diagram of a simple MD process

4.2.2   MD Membranes

Hydrophobic microfiltration (MF) membranes have been widely used as MD membranes. The most common MF membrane material used for MD include polypropylene (PP), polytetrafluroethylene (PTFE), and polyvinylidene fluoride (PVDF). The typical porosity, pore size and membrane thickness are 0.3–0.8%, 0.1–0.6 μm and 10–60 μm, respectively [22, 23]. To avoid wetting, MD membranes should have high liquid entry pressure (LEP). LEP is the pressure needed to force the liquid to penetrate the pores of the membrane and passes through to the other side. Membranes with high LEP are favorable in MD applications.

The membranes modules used in the MD system can be flat sheet, tubular or hollow fiber. Extensive research was done using plate and frame flat sheet membrane [24–29]. The main advantage of this configuration is that it is easy to clean and replace. However, the fact that it has a low packing density and requires a membrane support has limited its use beyond laboratory scale applications. Similarly, the tubular membrane has low packing density and high operating cost. Nevertheless, tubular can be attractive in certain cases because it is easy to clean, has a higher effective area, and has a higher resistance to fouling [30–32]. In the hollow fiber module, thousands of tiny hollow fibers are bundled and sealed inside a shell tube. This allows for a very high packing density, which makes this configuration a favorable commercial choice especially for the treatment of low fouling feed [33–37]. One key limitation of the hollow fiber configuration in MD is its relative low flux compared to tubular and flat sheet modules. The low flux is attributed to the poor flow dynamics that usually result in high temperature polarization [38–40].

4.2.3   MD Configurations

Different MD system configurations have been proposed in the last few decades. Table 4.1 compares the main four configurations highlighting their attributes and limitations [29, 41–50]. Direct contact membrane distillation (DCMD) is the most commonly used system configuration for lab scale research due to its simplicity where the condensation of water vapor is carried out inside the membrane module (see Fig. 4.2). Air gap membrane distillation (AGMD) is more widely used in commercial application due to its high-energy efficiency compared with other configuration. Both sweeping gas membrane distillation (SGMD) and vacuum membrane distillation (VMD) are suitable for treating feed that contains volatiles.

4.2.4   Challenges and Opportunities

The main challenges facing MD include wetting, low flux, scaling, heat loss across the membrane, and temperature polarization [52–57]. Scaling often occurs when the system operates at high recoveries. Scale layers can aggravate temperature polarization

Table 4.1 Comparison between the four well tested MD system configurations [29, 41–50]

System Configuration	Direct contact membrane distillation (DCMD)	Air gap membrane distillation (AGMD)	Sweeping gas membrane distillation (SGMD)	Vacuum membrane distillation (VMD)
Principle	The hot feed is passed on one side of a porous hydrophobic membrane while a cold pure water is flowing on the other side.	A cold condensing surface near the permeate side of the membrane condenses the passing water vapor directly. The air gap functions as an insulation layer.	Stripping gas is used to carry the water vapor in the permeate side to be condensed in an external condenser.	The permeate side is vacuum which allows water vapor to pass through the membrane and condenses elsewhere.
Advantages	– Simplest operation – Requires least equipment – Doesn't require an external condenser	– Highest energy efficiency – Latent heat can be recovered – Less complex than SGMD and VMD	– Higher mass transfer rates than AGMD – Suitable when volatiles are present in the feed	– Suitable when volatiles are present in the feed – Gives highest driving force compared to others at the same temperature
Disadvantages	– Highest heat conduction loss – Energy inefficient	– Lower mass transfer rates – Relatively low flux	– More complex system – Requires additional equipment (i.e. external condenser and air blower)	– More complex system – Higher capital cost due to the need for additional devices

effect and reduce the active membrane surface for evaporation, which will result in a significant reduction of water flux. Furthermore, scaling on the membrane surface can change the hydrophobicity of the membrane, leading to membrane wetting that will cause decline in product water quality. Membrane wetting can also occur due to unfavorable membrane material, poor operation of the MD system, and organic fouling on the membrane surface. Temperature polarization is a negative effect that occurs when the temperature at the membrane/feed interface is different from the bulk temperature on both sides of the channel. Since the flow in an MD system is typically laminar, the mixing in the channel is far from ideal. Therefore, temperature polarization becomes significant leading to lower actual driving force. This effect can be minimized by improving the hydrodynamics of the MD system and by using membranes that are relatively thick and have low thermal conductivity.

Most of the challenges facing MD can be addressed with suitable membrane material [22, 23]. Currently, the most common membranes used in MD include

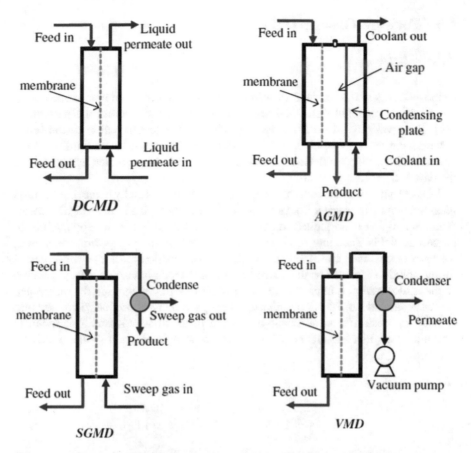

Fig. 4.2 The four main MD configurations. Reproduced with permission from Ref. [51]; Copyright 2011 @ Elsevier

PVDF, polyethersulfone (PES), PP and PTFE. In fact, these membranes were initially made as MF membranes. PVDF and PES are not very hydrophobic, which often lead to partial wetting. PP and PTFE, on the other hand, are hydrophobic but less porous with larger pore size. The cause of the low porosity of PP and PTFE come from its preparation method. These polymers are hardly soluble in common solvents at room temperature. Their preparation usually includes extrusion, followed by stretching. This naturally leads to membranes with larger pores and low porosity. Several new membranes materials have been researched and proposed. The objective is to look for a material that combines the characteristic of PVDF's solubility and PTFE's hydrophobicity. One example of a novel approach is the use of fluorinated polyoxadiazole (F-POD) membrane [58]. This membrane exhibits high hydrophobicity, high porosity and excellent thermal stability. Additionally, initial testing confirmed salt selectivity as high as 99.95% using DCMD configuration. Having said that, further long-term performance evaluation is needed to fully assess F-POD membrane before its commercial utilization in MD application.

4.3   Forward Osmosis (FO)

4.3.1   Principles

Osmosis refers to the transport of water through semipermeable membrane due to the chemical potential difference between the two regions across the membrane [59]. Pure water is driven from the high chemical potential region to the lower potential region. This results in a concentration of the feed side and dilution of the draw solution in the permeate side. FO technology has been studied extensively in the past two decades [4, 9, 60].

   FO is a membrane-based technology that utilizes the naturally occurring osmosis phenomenon to draw water from a saline feed water through a semipermeable membrane to a higher concentrated solution (See Fig. 4.3). It has found applications in numerous fields including food processing, pharmaceuticals, power generation, wastewater treatment and desalination. In a FO desalination process, a concentrated 'draw' solution is placed on the permeate side of the membrane that acts as a source of the osmotic driving force [61–63]. The water that passes through the membrane is then separated from the draw solution to get pure water as product. The literature sometimes uses different terminology to refer to the draw solution. This includes osmotic agent, osmotic engine, osmotic media, sample solution, brine or driving solution.

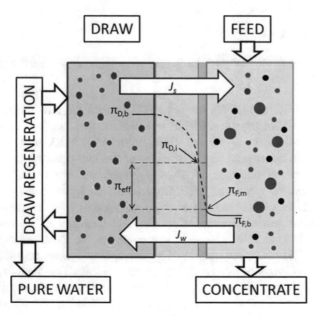

Fig. 4.3 Illustration of the FO process utilizing a draw solution with regeneration step. Reproduced with permission from Ref. [11]. Copyright 2018 @ Elsevier

The main benefits of FO technology are that it eliminates the need to produce desalinated water using high pressures (i.e. RO technology) or high temperature (i.e. MSF, MED, etc.). As a result, the energy requirement is relatively low, while the equipment used in the process is simple and cheap. Unlike RO where hydraulic pressure differential is needed to drive the desalination process, FO uses osmotic pressure differential to drive water across the membrane. Since there is no need for hydraulic pressure, the FO process to be less prone to membrane fouling compared with conventional RO [64–66].

4.3.2 FO System Components

4.3.2.1 Draw Solution

The concentrated solution in the permeate side is used to create the driving force to drive water through the membrane from the feed side to the permeate side. This concentrated solution is commonly referred to as a draw solution. Selection of suitable draw solution is critical for the success of the FO process for seawater desalination. The draw solutions can generally be classified into inorganic and organic solutions (Table 4.2).

4.3.2.2 Membrane Material

The development of enhanced membranes specific for FO application is critical for the advancement of FO technology. Key characteristics of FO membranes include high hydrophilicity, high solute rejection and minimum thickness of the support membrane layer for low internal concentration polarization. The first FO membrane was used in 1965 and was based on natural cellulose [68]. Soon after, cellulose acetate (CA) RO membranes were adapted for FO test trails [63, 83]. However, these membranes exhibit characteristics that make them not ideal for FO system.

Therefore, since the late 1990s researchers and developers have been proposing special materials for FO applications [24, 74, 84, 85]. In general, FO membrane materials can be classified into three main groups: cellulose triacetate (CTA), polybenzimidazole-based (PBI) and thin film composite (TFC) membranes.

The cellulose-based FO membranes have been used extensively by numerous researchers due to their low cost, high mechanical strength, and good chlorine resistance. However, CTA is susceptible to hydrolysis and showed poor resistance to organic and biological fouling. PBI based membrane was proposed by Wang et al. and showed promising results that include good water flux, high salt retention and good chemical and mechanical strength [86, 87]. TFC membranes that were developed especially for FO displayed high flux and excellent selectivity [88–90]. The high flux was helpful to achieve successful minimization of the structural support of the TFC membranes.

Table 4.2 Comparison between selected draw solution

Name	Chemical formula	Advantages	Disadvantages	Ref.
Sulfur dioxide	SO_2	High osmotic pressure	Costly and dangerous recovery method, bad odor and corrosive solution	[67–69]
Aluminum sulfate	$Al_2(SO_4)_3$	Energy-efficient recovery	Costly recovery (consumables), toxic by-products	[70]
Potassium nitrate	KNO_3	No need for regeneration as the diluted solution can be used for fertigation	Increased potential for biofouling, high reverse solute flux	[71]
Sugar	$C_6H_{12}O_6$	Low reverse solute flux, diluted draw solutions can have beneficial uses	Unavailability of efficient recovery method	[63, 72, 73]
Ammonia–carbondioxide	CO_2/NH_3	High osmotic pressure, low molecular weight	Volatile gas, requires heating to recover draw solution	[74, 75]
Magnesium sulfate, Cupper sulfate	$MgSO_4$, $CuSO_4$	Efficient recovery, high performance	High scaling tendency	[76–78]
Functionalized magnetic nanoparticles	PAA-NPs, PNIPAM/T, RI-NPs	High water flux, high reverse solubility	Expensive, high viscosity	[79–81]
Thermosensitive polyelectrolytes	PSSS-PNIPAM, nBu-AEA	Easy recovery, very high osmotic pressure	Expensive, complex synthesis	[69, 82]

4.3.3 Challenges and Opportunities

Despite the above-mentioned benefits of FO compared to conventional RO and thermal desalination processes, FO suffer from scale up challenge due to module size and mass transfer limitations [91]. Furthermore, since FO is a combined separation and mixing process, the theoretical minimal energy for desalination using FO is always higher than RO process [8, 92].

Despite recent advancements of FO technology, there remain few barriers to overcome for successful and wide commercial implementation of this technology. The main challenges of FO technology can be primarily attributed to the lack of effective FO membranes materials and draw solutions.

In terms of membrane material, Ideally, a FO membrane has to have low concentration polarization, high salt retention, high water flux, and strong mechanical and chemical stability. Most of the currently used membranes in the FO process are commercial membranes designed for pressure-driven applications [86, 93–97]. The conventional pressure-driven membranes are asymmetric and can aggravate the concentration polarization effect in the FO process.

Regarding the draw solution challenge, a perfect draw solution would have easy recovery method, very high osmotic pressure and relatively low cost. The solution has to be also completely non-toxic and compatible with the FO membrane [98–101]. Over the last decade, a number of promising draw solutions have been developed [70, 74, 102–106]. Still, none was able to address all the desired characteristics of a draw solution.

Although FO is unlikely to replace RO in the near future as a standalone replacement for seawater or brackish water desalination, hybrid FO process has a promising potential when used to desalinate brine from RO seawater desalination plant. For example, hybrid FO-distillation system has been successfully tested to treat high salinity water [107–110]. In this case, ammonium bicarbonate or ammonium carbon dioxide are dissolved in water to be used as a draw solution with very high osmotic pressure. Once the draw solution becomes diluted after passing through the FO membrane module, it is sent to a distillation column where upon moderate heating the ammonium salts are decomposed into ammonia and carbon dioxide gases. These gases can be recovered and used again to make a concentrated draw solution.

One interesting use of FO technology is for combined power generation and desalination. Chung et al. proposed the use of fresh surface water and seawater to generate power [79]. River water is sent to an osmotic membrane bioreactor where water passes across the submerged membrane into the other side where there is a seawater stream. Part of the pressurized diluted seawater is passed through a pressure exchanger to reuse its energy to increase the pressure of the raw feed seawater. The remaining majority of the pressurized diluted seawater is sent to a turbine to generate electrical power. The diluted seawater is then passed through a FO membrane where a draw solution from the other side of the membrane forces water to pass through the membrane. Finally, fresh water is recovered from the diluted draw solution.

4.4   Adsorption Desalination

4.4.1   Principles

Adsorption Desalination (AD), also known as Adsorption Desalination Cooling (ADC), is comparatively a new desalination technology. The concept of AD or ADC is based on the evaporation of saline water followed by the adsorption and desorption of the vapor using a low temperature waste heat available from sources like solar, geothermal and exhaust energies of industrial processes [111–116]. The evaporation is artificially induced by using an adsorbent inside a vacuum environment. There are two byproducts of this evaporation-adsorption process: (i) chilled water and (ii) brine. Once the adsorbent is saturated with water, the adsorbent bed is heated to around 80 °C to desorb the vapor and, then, the vapor is condensed to produce potable water. The most popular adsorbent used for this kind of desalination is silica gel.

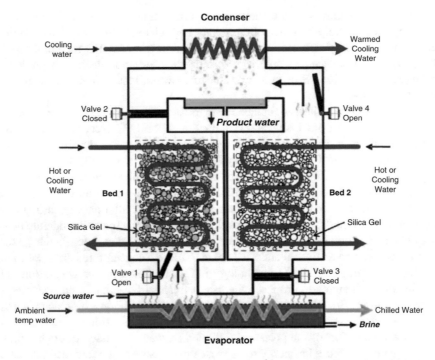

Fig. 4.4 Schematic presentation of AD process. Reproduced with permission from Ref. [114]; Copyright 2012 @ Elsevier

In general, an AD or ADC device is comprised of three major units: (i) evaporator, (ii) adsorbent (silica) bed, and (iii) condenser. The feed (sea or brackish water) is supplied to the evaporator, which is usually a constant pressure vacuum chamber consisting of a heat exchanger. In the course of the vaporization, energy is extracted from the room temperature water supplied to the heat exchanger and thereby a cooling capacity is produced. The evaporation process also produces brine (a concentrated solution of salt) and other contaminants. The vapor produced in the evaporator is transported to be adsorbed in a silica bed. When the adsorbent bed is saturated with water, it is isolated from the evaporator and heated to desorb the water in a batch AD process. The vapor produced in the evaporator unit during the desorption process can be adsorbed in a second similar bed operated in parallel to make the process continuous. The pristine vapor produced during the desorption is transported further to a condenser, where a heat exchanger is used to transform the gaseous vapor to liquid water.

The primary steps of the AD process can be described as follows, see Fig. 4.4, [117, 118]:

(i) Pre-processing

Prior to feeding the saline water stream into the evaporator, it is usually pre-treated. The important steps of the pre-treatment are degassing, filtration and pre-heating.

(ii) Evaporation

Pre-processed saline water is fed into an evaporator. Usually, a vacuum is maintained to facilitate the evaporation process. The heat exchanger inside the evaporation unit is supplied with ambient temperature water to provide the heat required for the vaporization of pure water from the saline water. As a result, the temperature of the water falls down significantly and chilled water comes out of the heat exchanger. That is how a cooling capacity is produced in this step. It can be used for refrigeration.

(iii) Adsorption

The evaporator is connected to an insulated chamber housing a silica bed, so that the produced vapor can be adsorbed. The beads of silica gel are usually placed around a heat exchanger. Cold water is supplied to the heat exchanger to cool the adsorbent bed, as the adsorption process is exothermic. It should be mentioned that silica gel selectively adsorbs only water. It is capable of separating pure water from all other contaminants.

(iv) Desorption

When the adsorbent is saturated, the silica bed is isolated from the evaporator. The cooling water in the heat exchanger is replaced with hot water to increase the temperature of the adsorbent bed. At higher temperature, the water is desorbed from the silica bed as vapor.

(v) Condensation

The chamber housing the adsorbent bed is connected to a condenser while desorbing water. The desorbed vapor is transported to the condenser, where cooling water is supplied through a heat exchanger. The cooling water extracts heat from the vapor to transform it into potable water.

(vi) Product collection and restarting the process

At the completion of the vapor desorption, the adsorbent bed is separated from the condenser and recoupled with the evaporator. The hot water in the bed is replaced with the cold water to cool down the adsorbent silica gel. The cold silica bed then restarts adsorbing evaporated water. If the desalination under consideration operates in batch mode, the products (potable water and brine) should be transported to a larger storage or other processing facility prior to restarting the process. If it is a continuous process, the products need to be pumped regularly for storing/packaging/selling.

The significant advantages of the AD process compared to other desalination technologies like RO, MD, MSF and MED can be identified as follows:

- Environment friendly
- Less energy demanding
- Does not involve any moving part
- Less susceptible to fouling and/or erosion
- Capable of removing not only salt but also biological contaminants from sea/brackish water.

4.4.2 Recent Advancements

Wang and Ng [111] developed a four adsorbent-bed AD plant with the specific water production (SWP) of 4.7 kg per kilogram silica gel at an operating cycle time of 180 s. They found the SWP to be sensitive to both cooling and chilled water temperatures. Increasing heat source temperatures were observed to increase both SWP and coefficient of performance (COP). Thu et al. [112] studied the AD process based on two and four adsorbent-beds. The four-bed system was capable of yielding the daily production of desalinated water or equivalent specific daily water production (SDWP) of 10 m^3 and a performance ratio (PR) of 0.61. Similar values for the two-bed system were 9.0 m^3 and 0.57 (hot water temperature, HWT $\approx$ 85 °C), respectively. Later, Thu et al. [113] suggested an improvement of the process design by recovering heat from the condenser and evaporator. They could obtain a SDWP of 9.24 m^3 and a PR of 0.77 (HWT $\approx$ 70 °C).

Ng et al. [115] proposed a waste heat-driven four-bed ADC system to achieve SDWP of 3.6 m^3 (HWT $\approx$ 85 °C) and a specific cooling power (SCP) of 23 ton of refrigeration (TR) per ton of silica gel. They studied the impact of chilled water temperature on the performance of the process. At a chilled water temperature of 10 °C, SDWP was 8 m^3 and SCP was 51.6 TR per ton of silica gel. Later, Ng et al. [116] presented two separate four-bed pilot-scale AD facilities. The AD plants were operated based on solar energy (KAUST, Saudi Arabia) and electric energy (NUS, Singapore). The plant in KAUST was built with a nominal SDWP of 12.5 m^3 per ton of silica gel. Its SCP was 24 TR. The temperatures of heat source (HWT), cooling water and chilled water were 85 °C, 30 °C and 7 °C, respectively. The NUS plant had a nominal SCP of 5 TR and could be operated in both two and four-bed configurations. The pilot was equipped with heat/mass recovery scheme and high-quality modern apparatuses.

A recent study investigated the hybridization of AD and Humidification Dehumidification (HDH) system to boost the production capability of the desalination process [119]. The proposal was assessed based on two alternative process designs. Both designs were consisted of a humidifier – dehumidifier pair, two adsorbent-beds, a condenser and an evaporator. The proposed AD-HDH hybrid schemes were found to produce economic outputs based on a validated process simulation. However, this kind of hybridization is yet to be implemented in practice.

Ma et al. [117] tried to identify the optimum AD operating conditions using silica gel as the adsorbent in a two-bed system. They could achieve SDWP and PR of 4.69 m^3 and 0.766, respectively, by increasing the temperature of the adsorbing bed from 30.8 °C to 44 °C and decreasing the desorbing bed temperature from 68.2 °C to 55 °C. Similarly, Rezk et al. [120] used a solar driven two-bed ADC system to know the optimum process conditions. It was possible to obtain a 70% increase in both SDWP and SCP with the optimization. The outputs were 6.9 m^3/day/ton desalinated water, 191 W/kg cooling capacity and 0.961 COP. The temperatures of heat source and cooling water were 90 °C and 15 °C, respectively. A swarm-based stochastic optimization method, radial movement optimization (RMO) was used in this study.

Ng et al. [116] modelled an AD system based on detailed coupling of heat and mass transfers (CHMT) across all process units. All balance equations were solved using the method of Gear's backward differentiation formula. FORTRAN PowerStation and IMSL library were used to write the simulation code. The solver utilized the concept of double-precision and a tolerance value of 10^{-6} for the simulation. The input array consisted of 22 variables. Later, Sadri et al. [117] advanced the CHMT modeling approach. The mass and energy balance equations were supplemented with a number of correlations, such as Toth isotherm, linear driving force (LDF) model, and a correlation for the water vapor saturation pressure. The series of non-linear equations were solved using a lumped parameter (LP) model. As the differential equations comprised an initial value problem, Range-Kutta method was employed for the solution. The coding was done with MATLAB using ODE45. The solutions were obtained on the basis of 7 input variables and 13 design parameters. This model was validated using the data available in [116]. The model was cross-checked with a thermodynamic analysis based on exergy calculations. Recently, Mohammed et al. [118] compared the performances of CHMT 1 and LP models. They identified that neither of these modeling approaches are accurate due to the respective limitations of addressing the adsorption dynamics. A modified version of CHMT, which is a combination of traditional CHMT and LP models, was developed as part of this study to improve the prediction accuracy.

The selection of an appropriate adsorbent is a critical step, as the efficiency of an AD process is subject to its characteristics. Although a large number of adsorbents are available, the qualification of an adsorbent for an AD or ADC system depends on the following considerations [121–123]:

- Hydrophilic nature
- Lower regeneration temperature
- Higher adsoption/desorption efficiency
- Latent heat of adsorption should be higher than the sensible heat
- Technical life time should be longer
- Non-toxic
- Non-corrosive
- Low cost and
- Easy availability

Based on the above-mentioned criteria, three adsorbents are used mostly for different AD processes: silica gel, activated alumina and zeolite. The regeneration temperatures of these adsorbents are compared in Table 4.3.

Due to the lowest regeneration temperature and easy availability, silica gel is the most popular AD adsorbent. A detailed analysis of the types and efficiencies of different silica gels are available in [116]. At present, multiple researches are underway to identify the best packing material for the silica gel [124, 125]. It should be mentioned that, even though silica gel is very popular, copper sulfate salt hydrate can also be used as an AD adsorbent [126].

Table 4.3 Regeneration temperatures of commonly used AD adsorbents. Reproduced with permission from Ref. [116]; Copyright 2013 @ Elsevier

Adsorbent	Regeneration temperature (°C)
Silica gel	55–140
Activated alumina	120–260
Zeolite	175–370

4.4.3 Challenges and Opportunities

It is evident from the literature review that the applications of AD process are not as extensively reported as other desalination technologies. The challenges and opportunities for further development of this particular desalination technology to make it more efficient and economical can be discussed as follows:

i) Most of the AD researches reported to date were lab-scale studies. That is, the desalination process is not ready for industrial-scale application. Further researches in scale up, system set up and regulatory/environmental compliance are required to ensure the commercial application of this technology.
ii) The desorption of adsorbed water and the regeneration of adsorbent are not investigated satisfactorily. Methods that are more efficient should be explored to reduce the regeneration temperature significantly.
iii) More large-scale investigations are necessary to figure out the optimum design and process conditions for an AD system.
iv) Although silica gel is considered a convenient adsorbent, comprehensive studies are required to ascertain the most economic and technologically sustainable solution for the AD process. In addition, the exhaustion of the adsorbent should also be studied. A reliable model that can predict the exhaustion time based on the system properties would greatly facilitate the industrial application of the AD technology.
v) One of the major challenges to the wider application of AD process is the unavailability of a reliable model to predict the output and performance. Thorough investigations are required not only to identify the limitations of current models but also to develop novel models capable of producing dependable results.

4.5 Freeze Desalination

4.5.1 Principles

The basis of Freeze Desalination (FD) technology is the fact that crystallization of saline water is a purification process that separates pure water from dissolved salts as ice crystals (Fig. 4.5). The phase of water changes from liquid to solid in the

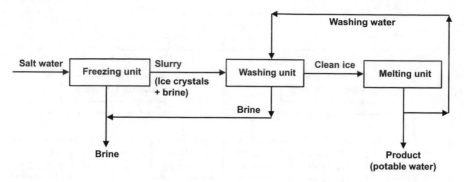

Fig. 4.5 Presentation of the flow diagram demonstrating the fundamental FD process

separation process. Salts and other dissolved solids in saline water behave as impurities during the growth of ice crystals in supersaturated brine solution. These impurities are excluded naturally in the crystal structure. There are many advantages of FD, such as [127–130]:

(a) 75% to 90% lower energy requirement compared to the evaporative desalination processes as the latent heat of ice is 333 kJ/kg and that of water vapor is 2500 kJ/kg
(b) negligible fouling or corrosion
(c) wider options to select materials of construction for process equipment
(d) requirement of no pre-treatment as the formation of ice crystals naturally rejects all impurities
(e) insensitive to feed-water contents or degree of salinity
(f) high removal efficiency of contaminants

The overall FD process is comprised of ice-formation, ice-cleaning and ice-melting. Three basic units, namely freezing unit, washing unit and melting unit, are used in the process. The freezing unit or freezer used in FD is actually a crystallizer. The ice crystals formed in this unit on the supersaturated brine are transferred to the washing unit or washer as a slurry. A minor fraction of the produced water is used in the washer to clean the crystals. The clean crystals are then conveyed to the melting unit or melter. In this unit, the heat removed from the freezer during crystallization is used to melt the brine free ice and produce potable water. A basic FD process is presented in Fig. 4.5 with a flow diagram [129].

As the freezing process is operated below ambient temperature, a heat removing system consisting of the evaporator, vapor compressor and heat exchanger is required to remove the heat continuously. It should be mentioned that the crystallization in the freezing unit could be achieved by adding refrigerant to the brine directly or indirectly. Based on the method of introducing refrigerant, the FD process can be categorized as follows (Fig. 4.6) [1–4]:

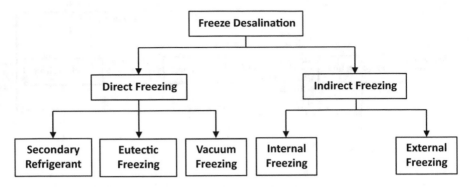

Fig. 4.6 Presentation of different categories of FD

4.5.2 Recent Advancements

Computational fluid dynamics (CFD) was used recently to model the dynamics of crystallization in an indirect FD process [129]. Primarily, the heat transfer phenomena was studied within a 2D rectangular enclosure. The movement of ice was modelled by assuming both brine and ice as two separate fluids. On the other hand, Jayakody et al. [131, 132] conducted a thorough CFD analysis of the crystallization process. They developed a 3D model based on the species transport, solidification/ melting and energy modules in ANSYS Fluent. The CFD model was validated with experimental results obtained using icemaker machine. The validated CFD model was further used to carry out a detailed parametric study.

To make the FD process less complex, Attia [133] proposed an indirect FD process using a vapor-compression heat cycle comprised of compressor, condenser, evaporator and throttling valve. The difluro-monochloromethane (R-22) was used in the cycle as the refrigerant. The most important feature of the proposed process was that it did not require using a complex mechanical system to remove the ice crystals, rather a single unit could be used for forming, washing and melting ice by reversing the flow of R-22 through the heat cycle. An economic analysis demonstrated the cost associated with the proposed FD process to be more that 50% less than other vapor-compression desalination processes. Similarly, Erlbeck et al. [134] developed a single step FD process with a specially fabricated crystallizer or freezing unit. Instead of washing and melting, the ice crystals were pressed to produce water directly. The crystallizer was equipped with a perforated cone and a screwing system for the purpose. The experimental results demonstrated that the proposed technology has the potential to be commercialized.

In an effort to reduce the production cost, Williams et al. [135] tested the application of an icemaker in an external FD process. They were successful in producing ice from different saline waters like sodium chloride solutions, Arabian Gulf seawater and RO brines. The parameters that influence the freezing process were identified as crystallization time, salt concentration of feed and brine flow rate. It was suggested that a modified icemaker can be used as a FD freezing unit. Mtombeni et al. [136] studied a pilot-scale application of a patented FD process (HybridICE™). The indirect cooling was achieved in the crystallizer with the refrigerant, R404a. It could achieve a salt removal efficiency of 96% and produce 1 m³ of potable water at cost of around 1 USD. Shin et al. [137] used a surface scraped freeze crystallizer (SSFC) as the freezing unit to develop a FD process capable of producing water for irrigation. No seeding was required for the crystallization. Ice crystals were artificially induced on the wall and a specially shaped agitator was used to scrap the ice.

Zambrano et al. [138] proposed a novel integration of falling film freeze concentration (FFFC), fractionated thawing (FT) and block freeze concentration (BFC) to design an indirect multi-stage FD process. The product ice crystals of a FFFC unit was fed into a FT unit, while the effluent brine was supplied to another FFFC unit. The outputs of a FT unit were thawed ice and dilute solution of salts occluded in the ice crystals. The solution was recycled back to the FFFC unit and thawed ice was fed into a BFC unit. The effluent of the BFC unit, which is a dilute salt solution, was recycled back to the initial FFFC unit. The output of the BFC unit was pure ice that could be melted into potable water. A combination of five sets of coupled FFFC-FT units and three BFC units were suggested to achieve a desalination efficiency of 98.5%. It should be mentioned that the proposal was supported with small-scale bench-top experiments.

Cao et al. [139] explored the feasibility of using the cooling capacity produced in the course of processing liquefied natural gas (LNG) for the crystallization in an indirect FD system. They proposed using a suitable intermediate refrigerant to transfer the cold energy from the LNG re-gasification. In this study, a flake icemaker was converted into a freezing unit as it could be utilized to produce and remove ice continuously. HYSYS and gPROMS software were used for process simulation and modeling. Later, Chang et al. [140] tested the proposal with a real-time experimental setup. The LNG regasification process was mimicked using Freon gas (R-23). A cylindrical freezing unit containing a stirrer was fabricated with glass or iron. The experimental investigation was used to identify the most important parameter as the washing procedure that might affect the FD process. Other investigated parameters were washing water requirement, intermediate refrigerant temperature, process duration, supercooling and stirring speed. Optimum values of these parameters were determined experimentally. Advancing this lab-scale investigation, Lin et al. [141] produced a prototype FD system by combining the processes of LNG vaporization and seawater crystallization. The intermediate refrigerant for this pilot-scale setup was R410A. A flake icemaker was used as the freezing unit and liquid nitrogen was utilized to simulate LNG. The effects of the parameters like refrigerant evaporating temperature, number of water spraying nozzles and seawater flowrate on the impurity of produced water were investigated. The prototype system could produce fresh water at a rate of 150 L/h with a cold energy efficiency of more than 2 kg fresh water/kg LNG. Recently, Ong and Chen [142] supported the application of a direct FD process for seawater desalination with a thorough economic analysis.

4.5.3 Challenges and Opportunities

The FD process is yet to gain a substantial attention in the research community. The primary reason for this lack of interest is most likely the capital cost and process complexity. A limited number of studies on this desalination technology are being reported annually in the open literature. The challenges and opportunities associated with the further development of FD technology can be discussed as follows [127, 128, 130]:

i) Most of the FD studies reported to date were theoretical proposals. Few of these proposals were supported with small-scale benchtop experiments. Only a handful of pilot-scale FD applications were reported in open literature. Extensive lab-scale investigations are necessary to ensure the scale up for industrial applications.

ii) The most important requirement for the further development of FD technology is an efficient process to separate ice crystals from brine. One of the probable solutions for this separation is Eutectic Freeze Desalination (EFD). In an EFD process, the temperature of brine is decreased to a eutectic point where the crystals of ice and salt are formed simultaneously. Denser salt sinks down and lighter ice floats on the solution. Thereby the salt and ice can be separated in a single step.

iii) Fundamental studies are necessary to clarify the complex process of crystal growth in brines under various process conditions. Validated CFD modeling can be an effective tool for the purpose. This kind of investigations would assist to design novel crystallizer, which could reduce the cost and improve the efficiency of FD process.

iv) Dedicated analyses are necessary about the heat cycle with the objective to reduce energy consumption. The studies in this field should include cooling system, refrigeration technology, integrating hot and cold process streams, and utilization of renewable energy in the FD process.

v) Efforts should be undertaken to integrate FD to other desalination technologies like RO, Multistage Evaporation and AD. This kind of hybridization would enhance the process efficiency by increasing yield.

vi) The technique of ice crystallization has been applied to a considerable extent in food and pharmaceutical industry. The commercial processes applied in these industries should be investigated for their applications in desalination industry.

4.6 Conclusions and Outlook

Desalination of seawater is a viable option to produce potable water. The technology available for desalination can provide extensive amount of water for human usage irrespective of the climate. Further development of this technology is becoming progressively more important to address the water scarcity around the world. This is because the gap between global supply and demand of water is increasing on an annual basis due to the rising population, pollution, industrial development and the climate change. The prominent conventional technologies for extracting pure water from seawater are membrane separation (RO), thermal evaporation (MSF, MEF and VC) and chemical separation (IX and ED). Eventhough it is possible to produce high-quality water with these separation mechanisms, there are challenges to their specific applications like high capital and/or operating cost, and soaring energy requirement. A number of unconventional desalination methodologies, such as MD, FO, AD and FD, have emerged for the purpose of addressing some of the existing limitations of the conventional desalination processes.

The MD is a developing thermal based technology. While the technology can utilize low grade heat to drive the desalination process, it still faces few challenges including low flux and temperature polarization. Further development of this technology requires substantial improvement of system hydrodynamics and membrane materials. The FO is developed based on the natural phenomenon of osmosis to separate the pure water from the seawater through a suitable membrane with a draw solution. This desalination technology, which is a combination of separation and mixing processes, does not require high temperature or pressure. As a result, the potential of membrane fouling is far less than traditional RO process. However, the FO methodology suffers from the shortage of appropriate draw solution and membrane material. It is also challenging to scale up a FO process. The AD technology uses low temperature waste heat to produce water vapor that is adsorbed and, then, desorbed to produce potable water. Although a low energy solution, successful scale up of a laboratory-based AD system is yet to be reported in literature. It is also challenging to model an AD process. In the FD technology, seawater or brine is desalinated by crystallizing water and melting the separated ice crystals into pure water. The complexity of crystal formation and separation is the most challenging part to its commercial application. Comparatively higher capital cost is also an impending block.

The apposite solution for the preemptive challenges is to treat these as opportunities of research and development (R & D). Currently, desalination is primarily used to produce water for municipal and industrial usage in economically developed countries. However, it is expected to be used worldwide in many other sectors, such as, power and irrigation in future [143]. Desalination is likely to be a critical tool to ensure sustainable water supply around the world. Unconventional desalination technologies discussed in this chapter are likely to play a major role in the process. Continual R&D in this sector will lead to the advancement of the desalination technology for its global application. Intensive scientific interests and long-term investments in R&D are necessary to ensure affordable, sustainable and secure supply of water to our human race in the resource-constrained future world.

References

1. I.A. Shiklomanov, World water resources – A new appraisal and assessment for the 21st Century. 1998, https://unesdoc.unesco.org/ark:/48223/pf0000112671
2. D. Seckler, U. Amarasinghe, D. Molden, R. de Silva, R. Barker, World water demand and supply, 1990 to 2025: Scenarios and issues. Colombo, Sri Lanka: International Irrigation Management Institute (IIMI). vi, p. 40 (IWMI Research Report 019 / IIMI Research Report 019), (1998). https://doi.org/10.3910/2009.019
3. T.-S. Chung, S. Zhang, K.Y. Wang, J. Su, M.M. Ling, Forward osmosis processes: Yesterday, today and tomorrow. Desalination 287, 78–81 (2012)
4. S. Zhao, L. Zou, C.Y. Tang, D. Mulcahy, Recent developments in forward osmosis: Opportunities and challenges. J. Membr. Sci. 396, 1–21 (2012)
5. J.-J. Qin, W.C.L. Lay, K.A. Kekre, Recent developments and future challenges of forward osmosis for desalination: A review. Desalin. Water Treat. 39, 123–136 (2012)
6. H. Luo, Q. Wang, T.C. Zhang, T. Tao, A. Zhou, L. Chen, X. Bie, A review on the recovery methods of draw solutes in forward osmosis. J. Water Process Eng 4, 212–223 (2014)
7. K. Lutchmiah, Reclaiming water from wastewater using forward osmosis. Mannheim University of Applied Sciences, (2014) https://doi.org/10.4233/uuid:b9b25f74-5999-43c1-a61d-c36e3062f6f3
8. D.L. Shaffer, J.R. Werber, H. Jaramillo, S. Lin, M. Elimelech, Forward osmosis: Where are we now? Desalination 356, 271–284 (2015)
9. N. Akther, A. Sodiq, A. Giwa, S. Daer, H. Arafat, S. Hasan, Recent advancements in forward osmosis desalination: A review. Chem. Eng. J 281, 502–522 (2015)
10. L. Chekli, S. Phuntsho, J.E. Kim, J. Kim, J.Y. Choi, J.-S. Choi, S. Kim, J.H. Kim, S. Hong, J. Sohn, A comprehensive review of hybrid forward osmosis systems: Performance, applications and future prospects. J. Membr. Sci 497, 430–449 (2016)
11. D.J. Johnson, W.A. Suwaileh, A.W. Mohammed, N. Hilal, Osmotic's potential: An overview of draw solutes for forward osmosis. Desalination 434, 100–120 (2018)
12. N. Hilal, V. Kochkodan, H. Al Abdulgader, D. Johnson, A combined ion exchange–nanofiltration process for water desalination: II. Membrane selection. Desalination 363, 51–57 (2015)
13. A. Venkatesan, P.C. Wankat, Simulation of ion exchange water softening pretreatment for reverse osmosis desalination of brackish water. Desalination 271, 122–131 (2011)
14. O.A. Hamed, Overview of hybrid desalination systems – current status and future prospects. Desalination 186, 207–214 (2005)
15. M.W. Shahzad, K.C. Ng, K. Thu, B.B. Saha, W.G. Chun, Multi effect desalination and adsorption desalination (MEDAD): A hybrid desalination method. Appl. Therm. Eng 72, 289–297 (2014)
16. A.D. Khawaji, I.K. Kutubkhanah, J.-M. Wie, Advances in seawater desalination technologies. Desalination 221, 47–69 (2008)
17. T. Matsuura, Progress in membrane science and technology for seawater desalination – A review. Desalination 134, 47–54 (2001)
18. M.A. Shannon, P.W. Bohn, M. Elimelech, J.G. Georgiadis, B.J. Marinas, A.M. Mayes, Science and technology for water purification in the coming decades, in Nanoscience and technology: A collection of reviews from nature Journals, (World Scientific, 2010), pp. 337–346
19. P.K. Weyl, Recovery of demineralized water from saline waters. US3340186A. (1967)
20. M. Findley, Vaporization through porous membranes. Ind. Eng. Chem. Process Des Dev 6, 226–230 (1967)
21. W. Hanbury, T. Hodgkiess, Membrane distillation-an assessment. Desalination 56, 287–297 (1985)
22. M. Khayet, Membranes and theoretical modeling of membrane distillation: A review. Adv. Colloid Interfac. Sci 164, 56–88 (2011)

23. L. Eykens, K. De Sitter, C. Dotremont, L. Pinoy, B. Van der Bruggen, Membrane synthesis for membrane distillation: A review. Sep. Purif. Technol **182**, 36–51 (2017)
24. S. Hsu, K. Cheng, J.-S. Chiou, Seawater desalination by direct contact membrane distillation. Desalination **143**, 279–287 (2002)
25. P. Termpiyakul, R. Jiraratananon, S. Srisurichan, Heat and mass transfer characteristics of a direct contact membrane distillation process for desalination. Desalination **177**, 133–141 (2005)
26. Y. Yun, R. Ma, W. Zhang, A. Fane, J. Li, Direct contact membrane distillation mechanism for high concentration NaCl solutions. Desalination **188**, 251–262 (2006)
27. S. Srisurichan, R. Jiraratananon, A. Fane, Humic acid fouling in the membrane distillation process. Desalination **174**, 63–72 (2005)
28. R. Schofield, A. Fane, C. Fell, R. Macoun, Factors affecting flux in membrane distillation. Desalination **77**, 279–294 (1990)
29. M.M.A. Shirazi, A. Kargari, M.J.A. Shirazi, Direct contact membrane distillation for seawater desalination. Desalin. Water Treat **49**, 368–375 (2012)
30. S. Cerneaux, I. Strużyńska, W.M. Kujawski, M. Persin, A. Larbot, Comparison of various membrane distillation methods for desalination using hydrophobic ceramic membranes. J. Membr. Sci **337**, 55–60 (2009)
31. A. Larbot, L. Gazagnes, S. Krajewski, M. Bukowska, W. Kujawski, Water desalination using ceramic membrane distillation. Desalination **168**, 367–372 (2004)
32. X. Chen, X. Gao, K. Fu, M. Qiu, F. Xiong, D. Ding, Z. Cui, Z. Wang, Y. Fan, E. Drioli, Tubular hydrophobic ceramic membrane with asymmetric structure for water desalination via vacuum membrane distillation process. Desalination **443**, 212–220 (2018)
33. M. Khayet, M. Essalhi, M. Qtaishat, T. Matsuura, Robust surface modified polyetherimide hollow fiber membrane for long-term desalination by membrane distillation. Desalination **466**, 107–117 (2019)
34. L. García-Fernández, B. Wang, M. García-Payo, K. Li, M. Khayet, Morphological design of alumina hollow fiber membranes for desalination by air gap membrane distillation. Desalination **420**, 226–240 (2017)
35. S.K. Hubadillah, M.H.D. Othman, T. Matsuura, M.A. Rahman, J. Jaafar, A. Ismail, S.Z.M. Amin, Green silica-based ceramic hollow fiber membrane for seawater desalination via direct contact membrane distillation. Sep. Purif. Technol **205**, 22–31 (2018)
36. L. Francis, N. Ghaffour, A.S. Al-Saadi, G. Amy, Performance of different hollow fiber membranes for seawater desalination using membrane distillation. Desalin. Water Treat **55**, 2786–2791 (2015)
37. M. Gryta, Concentration of saline wastewater from the production of heparin. Desalination **129**, 35–44 (2000)
38. S. Bonyadi, T.S. Chung, Flux enhancement in membrane distillation by fabrication of dual layer hydrophilic–hydrophobic hollow fiber membranes. J. Membr. Sci **306**, 134–146 (2007)
39. L.-H. Cheng, P.-C. Wu, J. Chen, Modeling and optimization of hollow fiber DCMD module for desalination. J. Membr. Sci **318**, 154–166 (2008)
40. D. Cheng, W. Gong, N. Li, Response surface modeling and optimization of direct contact membrane distillation for water desalination. Desalination **394**, 108–122 (2016)
41. L. Martinez, F. Florido-Diaz, Theoretical and experimental studies on desalination using membrane distillation. Desalination **139**, 373–379 (2001)
42. J. Phattaranawik, R. Jiraratananon, Direct contact membrane distillation: Effect of mass transfer on heat transfer. J. Membr. Sci **188**, 137–143 (2001)
43. K.W. Lawson, D.R. Lloyd, Membrane distillation. I. Module design and performance evaluation using vacuum membrane distillation. J. Membr. Sci **120**, 111–121 (1996)
44. S. Bandini, C. Gostoli, G. Sarti, Separation efficiency in vacuum membrane distillation. J. Membr. Sci **73**, 217–229 (1992)
45. L. Basini, G. D'Angelo, M. Gobbi, G. Sarti, C. Gostoli, A desalination process through sweeping gas membrane distillation. Desalination **64**, 245–257 (1987)

46. M. Khayet, P. Godino, J.I. Mengual, Theory and experiments on sweeping gas membrane distillation. J. Membr. Sci **165**, 261–272 (2000)
47. M. Chernyshov, G. Meindersma, A. De Haan, Comparison of spacers for temperature polarization reduction in air gap membrane distillation. Desalination **183**, 363–374 (2005)
48. Y. Xu, B.-K. Zhu, Y.-Y. Xu, Pilot test of vacuum membrane distillation for seawater desalination on a ship. Desalination **189**, 165–169 (2006)
49. M.M.A. Shirazi, A. Kargari, D. Bastani, L. Fatehi, Production of drinking water from seawater using membrane distillation (MD) alternative: Direct contact MD and sweeping gas MD approaches. Desalin. Water Treat **52**, 2372–2381 (2014)
50. M. Khayet, C. Cojocaru, Air gap membrane distillation: Desalination, modeling and optimization. Desalination **287**, 138–145 (2012)
51. M. Khayet, T. Matsuura, *Membrane Distillation: Principles and Applications*. (Elsevier, 2011), ISBN: 978-0-444-53126-1
52. R. Schofield, A. Fane, C. Fell, Heat and mass transfer in membrane distillation. J. Membr. Sci **33**, 299–313 (1987)
53. J. Phattaranawik, R. Jiraratananon, A.G. Fane, Heat transport and membrane distillation coefficients in direct contact membrane distillation. J. Membr. Sci **212**, 177–193 (2003)
54. M. Qtaishat, T. Matsuura, B. Kruczek, M. Khayet, Heat and mass transfer analysis in direct contact membrane distillation. Desalination **219**, 272–292 (2008)
55. S. Srisurichan, R. Jiraratananon, A. Fane, Mass transfer mechanisms and transport resistances in direct contact membrane distillation process. J. Membr. Sci **277**, 186–194 (2006)
56. A. Franken, J. Nolten, M. Mulder, D. Bargeman, C. Smolders, Wetting criteria for the applicability of membrane distillation. J. Membr. Sci **33**, 315–328 (1987)
57. A. Alkhudhiri, N. Darwish, N. Hilal, Membrane distillation: A comprehensive review. Desalination **287**, 2–18 (2012)
58. H. Maab, L. Francis, A. Al-Saadi, C. Aubry, N. Ghaffour, G. Amy, S.P. Nunes, Synthesis and fabrication of nanostructured hydrophobic polyazole membranes for low-energy water recovery. J. Membr. Sci **423**, 11–19 (2012)
59. T.Y. Cath, A.E. Childress, M. Elimelech, Forward osmosis: Principles, applications, and recent developments. J. Membr. Sci **281**, 70–87 (2006)
60. M. Qasim, N.A. Darwish, S. Sarp, N. Hilal, Water desalination by forward (direct) osmosis phenomenon: A comprehensive review. Desalination **374**, 47–69 (2015)
61. R. Pattle, Production of electric power by mixing fresh and salt water in the hydroelectric pile. Nature **174**, 660 (1954)
62. R.S. Norman, Water salination: A source of energy. Science **186**, 350–352 (1974)
63. R.E. Kravath, J.A. Davis, Desalination of sea water by direct osmosis. Desalination **16**, 151–155 (1975)
64. J.R. McCutcheon, M. Elimelech, Influence of concentrative and dilutive internal concentration polarization on flux behavior in forward osmosis. J. Membr. Sci **284**, 237–247 (2006)
65. Y. Kim, M. Elimelech, H.K. Shon, S. Hong, Combined organic and colloidal fouling in forward osmosis: Fouling reversibility and the role of applied pressure. J. Membr. Sci **460**, 206–212 (2014)
66. S. Lee, C. Boo, M. Elimelech, S. Hong, Comparison of fouling behavior in Forward Osmosis (FO) and Reverse Osmosis (RO). J. Membr. Sci **365**, 34–39 (2010)
67. D.N. Glew, Process for liquid recovery and solution concentration. US24930763A. (1965)
68. G.W. Batchelder, Process for the demineralization of water. 1965, Google Patents
69. R. Mc Ginnis, Osmotic desalination process. US 8753514, 2009
70. B.S. Frank, Desalination of sea water. US 3670897, 1972
71. R.L. McGinnis, Osmotic desalinization process. US 8753514, 2002
72. J. Kessler, C. Moody, Drinking water from sea water by forward osmosis. Desalination **18**, 297–306 (1976)
73. K. Stache, Apparatus for transforming sea water, brackish water, polluted water or the like into a nutrious drink by means of osmosis. US 4879030, 1989
74. J.R. McCutcheon, R.L. McGinnis, M. Elimelech, A novel ammonia-carbon dioxide forward (direct) osmosis desalination process. Desalination **174**, 1–11 (2005)

75. R.L. McGinnis, M. Elimelech, Global challenges in energy and water supply: The promise of engineered osmosis. Environ. Sci. Technol **42**, 8625–8629 (2008)
76. Z. Liu, H. Bai, J. Lee, D.D. Sun, A low-energy forward osmosis process to produce drinking water. Energy Environ. Sci **4**, 2582–2585 (2011)
77. R. Alnaizy, A. Aidan, M. Qasim, Copper sulfate as draw solute in forward osmosis desalination. J. Environ. Chem. Eng **1**, 424–430 (2013)
78. R. Alnaizy, A. Aidan, M. Qasim, Draw solute recovery by metathesis precipitation in forward osmosis desalination. Desalin. Water Treat **51**, 5516–5525 (2013)
79. M.M. Ling, T.-S. Chung, Surface-dissociated nanoparticle draw solutions in forward osmosis and the regeneration in an integrated electric field and nanofiltration system. Ind. Eng. Chem. Res **51**, 15463–15471 (2012)
80. M.M. Ling, T.-S. Chung, Desalination process using super hydrophilic nanoparticles via forward osmosis integrated with ultrafiltration regeneration. Desalination **278**, 194–202 (2011)
81. M. Mingá Ling, Facile synthesis of thermosensitive magnetic nanoparticles as "smart" draw solutes in forward osmosis. Chem. Commun **47**, 10788–10790 (2011)
82. M. Noh, Y. Mok, S. Lee, H. Kim, S.H. Lee, G.-W. Jin, J.-H. Seo, H. Koo, T.H. Park, Y. Lee, Novel lower critical solution temperature phase transition materials effectively control osmosis by mild temperature changes. Chem. Commun **48**, 3845–3847 (2012)
83. I. Goosens, A. Van Haute, The use of direct osmosis tests as complementary experiments to determine the water and salt permeabilities of reinforced cellulose acetate membranes. Desalination **26**, 299–308 (1978)
84. E. Beaudry, K. Lampi, Membrane technology for direct-osmosis concentration of fruit juices. Food Technol **44**, 121 (1990)
85. E.G. Beaudry, J.R. Herron, Direct osmosis for concentrating wastewater. SAE Trans **106**, 460–466 (1997)
86. K.Y. Wang, T.-S. Chung, J.-J. Qin, Polybenzimidazole (PBI) nanofiltration hollow fiber membranes applied in forward osmosis process. J. Membr. Sci **300**, 6–12 (2007)
87. K.Y. Wang, Q. Yang, T.-S. Chung, R. Rajagopalan, Enhanced forward osmosis from chemically modified polybenzimidazole (PBI) nanofiltration hollow fiber membranes with a thin wall. Chem. Eng. Sci **64**, 1577–1584 (2009)
88. N. Widjojo, T.-S. Chung, M. Weber, C. Maletzko, V. Warzelhan, The role of sulphonated polymer and macrovoid-free structure in the support layer for thin-film composite (TFC) forward osmosis (FO) membranes. J. Membr. Sci **383**, 214–223 (2011)
89. N.-N. Bui, M.L. Lind, E.M. Hoek, J.R. McCutcheon, Electrospun nanofiber supported thin film composite membranes for engineered osmosis. J. Membr. Sci **385**, 10–19 (2011)
90. N.-N. Bui, J.R. McCutcheon, Hydrophilic nanofibers as new supports for thin film composite membranes for engineered osmosis. Environ. Sci. Technol **47**, 1761–1769 (2013)
91. R.W. Field, J.J. Wu, Mass transfer limitations in forward osmosis: Are some potential applications overhyped? Desalination **318**, 118–124 (2013)
92. M. Elimelech, W.A. Phillip, The future of seawater desalination: Energy, technology, and the environment. Science **333**, 712–717 (2011)
93. J.R. McCutcheon, M. Elimelech, Modeling water flux in forward osmosis: Implications for improved membrane design. AICHE J **53**, 1736–1744 (2007)
94. J.R. McCutcheon, M. Elimelech, Influence of membrane support layer hydrophobicity on water flux in osmotically driven membrane processes. J. Membr. Sci **318**, 458–466 (2008)
95. J. Su, Q. Yang, J.F. Teo, T.-S. Chung, Cellulose acetate nanofiltration hollow fiber membranes for forward osmosis processes. J. Membr. Sci **355**, 36–44 (2010)
96. S. Zhang, K.Y. Wang, T.-S. Chung, H. Chen, Y. Jean, G. Amy, Well-constructed cellulose acetate membranes for forward osmosis: Minimized internal concentration polarization with an ultra-thin selective layer. J. Membr. Sci **360**, 522–535 (2010)
97. I. Alsvik, M.-B. Hägg, Pressure retarded osmosis and forward osmosis membranes: Materials and methods. Polymers **5**, 303–327 (2013)
98. T.-S. Chung, X. Li, R.C. Ong, Q. Ge, H. Wang, G. Han, Emerging forward osmosis (FO) technologies and challenges ahead for clean water and clean energy applications. Curr. Opin. Chem. Eng **1**, 246–257 (2012)

99. Q. Ge, M. Ling, T.-S. Chung, Draw solutions for forward osmosis processes: Developments, challenges, and prospects for the future. J. Membr. Sci. **442**, 225–237 (2013)
100. D. Li, H. Wang, Smart draw agents for emerging forward osmosis application. J. Mater. Chem. A **1**, 14049–14060 (2013)
101. L. Chekli, S. Phuntsho, H.K. Shon, S. Vigneswaran, J. Kandasamy, A. Chanan, A review of draw solutes in forward osmosis process and their use in modern applications. Desalin. Water Treat **43**, 167–184 (2012)
102. N.T. Hancock, T.Y. Cath, Solute coupled diffusion in osmotically driven membrane processes. Environ. Sci. Technol **43**, 6769–6775 (2009)
103. C.R. Martinetti, A.E. Childress, T.Y. Cath, High recovery of concentrated RO brines using forward osmosis and membrane distillation. J. Membr. Sci **331**, 31–39 (2009)
104. M.M. Ling, K.Y. Wang, T.-S. Chung, Highly water-soluble magnetic nanoparticles as novel draw solutes in forward osmosis for water reuse. Ind. Eng. Chem. Res **49**, 5869–5876 (2010)
105. Q. Ge, J. Su, T.-S. Chung, G. Amy, Hydrophilic superparamagnetic nanoparticles: Synthesis, characterization, and performance in forward osmosis processes. Ind. Eng. Chem. Res **50**, 382–388 (2010)
106. H. Bai, Z. Liu, D.D. Sun, Highly water soluble and recovered dextran coated Fe3O4 magnetic nanoparticles for brackish water desalination. Sep. Purif. Technol **81**, 392–399 (2011)
107. D.L. Shaffer, L.H. Arias Chavez, M. Ben-Sasson, S. Romero-Vargas Castrillón, N.Y. Yip, M. Elimelech, Desalination and reuse of high-salinity shale gas produced water: Drivers, technologies, and future directions. Environ. Sci. Technol **47**, 9569–9583 (2013)
108. J.R. McCutcheon, R.L. McGinnis, M. Elimelech, Desalination by ammonia–carbon dioxide forward osmosis: Influence of draw and feed solution concentrations on process performance. J. Membr. Sci **278**, 114–123 (2006)
109. R.L. McGinnis, N.T. Hancock, M.S. Nowosielski-Slepowron, G.D. McGurgan, Pilot demonstration of the NH$_3$/CO$_2$ forward osmosis desalination process on high salinity brines. Desalination **312**, 67–74 (2013)
110. B.D. Coday, P. Xu, E.G. Beaudry, J. Herron, K. Lampi, N.T. Hancock, T.Y. Cath, The sweet spot of forward osmosis: Treatment of produced water, drilling wastewater, and other complex and difficult liquid streams. Desalination **333**, 23–35 (2014)
111. X. Wang, K.C. Ng, Experimental investigation of an adsorption desalination plant using low-temperature waste heat. Appl. Therm. Eng **25**, 2780–2789 (2005)
112. K. Thu, K.C. Ng, B.B. Saha, A. Chakraborty, S. Koyama, Operational strategy of adsorption desalination systems. Intl. J. Heat Mass Transfer **52**, 1811–1816 (2009)
113. K. Thu, B.B. Saha, A. Chakraborty, W.G. Chun, K.C. Ng, Study on an advanced adsorption desalination cycle with evaporator–condenser heat recovery circuit. Intl. J. Heat Mass Transfer **54**, 43–51 (2011)
114. J.W. Wu, E.J. Hu, M.J. Biggs, Thermodynamic cycles of adsorption desalination system. Appl. Energy **90**, 316–322 (2012)
115. K.C. Ng, K. Thu, B.B. Saha, A. Chakraborty, Study on a waste heat-driven adsorption cooling cum desalination cycle. Intl. J. Refrig **35**, 685–693 (2012)
116. K.C. Ng, K. Thu, Y. Kim, A. Chakraborty, G. Amy, Adsorption desalination: An emerging low-cost thermal desalination method. Desalination **308**, 161–179 (2013)
117. H. Ma, J. Zhang, C. Liu, X. Lin, Y. Sun, Experimental investigation on an adsorption desalination system with heat and mass recovery between adsorber and desorber beds. Desalination **446**, 42–50 (2018)
118. R.H. Mohammed, O. Mesalhy, M.L. Elsayed, L.C. Chow, Assessment of numerical models in the evaluation of adsorption cooling system performance. Intl. J. Refrig **99**, 166–175 (2019)
119. N.A. Qasem, S.M. Zubair, Performance evaluation of a novel hybrid humidification-dehumidification (air-heated) system with an adsorption desalination system. Desalination **461**, 37–54 (2019)
120. H. Rezk, A.S. Alsaman, M. Al-Dhaifallah, A.A. Askalany, M.A. Abdelkareem, A.M. Nassef, Identifying optimal operating conditions of solar-driven silica gel based adsorption desalination cooling system via modern optimization. Sol. Energy **181**, 475–489 (2019)

121. N. Lior, Advances in water desalination. (Wiley, 2013) ISBN:9780470054598
122. M. Alghoul, M. Sulaiman, B. Azmi, M.A. Wahab, Advances on multi-purpose solar adsorption systems for domestic refrigeration and water heating. Appl. Therm. Eng **27**, 813–822 (2007)
123. A.A. Askalany, M. Salem, I. Ismail, A.H.H. Ali, M. Morsy, A review on adsorption cooling systems with adsorbent carbon. Renew. Sust. Energy Rev **16**, 493–500 (2012)
124. M.M. Younes, I.I. El-sharkawy, A. Kabeel, K. Uddin, A. Pal, S. Mitra, K. Thu, B.B. Saha, Synthesis and characterization of silica gel composite with polymer binders for adsorption cooling applications. Intl. J. Refrig **98**, 161–170 (2019)
125. R.H. Mohammed, O. Mesalhy, M.L. Elsayed, L.C. Chow, Adsorption cooling cycle using silica-gel packed in open-cell aluminum foams. Intl. J. Refrig **104**, 201–212 (2019)
126. E.S. Ali, A.A. Askalany, K. Harby, M.R. Diab, A.S. Alsaman, Adsorption desalination-cooling system employing copper sulfate driven by low grade heat sources. Appl. Therm. Eng **136**, 169–176 (2018)
127. D. Randall, J. Nathoo, A succinct review of the treatment of reverse osmosis brines using freeze crystallization. J. Water Process Eng **8**, 186–194 (2015)
128. P.M. Williams, M. Ahmad, B.S. Connolly, D.L. Oatley-Radcliffe, Technology for freeze concentration in the desalination industry. Desalination **356**, 314–327 (2015)
129. K. El Kadi, I. Janajreh, Desalination by freeze crystallization: An overview. Int. J. Therm. Environ. Eng **15**, 103–110 (2017)
130. B. Kalista, H. Shin, J. Cho, A. Jang, Current development and future prospect review of freeze desalination. Desalination **447**, 167–181 (2018)
131. H. Jayakody, R. Al-Dadah, S. Mahmoud, Computational fluid dynamics investigation on indirect contact freeze desalination. Desalination **420**, 21–33 (2017)
132. H. Jayakody, R. Al-Dadah, S. Mahmoud, Numerical investigation of indirect freeze desalination using an ice maker machine. Energy Convers. Manag **168**, 407–420 (2018)
133. A.A. Attia, New proposed system for freeze water desalination using auto reversed R-22 vapor compression heat pump. Desalination **254**, 179–184 (2010)
134. L. Erlbeck, D. Wössner, T. Kunz, M. Rädle, F.-J. Methner, Investigation of freeze crystallization and ice pressing in a semi-batch process for the development of a novel single-step desalination plant. Desalination **448**, 76–86 (2018)
135. P. Williams, M. Ahmad, B. Connolly, Freeze desalination: An assessment of an ice maker machine for desalting brines. Desalination **308**, 219–224 (2013)
136. T. Mtombeni, J. Maree, C. Zvinowanda, J. Asante, F. Oosthuizen, W. Louw, Evaluation of the performance of a new freeze desalination technology. Intl. J. Environ. Sci. Technol **10**, 545–550 (2013)
137. H. Shin, B. Kalista, S. Jeong, A. Jang, Optimization of simplified freeze desalination with surface scraped freeze crystallizer for producing irrigation water without seeding. Desalination **452**, 68–74 (2019)
138. A. Zambrano, Y. Ruiz, E. Hernández, M. Raventós, F. Moreno, Freeze desalination by the integration of falling film and block freeze-concentration techniques. Desalination **436**, 56–62 (2018)
139. W. Cao, C. Beggs, I.M. Mujtaba, Theoretical approach of freeze seawater desalination on flake ice maker utilizing LNG cold energy. Desalination **355**, 22–32 (2015)
140. J. Chang, J. Zuo, K.-J. Lu, T.-S. Chung, Freeze desalination of seawater using LNG cold energy. Water Res **102**, 282–293 (2016)
141. W. Lin, M. Huang, A. Gu, A seawater freeze desalination prototype system utilizing LNG cold energy. Intl. J. Hydrog. Energy **42**, 18691–18698 (2017)
142. C.-W. Ong, C.-L. Chen, Technical and economic evaluation of seawater freezing desalination using liquefied natural gas. Energy **181**, 429–439 (2019)
143. A. Silber, Y. Israeli, I. Elingold, M. Levi, I. Levkovitch, D. Russo, S. Assouline, Irrigation with desalinated water: A step toward increasing water saving and crop yields. Water Resour. Res **51**, 450–464 (2015)

Part II
Corrosion in Desalination

Chapter 5
Corrosion in Thermal Desalination Processes: Forms and Mitigation Practices

Abdelkader A. Meroufel

5.1 Introduction

Thermal seawater desalination was competing with membrane-based desalination such as Reverse Osmosis (RO) up to the mid-1990s where RO started to dominate globally (Fig. 5.1). However, Gulf Cooperation Council (GCC) still relying on the robust and reliable thermal desalination technologies due to many reasons. This include the subsidized fossil fuel cost, the huge water demand and the challenging water quality for the sensitive RO desalination process.

The description of thermal desalination technologies is covered in the first and second chapters of this book. As one of their major challenges, corrosion attracted a lot of research and engineering interest. This is due to the severe conditions where the desalination industry accumulated an important level of learned lessons and practical experience particularly in the mature brine recycle Multiple-Stage-Flashing (MSF) process with more than 50 years of record. Consequently, the MSF plant design life increased from 25 years to 40 years between 1980 and 2005. Nevertheless, Multiple Effect Desalination (MED) still facing some challenges making its market penetration difficult and corrosion studies very limited. The cost-effectiveness and reliability of these two processes are dependent among material performance in the different process streams and outdoor conditions. Indeed, special care is considered for equipment in desalination plants where double corrosion threats are expected i.e. internal and external. While the external corrosion risks are similar to other marine atmosphere industries, the internal ones are specific in terms of corrosion factors such as temperature, oxygen concentration and flowing conditions.

Although corrosion was pointed as a source of the environmental impact of desalination, this chapter aims to discuss the main corrosion forms and their

A. A. Meroufel (✉)
Desalination Technologies Research Institute, Saline Water Conversion Corporation,
Jubail, Saudi Arabia
e-mail: ameroufel@swcc.gov.sa

© Springer Nature Switzerland AG 2020
V. S. Saji et al. (eds.), *Corrosion and Fouling Control in Desalination Industry*,
https://doi.org/10.1007/978-3-030-34284-5_5

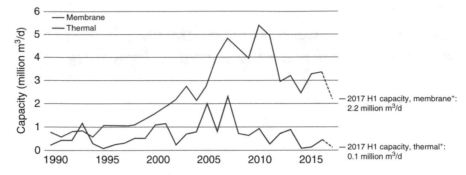

Fig. 5.1 World trend of desalination technologies since 1990. (Source: GWI Desaldata IDA) [1]

engineering mitigation practices that contribute to the limitation of this unwanted phenomenon. The chapter considers two major parts of the process i.e. pretreatment and desalting system. Corrosion in utilities (steam/power plant) and post-treatment (water transmission section) sections are out of the scope of the discussion. Finally, the corrosion in thermal desalination in this chapter goes beyond the practical experience to outline the standardization issues and research niches for future efforts.

5.2   Economics of Equipment Integrity in Thermal Desalination Industry

The economics of corrosion in thermal desalination still an aspect with many uncertainties. Indeed, such data are plant-specific where site conditions, plant design are so variable making the accurate estimation of corrosion cost a challenge. In addition, the absence of standardization of corrosion engineering in the desalination industry contributes to the scattering of the cost figure. However, there is a general agreement on the materials cost at least in terms of Capital Expenditure (CAPEX). Figures 5.2a and 5.2b show the contribution of corrosion-related items for MSF and MED desalination processes for a plant with 450,000 m³/d production capacity [1]. The values presented in the figures should be considered as percentage relative to the water production cost. Then, it is important to notice the close capital cost of materials between MSF and MED despite the difference in process design and conditions.

For operational expenditure (OPEX) cost, the corrosion cost represented through indirect factors such as parts and chemicals is shown in Figs. 5.3a and 5.3b. These contributions interfere with other functionalities such as mechanically failed parts, disinfection and anti-scaling activities. The two chemicals related to the corrosion cost are biocides and oxygen scavenger. While the former is adopted for both processes, oxygen scavenger is present only in MSF desalination. The higher cost of chemicals in MED is usually attributed to the higher dozing of antiscalant.

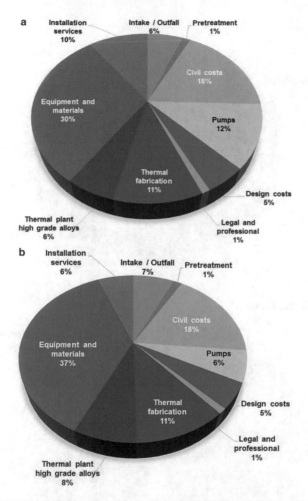

Fig. 5.2 (**a**) CAPEX breakdown of MSF desalination plant, (**b**) CAPEX breakdown of MED desalination plant [1]

Unfortunately, other corrosion cost items are not included in most of the calculations such as corrosion monitoring (inspection/life assessment) and maintenance (preventive or corrective). This is mainly due to the scattering of investment between thermal desalination owners.

Some authors discussed CAPEX and OPEX of thermal desalination processes from a perspective of water production cost [2–4]. They aimed to find opportunities to reduce the CAPEX especially through materials selection and process configuration choice. This kind of exercise fit to the Front End Engineering Design (FEED) phase where corrosion can be tackled early before the plant construction. Materials selection in thermal desalination plants will be discussed later in this chapter. In the following section, the desalination philosophy in terms of equipment reliability will be described briefly especially as tactics.

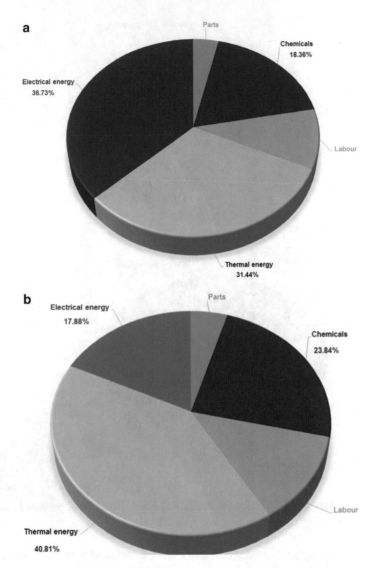

Fig. 5.3 (**a**) OPEX breakdown of MSF desalination plant, (**b**) OPEX breakdown of MED desalination plant [1]

5.3 Reliability Philosophy Within Thermal Desalination Industry

From an engineering point of view, the equipment integrity in the desalination industry can be classified into two groups; static and rotating equipment. The differences in corrosion or failure risks between these two equipment groups define the investment level of the desalination plant owner. For instance, due to the severe operating conditions, rotating equipment are subject to duplication and alternating operation. The main equipment concerned by this practice are the pumps handling the different waters (feed seawater, brine and distilled water).

Unlike oil and gas industry where professional engineering practices are well spread (RBI,[1] predictive maintenance, etc.), desalination still dealing with corrosion challenge based on manufacturer recommendations through two maintenance practices i.e. preventive and corrective maintenance with huge financial and time investments. With experience accumulation, some manufacturer recommendations are modified to optimize the maintenance cost.

Recently, an initiative to move for a condition-based maintenance was raised to reduce the water production cost. However, this asset management tool for high production capacity plants can jump to higher values if this practice is not optimized. In addition, this is not the end of the story where reliability initiatives do have a maturation process. For reliability initiatives with condition-based maintenance, more assessment structures are needed. The advantages are the ability to sense operational changes, the impact on the performance increase and more proactive maintenance strategy [5].

5.4   Corrosion Forms in Thermal Desalination Plants

For simplicity, corrosion forms encountered in thermal desalination plants will be detailed by environment. Under each environment, the discussion are organized by the material type; and the corrosion forms with each material type are explained with supportive evidences from real desalination plants. It is important to notice that, due to their widespread implementation, we are considering MSF desalination plants operating by using high-temperature additives (anti-scalant). Acid dozing MSF plants are out of the scope of this chapter. Similarly, few cases from MED desalination plants are considered because of the limited number of MED plants.

5.4.1   Saline Water Corrosion

Seawater is a common challenge between desalination and other sectors such as oil and gas, power plants, shipping, etc. However, this challenge is more pronounced in desalination where the water is heated and concentrated producing highly saline brine. Depending on operation parameters and the metal/alloy in contact with the seawater, different corrosion impacts can be expected. Corrosion due to acidic saline waters is described in Chap. 10.

---

[1] Risk Based Inspection

Despite the high number of published results on marine corrosion, failures still occur [6]. From general to localized corrosion, the spectrum is wide when the equipment is fully immersed in this aggressive medium. Through desalination industry history, many authors studied and updated the knowledge about the different saline corrosion forms that can be expected in thermal desalination plants [7–13]. Indeed, Giuliani et al. [8], summarized the reasons of the unexpected desalination corrosion failures including corrosive transitory conditions, and malfunctioning of control instrumentation. Due to the variety of used alloys, the information on thermal desalination corrosion in the literature is highly dispersed. In the present section, an attempt is made to organize and summarize this information. Starting from the industry workhorse material which is the carbon steel, desalination materials moved toward corrosion resistant alloys (CRAs) seeking maintenance-free plant.

Table 5.1 summarizes the expected corrosion forms of metals/alloys in saline waters and their accelerating factors. The factors affecting the corrosion of materials in saline environments are well documented [14, 15].

In the following, corrosion risks will be discussed by the material type including for MSF and MED desalination processes and considering the two main types of corrosion (general and localized).

Despite the variation in seawater characteristics (salinity, temperature, microbial activity) around the world, differences in the general corrosion rate of submerged steel in seawater were found relatively small. This was well-explained by a compensation relation between controlling factors as discussed by LaQue [16]. For instance, the temperature increase induces a decrease of dissolved oxygen concentration, increase of its diffusion and water salinity, and enhance the development of protective surface products such as inorganic (calcium/magnesium-based) deposits. In the end, it was found that steel corrosion when immersed continuously in seawater

Table 5.1 Alloys used in thermal desalination and their corrosion risks in saline waters

Material	Corrosion risk	Critical factors
C-Steels	General	1,2,3,7
Ductile iron: ferritic ASTM A536	General	1,2,3,7
Ductile iron: austenitic Ni-resist D2/D2W	SCC[a]	1,2,8
SS: austenitic 304 / 316 L	Pitting/crevice	1,2,3,4,6
SS: Austenitic Nitronic 50	Pitting/Crevice	1,2,3,4,6
SS: duplex CD4MCu	Crevice	1,2,3,4,6
SS: duplex 2205	Crevice	1,2,3,4,6
CuNi: 90/10, 70/30 and modified 70/30[b]	VSC[c]- Pitting	1,3,7
Al Brass/AlBr / NiAlBr	Erosion-corrosion	1,2,3,5
NiCu: Monel 400/500	MIC[d]	1,3,4
Titanium Gr.2	Crevice	1,3,6

1: Temperature/2: Salinity/3: Oxidants/4: Biofouling/5: Velocity/6: Crevice/7: pH/8: Stress
[a]Stress corrosion cracking
[b]66Cu30Ni2Fe2Mn
[c]Vapor Side Corrosion
[d]Microbiological induced corrosion

Fig. 5.4 General corrosion of MSF flash chamber bottom after 35 years of operation

tends to decline with time [17]. In thermal MSF desalination plants, the actual carbon steel corrosion rate in the submerged area for different MSF stages was estimated based on thickness loss measurements and found less than 0.1 mm/yr. after 35 years of operation at TBT[2] that does not exceed 98 °C (Fig. 5.4). This could be explained by a combination of factors including the formation of protective calcareous deposit, reduced dissolved oxygen concentration and closed evaporator system. One of the consequences of this very low corrosion rate is the necessity to revise the corrosion allowance during the design phase of MSF evaporators.

For localized corrosion of carbon steels in saline waters, the common cause is the non-uniformity of surface products (partial surface coverage) causing a differential cell. In a similar way to the evolution of general corrosion, the localized attack also decreases with time as found by many authors [16, 18]. Although passivity is never reached in flowing seawater for carbon steels due to the high chlorides concentration, corrosion rate get stabilized beyond a critical velocity of seawater [19]. However, erosion-corrosion can proceed above critical velocity or even at low velocity if the area suffers from a flow disturbance as occurred on vertical non-cladded carbon steel pillars shown in Fig. 5.5. Later on, MSF designer specified the necessity to either clad carbon steel or adopt solid stainless steel to resist to this kind of attack.

With their carbon content in excess to 1.7 wt.% and presence of more corrosion resistant phase (graphite) in their microstructure, cast iron has some differences in seawater corrosion behavior compared to conventional carbon steels. Cast iron alloys are present in the list of thermal desalination materials mainly in an alloyed form such as austenitic ductile iron called Ni-resist D2W grade. This grade is well-recognized for its superior corrosion and erosion-corrosion resistances in seawater compared to the normal and low alloy cast irons [20, 21]. In terms of localized

---

[2] Top Brine Temperature

Fig. 5.5 Severe erosion-corrosion of carbon steel in MSF flash chamber structure

Fig. 5.6 SCC of Ni-resist D2 in brine recycle pump column

corrosion, Ni-resist seldom has pitting, crevice or stress corrosion cracking (SCC) [21]. However, SCC failures were reported in the early 1990s for this alloy as brine recycle pump component within MSF desalination plant (Fig. 5.6) [22]. Microscopic observation revealed the typical SCC cracking mode as shown in Fig. 5.7. These failures were attributed to manufacturing residual stresses combined with brine salinity circulating at temperature reaching 40 °C. The absence of residual stresses for this alloy in some other desalination plant locations was explained by a post-heating treatment conducted by the manufacturer; which raises the material capital cost.

The next family of materials in thermal desalination is the stainless steel (SS) with three metallurgies i.e. austenitic, duplex and martensitic. Stainless steels are known to be immune to general corrosion in saline waters with a corrosion rate far

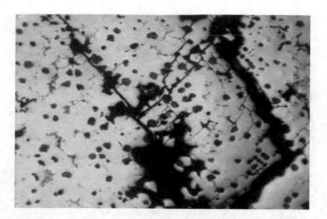

Fig. 5.7 Transgranular cracking revealed on Ni-resist D2 failed pump column (×50)

lower than 0.1 mm/yr. However, they are prone especially to localized corrosion in saline waters such as pitting, crevice, and environmental assisted cracking (EAC) depending on critical alloying elements, metallurgical features and association of extrinsic and intrinsic stresses.

In terms of critical alloying elements, lot of works were conducted to illustrate the beneficial role of chromium (Cr), molybdenum (Mo), nitrogen (N), copper (Cu) and tungsten (W). Then, an empirical formula was introduced to rank the different stainless steels in terms of localized corrosion susceptibility and degree i.e. Pitting Resistance Equivalent Number (PREN) [23]. Different formulas of PREN are cited in the literature that are function of the stainless steel composition, microstructure and the experimental findings. Bauernfeind et al. [24] reviewed these formulas and concluded that the most used formulas for standard austenitic stainless steels are $PREN_{20N}$ and $PREN_{30N}$. Due to the similarity between pitting and crevice corrosion mechanisms, the same formula is used to rank stainless steel alloys in terms of crevice corrosion resistance as well.

While stainless steel manufacturers usually present their developed products after conducting standard tests for pitting (ASTM G46-methods A and E) and crevice (ASTM G46-methods B and F), long-term tests in seawater conditions are the most reliable information for desalination plant owners. Indeed, laboratory standard tests are in most of the time over-conservative and difficult to be correlated with real conditions for certain important factors. Cases of pitting corrosion on stainless steels in thermal desalination are very rare in the open literature. This could be due to the formation of tightly adherent calcareous deposit protecting the underneath substrate (Fig. 5.8). Nevertheless, one case study published by Malik et al. [25] concerns 316 L cladding of MSF flash chamber bottom as shown in Fig. 5.9. This unusual failure initiated during a long period of shut-down combined with incomplete draining of the unit. In a similar way, Elshawesh et al. [26] discussed a case of multiple pitting of MSF flash chamber walls made of 316 L after 3 months of continuous operation. The corrosion attack was confined to the mechanical

Fig. 5.8 Calcareous deposit covering most of the 1st stage flash chamber in MSF desalination plant

Fig. 5.9 Pitting corrosion on MSF flash chamber bottom cladded by stainless steel 316 L

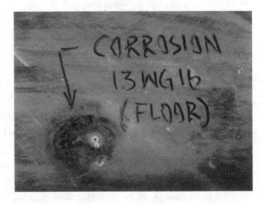

damaged and iron contaminated area. The failure to keep the oxygen level below the requested limit accelerated the localized pitting attack. Such kind of cases can be avoided by some pre-service practices centered on the quality of the material discussed by Giuliani [27].

Many authors studied the effect of metallurgical features of stainless steel grades on the three considered degradation modes. Indeed, the presence of inclusions, residual stresses, and the partition of elements (for duplex stainless steels) are the main factors that enhance the stainless steel susceptibility to localized degradation [28–30]. The role of MnS inclusion was more debated for pitting and crevice corrosion where the negative effect of Mn was included in the PREN equation proposed by Bauernfeind et al. [24]. Similarly, the precipitation of undesirable phases in duplex stainless steels such as sigma phase was found to decrease both localized corrosion and mechanical resistance of these alloys [29].

On the other hand, the manufacturing process of the part seems to affect the metallurgy of the material inducing variation in the corrosion resistance. Many authors observed that cast stainless steels suffered from crevice corrosion in seawater to a higher degree compared to the forged ones [31, 32]. This can be explained by the better uniformity of forged alloys in terms of composition and structure providing fewer sites for crevice corrosion attack.

The history of crevice corrosion of stainless steels in thermal desalination does not include significant published cases. Mohammed et al. [30] reported the crevice corrosion in MSF plant of seawater drain valve with a body made of 316 L SS after 5 years of service. All drain valves suffered from the same crevice corrosion causing an enormous replacement cost. It is important to mention that crevice corrosion occurrence is not limited to the presence of susceptible alloys such as 316 L or duplex 2205. The presence of crevice itself is the main factor combined with other environmental parameters such as presence of oxidants (oxygen, chlorine). For instance, Lee et al. [33] found that the corrosion of 316 L SS is reduced by a factor of 20 on deaeration to 25 ppb oxygen, even though the temperature has increased from ambient to 105 °C. The presence of calcareous deposit inhibiting the cathodic process limits the possibility of crevice corrosion occurrence in thermal desalination plants.

Due to their double microstructure (ferrite/austenite), duplex stainless steels received important research interest. Especially in the role of partitioning elements between constituent phases on the pitting/crevice susceptibility in saline waters [34]. Duplex stainless steel use within thermal desalination is limited to seawater pumps with the successful performance of the older grade CD4MCuN containing nitrogen. Saithala et al. [34] introduced a PREN for each constituent phases of duplex stainless steel to explain the selective attack. Olsson et al. [35] suggested the use of solid duplex stainless steel as evaporator parts including shell, floor, and roof. This material choice will be further detailed under materials selection section.

When residual stresses are present in the alloy, either from manufacturing or from welding activity and associated with corrosive chlorides as in thermal desalination, the most dominant failure mode is either chloride SCC or corrosion fatigue. These degradation modes interfere with other failure mechanisms such as intergranular corrosion as it will be discussed in Chap. 7. According to Streitcher [36], the conclusion made by Staehle in 1969 on the absence of a unified mechanism of SCC for austenitic stainless steels is still valid. For duplex stainless steels, it is well accepted that these alloys have better resistance to SCC and corrosion fatigue than austenitic and precipitation hardening stainless steel grades [37].

Copper alloys are essential materials in thermal desalination plants. Depending on the equipment function (valve, pump, heat exchanger and water box), the most widely used alloys are cupronickel, bronze and brass. Their general corrosion in unpolluted seawater is known to be very low due to the thick and compact protective corrosion film based on cupric and cuprous oxides. However, Shams El-Din et al. [38] reported some unexpected pitting corrosion on the internal surface of cupronickel. The authors attributed this failure to the galvanic corrosion between cupronickel tube and carbon particles present at its surface. The source of this carbon was claimed from water box gasket material that was a carbon-based rubber.

However, carbon-based rubber is usually insoluble in seawater that makes this hypothesis unsupported. On the other hand, Giuliani et al. [8] discussed the presence of carbonaceous material as dirt on the surface of cupronickel tubes before installation (during manufacturing) which requires sandblasting. This is in line with practical recommendations in terms of tube quality discussed by many authors [39, 40]. Pitting corrosion of cupronickel can also be expected if the iron is present on their surface acting as a potential anodic site.

Under copper base alloys, aluminum bronze and nickel aluminum bronze alloys are well recognized in the list of thermal desalination cladding materials as mainly for tube sheets and water boxes in heat rejection systems of MSF. Their corrosion resistance in seawater relies on the protective alumina-rich corrosion film with an erosion resistance that exceeds that of 70Cu30Ni alloy [41]. One of the most cited corrosion risks on these alloys is the pitting due to the presence of a significant concentration of sulfide compounds (polluted seawater or microbial activity product).

In a similar way, aluminum brass specified in early design for MED heat exchanger tubes is sensitive to sulfide-polluted seawater. In addition, this alloy is known for its high susceptibility to erosion-corrosion when the velocity exceeds 1.8 m/s [42]. Malik et al. [43] reported its pitting corrosion due to an erosion of spraying seawater in MED plant (Fig. 5.10).

Aluminum bronze is also specified for valves handling seawater and brine with excellent record except for a few failure cases. Mohammed et al. [30] reported erosion-corrosion on the body of a seawater valve in MSF plant after 3 years of operation. The seawater temperature was 42 °C and the minimum linear velocity was 4.8 m/s (which is accepted for this alloy). However, in such cases, the maximum velocity should be considered along with the sand particle's size and medium temperature. Indeed, at seawater solids content higher than 100 mg/L, the particle's size and velocity start to affect the corrosion rate of copper alloys [40].

Titanium is considered as the seawater corrosion immune material in thermal desalination plants including MSF and MED. However, one of the exceptions is the crevice corrosion of titanium tube after 4 months of operation as discussed by Kido et al. [44]. The corrosion occurred between roll-expanded tube and tube sheet at the temperature range of 88–110 °C causing Ti-hydride formation near the corroded surface. Indeed, titanium is known for its immunity to crevice corrosion in seawater up to 82 °C or somewhat above as reported by Mountford [45]. In brine environment,

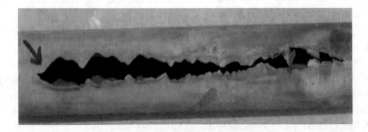

Fig. 5.10 Erosion-corrosion of Al brass tube from a MED plant

Been et al. [46] reported an upper-temperature limit of 80 °C as safe and conservative with a brine pH equal to or less than 9. However, this temperature limit is directly dependent on the crevice presence/geometry where the same authors mention the necessity of very narrow and much deeper crevice than most of stainless steels [45]. The rarity of crevice corrosion cases in desalination may be explained by the absence of crevice due to better junction design between tube and tube sheet. This can be explained mostly by the materials selection and environment control where the oxygen concentration is kept below 20 ppb by mechanical deaeration compensated by the injection of oxygen scavengers.

5.4.2 Vapor Phase Corrosion

One of the most studied corrosion modes in thermal desalination is the vapor phase corrosion. Indeed, this type of corrosion is critical to a certain level, which determines the lifetime of evaporator condenser parts [47]. Most of the publications focused on brine recycle MSF process with more susceptible alloys and conditions [47–51]. Three alloys are concerned by this corrosion form i.e. carbon steel, cupronickel and stainless steel 316 L. Two main types of corrosion can be expected; general and localized. Non-condensable gases and particularly CO_2 aggravate these corrosion forms. The contribution of oxygen ingress (air-in leakage), bromine gas, and sulfur-containing compounds was found to affect cupronickel, carbon steel and 316 L alloys.

Up to now, sulfur-containing compounds were never discussed in thermal desalination heat exchangers failures. Indeed, biogenic sulfur compounds cannot be excluded in these conditions where thermophilic bacteria could exist with microbial activity. Many authors discussed MIC cases for seawater cooled heat exchangers causing different kinds of localized corrosion attack, demonstrated for all alloys except titanium [52, 53].

While the general corrosion concerns carbon steel and cupronickel alloys with minor impact, the localized corrosion forms are the most critical affecting especially cupronickel and stainless steels. The vapor side corrosion can appear in different forms i.e. thinning, pitting, crevice, galvanic or SCC. SCC is usually limited to 316 L as venting pipe (see Chap. 7). Pitting corrosion was observed on 90Cu10Ni cupronickel tubes from stage 5 from a longitudinal tube configuration MSF plant as shown in Fig. 5.11. The investigation revealed pitting corrosion enhanced by low-performance venting combined to the presence of bromine gas [51].

Following these cases, Hodgkiess et al. [54] studied the pitting corrosion susceptibility of cupronickel alloys (70Cu30Ni and 90Cu10Ni) in hot distilled water in the presence of CO_2 leading to a pH of 5.3–5.7 to mimic vapor phase in MSF plants. In addition to the pitting, both alloys suffered from crevice corrosion after long exposure (20 days) at 80 °C.

Galvanic corrosion was suggested as a failure mechanism reported by Malik et al. [43] between 90Cu10Ni alloy as tubing material, 316 L SS as product water

Fig. 5.11 Pitting corrosion on MSF condenser tube from 5th stage after 13 years of operation

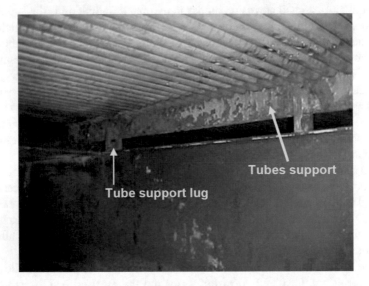

Fig. 5.12 Galvanic corrosion of tube support in MSF plant

tray and carbon steel as tube support material (Fig. 5.12). Whereas the distillate conductivity is very low which would limit the corrosion to the junction area, the extent of this kind of corrosion under near acidic pH and high temperature should be investigated for different area ratio to confirm the galavanic risk extent.

Clear galvanic corrosion was observed between cupronickel tube and tube baffle in heat recovery of an MSF brine recycle plant as shown in Fig. 5.13. Such kind of attack will increase the clearance gap between tube and baffles making a suitable condition for tube damage in the form of fretting corrosion when vibrating due to the high velocity flowing brine inside tubes.

Malik et al. [43] reported a crevice corrosion failure mechanism accelerated by the galvanic coupling between 90Cu10Ni tubes and 316 L tube support in the first stage inducing huge corrosion after 14 years of operation as detected by eddy

Fig. 5.13 Galvanic corrosion between tube and tube baffle in MSF plant

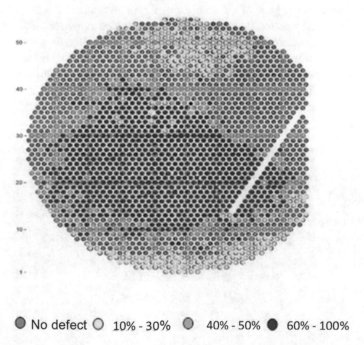

○ No defect ○ 10% - 30% ○ 40% - 50% ● 60% - 100%

Fig. 5.14 1st stage MSF condenser tubes corrosion after 14 years of service revealed by Eddy current technique

current testing (Fig. 5.14). Authors attributed the extensive failure degree to accelerating conditions including the galvanic action between 316 SS and 90/10 Cu/Ni, the low pH due to the accumulation of CO_2 and vibration of tubes following the gap created between tube and tube support.

Few works have been conducted to determine the extent of galvanic corrosion in acidic and low conductive media. In a recent work, Finšgar tried to determine the galvanic series in different acids including mineral and organic acids [55]. The author obtained a different order between stainless steel grades and copper and aluminum-based alloys in each acid and conclude that it is advisable to avoid galvanic coupling between stainless steel grades and copper alloys in all the tested acids.

5.4.3 Corrosion Under Insulation (CUI)

Corrosion under insulation is not specific to the desalination industry and it is a widespread corrosion form in all industries. Unfortunately, the absence of published cases and/or the related cost make the appreciation of this corrosion form difficult. Figure 5.15 shows one of the few cases published in the open literature where an MSF evaporator suffered from severe CUI after 25 years of service [56]. The root cause of such unexpected case was the stagnancy of significant and contaminated rainwater that penetrate through gaps of the insulation. It was found that the calcium silicate insulation material was totally soaked with contaminated rainwater. This raised the concern of roof structure design with limited draining capability. As a root cause, it is well-known that when aluminum cladding is removed, it is never re-fitted in the original way. Then, poorly fitted cladding particularly on a horizontal surface such as an evaporator roof frequently leads to roof corrosion problems due to moisture ingress. This was detected on a number of desalination units as shown in Fig. 5.15.

Corrosion condition of roof members

Fig. 5.15 CUI on MSF evaporator roof after 25 years of service

5.4.4   Microbial Induced Corrosion (MIC)

MIC corrosion is rarely reported in thermal desalination as discussed by Meroufel et al. [57]. Indeed, the high temperature and salinity contribute to the limitation of biofilm that excludes the possibility to develop an MIC scenario. It is well-established that marine biofilm development is stopped above a critical temperature. This temperature depends on the worldwide seawater location [58, 59]. On the other hand, the higher salinity reaching saturation induces exclusion of oxygen which is a vital element in biofilm development.

However, in some locations of thermal desalination plants, the possibility of MIC occurrence remains such as in MED condensers and MSF intake systems. An example of MIC case occurred in MED condenser where the cooling seawater is below 35 °C causing severe pitting on 70Cu30Ni tubes at 6 O'clock position after 6 years of operation is shown in Fig. 5.16. Meroufel et al. [60] discussed the root causes of this case including the use of water velocity below the acceptable limit and an inefficient chlorination system. These factors are favorable to the development of the microbial activity under deposit. The low velocity was adopted during low production period based on reduced water demand.

Another MIC failure occurred on seawater intake pump casing bolts that were in service for 10 years. These bolts made of Monel 400 developed circumferential cracks as shown in Fig. 5.17. At the microscopic level, the cracks appear to be intergranular filled with sulfur-containing compounds. These sulfur-rich compounds detected at grain boundaries support a mechanism of intergranular corrosion initiated by MIC.

Fig. 5.16   MIC on MED condenser tubes after 6 years of operation

Fig. 5.17   MIC on seawater pump bolt in desalination intake

Monel is a nickel-copper alloy that is susceptible to MIC corrosion as studied by Gouda et al. [61]. However, many authors raised the unclear mechanisms of MIC interaction with intergranular corrosion [62, 63].

5.4.5 Miscellaneous Corrosion

Under this category, we gather corrosion cases with a low negative impact on the operation of thermal desalination plants. Splash corrosion for instance, is an aggressive corrosion form that can affect thermal desalination especially in intake structures [64, 65]. With tide action, early designed plants adopted painted carbon steel as protection. However, mechanical impact on the paint would create a pathway to the aggressive seawater and lead to accelerated localized splash zone corrosion. A more reliable approach is the use of stainless steel grade 316 L along with cathodic protection by sacrificial anodes. However, the efficiency of sacrificial anodes to overcome the splash zone localized corrosion stay questionable. This is mainly due to the instability of the electrolyte in this location as discussed by Chen et al. [66]. These authors suggested wrapping carbon steel rods subject to splash corrosion with good seawater absorbent material (linen). The measurements of cathodic current showed good cathodic protection on carbon steel rods in seawater using zinc sacrificial anodes.

On the other hand, in all cases, macro-fouling attachment cannot be avoided particularly when chlorination system performance is unstable. Figure 5.18 shows an example of a bar screen painted and fouled with barnacles at the intake of a

Fig. 5.18 Splash and crevice corrosion caused by barnacles attachment on bar screen in desalination intake

thermal desalination plant. In addition to the crevice environment under barnacles which induces paint degradation, the chloride salts concentration at splash zone is well clear associated with differential aeration cell inducing accelerated corrosion.

During the late 1970s, Hodgkiess et al. [67] investigated mild steel and stainless steel welds corrosion in aerated seawater up to 60 °C. The authors confirmed the sensitivity of the heat-affected zone (HAZ) in mild steel welds to accelerated corrosion. Post-weld heat treatment provided a reduction of this corrosion but it was not enough to suppress it. On the other side, austenitic stainless steel welds were unaffected. Eid studied later the effect of heat treatment, testing media and its temperature [68]. The author gave some recommendations to mitigate the corrosion and/or cracking of austenitic stainless steel welds in contact with hot seawater/brine solutions. Similarly, Saricimen et al. [69] investigated the austenitic stainless steel weld corrosion in MSF desalination plant after only 6 years of operation. The root cause of the failure was the exposure to stagnant brine and a potential sensitization during welding/stress relieving operations. For duplex stainless steel welds, Petersson et al. [70] outlines the importance to ensure the right proportion of austenite phase in the weld metal and HAZ to respect the proportion of the phases and avoid intermetallics precipitation. In addition, the importance of good post-weld cleaning to ensure good corrosion properties cannot be over-emphasized. Figure 5.19 shows distillate tray cracks on the welds of the bottom stiffeners. After repetitive previous unsuccessful weld repairs, the cracks have propagated on to the tray itself. Some distortion of the distillate tray occurred due to the thermal stresses set up locally where the bottom stiffeners and the tube support plates were welded on either side of the tray.

Cavitation of pump impellers in thermal desalination was rarely reported in the open literature. Figure 5.20 shows a case of cavitation on brine recycle pump impeller made of austenitic stainless steel 316 L. The pump was in service for 19 years when cavitation was detected with hole formation which should affect the pump efficiency.

Fig. 5.19 Cracks on the welds of the bottom stiffeners from distillate tray

Fig. 5.20 Cavitation on brine recycle pump impeller after 19 years of service

5.5 Corrosion Control Practices

Depending on the corrosion forms and impact, corrosion control practices can vary upon cost and simplicity versus the operation and maintenance activities. In the following, three main corrosion control practices practiced in thermal desalination plants; viz. environment control, material selection and corrosion monitoring are described.

5.5.1 Environment Control

Seawater handling equipment is usually pretreated physically and chemically to minimize its aggressive action. Under physical treatment, macro and sub-macro screening are adopted to prevent or reduce the penetration of macro-fouling and solid particles. The former (macro-fouling) represents an obstacle to the stream flow and are a potential site for crevice corrosion when attached to the surface of the alloy. Crevice corrosion under barnacles was studied by some authors and is highly expected if the pretreatment is inefficient [71]. The mitigation of this risk is ensured by bar screens (trash racks) with an opening of 20 mm.

For solid particles, travelling band screen is used associated with a washing system to sustain the filtering performance. However, micrometer particles will penetrate to the process and usually accumulate with inorganic scale formed on heat transfer surfaces. Then, the thermal insulation properties of the scale increases. On the other hand, travelling band screen may encounter the risk of re-injecting debris on the clean waterside if the spray nozzles are unable to remove the debris clogging the screen. Travelling screens are usually under cathodic protection system that needs routine maintenance to avoid its corrosion, which can release corrosion product in the filtered feedwater.

For the chemical pretreatment, two main chemicals are used i.e. biocides and oxygen scavengers. Biocide is used to control biofouling through the chlorination by the injection of sodium hypochlorite ($NaClO_4$) which is often produced by electrolysis of the seawater itself. The treatment is managed to obtain a residual chlorine that does not exceed 0.25 ppm in the entering seawater either as cooling or to the heat rejection feed. Higher residual chlorine concentrations represent a risk for many ferrous and non-ferrous alloys [41, 72]. Biofouling control is well detailed from a microbiological perspective in Chap. 16.

To ensure deaerated stream, oxygen scavenger is dozed after mechanical deaeration of the seawater feed to compensate the oxygen reduction up to 20 ppb. Sodium sulfite is the most used with excellent performance. Contrary to passive alloys, a low oxygen level is beneficial for active alloys such as carbon steel [7]. This would enhance the formation of well protective magnetite corrosion layer and slow down the corrosion kinetics [15].

5.5.2 Material Selection

Materials selection for desalination plant attracted a lot of interest of desalination professionals. Taking benefit from materials development, offshore industry experience, industrial standards and desalination operating plants feedback, many authors studied materials selection in desalination plants [73–77]. Material selection should reflect the industry philosophy in terms of design lifetime, cost profile, monitoring and maintenance philosophy, etc.

However, most of the efforts were focused on MSF due to its implementation dominance compared to the MED. We can split materials selection in MSF into three major parts i.e. shell & evaporator internals, heat exchangers and pumps. Historically, materials selected for MSF evaporator shell and internals passed through different phases: totally made of carbon steel, epoxy coated carbon steel, and CRA cladding to the recent solid duplex stainless steel. Through this history, some non-successful idea of non-metallic evaporators were considered with limited use including concrete and polymer composite evaporators [78, 79]. In terms of standardization, Oldfield tried to suggest a standard guide for MSF evaporator materials based on 25 years of experience of 41 MSF desalination plant surveyed in MENA[3] region [80].

The idea of solid stainless steel as MSF evaporator material was discussed for the first time through the work of Olsson et al. [81]. Practically, this introduced in the market the concept of dual-duplex™ (developed by Outokoumpu) where duplex stainless steel 2205 is used for areas in contact with brine and seawater while the lean duplex 2304 is used for areas in contact with vapors and produced distilled

---

[3] MENA stands for Middle-East and North Africa

water. This concept was extrapolated to the low-temperature MED desalination plants [82]. In such a way, lower life cycle cost is claimed compared to the traditional carbon steel cladded by stainless steel 316 L for high-temperature stages. However, welding quality and acid cleaning operations are two main concerns that need to be considered for this choice. Indeed, due to the sensitivity of duplex stainless steels during welding, undesirable phases could precipitate creating corrosion risk areas and great care should be paid to the manufacturing of the evaporator to avoid failures at welded joints [83]. The corrosion resistance of duplex stainless steel to the acids used for descaling should be well-controlled [84].

Water boxes handling seawater or brine are usually made of carbon steel which is internally cladded by either cupronickel or nickel aluminum bronze (NAB) and painted externally. Sacrificial anodes are installed inside water boxes to avoid the impact of galvanic coupling with titanium. The galvanic corrosion of NAB due to the coupling with titanium was found dependent on the presence of oxygen [85]. This situation can be expected in heat rejection where aerated seawater circulate.

Whereas some authors recommended the use of Glass Fiber Polymers (GRP) for water boxes with successful performance and reduced cost, it seems that these materials are still questionable to be considered in projects specifications [86].

Heat exchanger materials within thermal desalination plants attracted lot of research interest. Due to their highest capital cost, it is well-established that heat exchangers determine the lifetime of a thermal desalination plant. While the cupronickel and titanium alloys dominate the tubing material, the tube sheet, baffles and tube supports can be made of cladded carbon steel, carbon steel and austenitic stainless steel 316 L respectively. Depending on operating conditions and galvanic compatibility, titanium grade 2 or 70Cu30Ni as tube and 316 L and 90Cu10Ni as cladding are reserved for the most aggressive conditions. Modified 70Cu30Ni with 66% of copper, 30% of nickel and 2% for iron and manganese elements is also used. This alloy showed the highest performance, which justifies its highest cost. Except the vapor side corrosion discussed above, the inlet erosion-corrosion has been well controlled in most of the plants. This was achieved either using the modified 70Cu30Ni or by the insertion of super-thin walls inserts that can be metallic (superior copper alloy grade, stainless steel or titanium). Therefore, the heat exchanger lifetime extends which avoid the huge retubing investment as discussed by Tallman [87]. Aluminum brass was suggested in the early stages of thermal MED desalination plants. However, this material was quickly abandoned after erosion-corrosion failures [43, 88]. In a similar way to the evaporator materials, Oldfield suggested a standard guide for heat exchangers in MSF desalination plants [80].

Stainless steels and particularly superferritic grades were also suggested for heat exchanger tubing in thermal desalination with limited interest [89, 90]. Despite the achieved good thermal conduction properties, their main drawback is the crevice corrosion risk. Alternatively, recent interest to the implementation of conductive polymer composites for tubing material is still in the early stages of validation for low-temperature MED process [91]. Due to the lack of long-term performance and difficulty to predict their lifetime, this technology needs more elaboration and qualification tests. In parallel, many efforts are made using additive manufacturing technologies to develop highly conductive polymer composites for heat transfer applications. Deisenroth et al. [92] discussed all the aspects and challenges in this research area. The absence of a database on different additive manufacturing techniques/products, and the low manufacturing speed are the challenges for the large scale production.

Pumps handling saline waters including seawater and concentrated brine are the most critical due to the coupling of the corrosive environment with mechanical stresses. For this reason, desalination plant pays more attention to this equipment by duplicating their numbers and investing a lot on their maintenance/monitoring to ensure a high level of plant reliability. Materials used for this equipment has to show good corrosion resistance, cavitation/erosion resistance and galvanic compatibilities in saline waters as well as in acidic condition for acid cleaning operation. Besides, standby conditions where stagnant waters should not induce corrosion. Francis et al. [93] considered also manufacturing/joining aspects (machinability, castability, weldability) and mechanical strength of pump alloys that follow the minimum requirements mentioned in industrial standards [94, 95].

Many references can be considered for the discussion of materials selection for pumps handling seawater and brine solutions [96–98]. For instance, Todd discussed the service experience and optimum material choice among the four main alloy categories i.e. copper-based, stainless steels, austenitic cast irons and nickel-based alloys [96]. The same author reported cavitation tests on different pump materials where stainless steel and nickel aluminum bronze are the most resistant to this kind of severe attack. That is why their use is widespread with successful performance. Previous to this, Morrow et al. [97] reported a table from Gutierez suggesting seawater materials based on performance and capital cost (Table 5.2).

Conventional austenitic stainless steels such as 316 L are still in seawater service with good performance for more than 40 years. However, the flowing condition should be maintained with a lower accepted velocity of 0.9–1.5 m/s and stagnancy should be strictly avoided. Recently, superduplex stainless steel is proposed for both seawater and brine solutions showing cost-effectiveness (Table 5.3) where 1 correspond to poor and 4 to excellent [93].

Conventional duplex stainless steels such as the CDM4Cu grade was suggested for pump's impeller and used since the 1980s [98]. In a similar way, industrial standards suggest duplex stainless steel 2205 for deaerated seawater and superduplex for the aerated condition [94]. However, it is necessary to consider the susceptibility of conventional duplex 2205 to the crevice corrosion in seawater that is difficult to predict due to the absence of a reliable method to assess the crevice geometry in the field.

As discussed by Skar et al. [99], material selection defined in industrial standards has to be safe and conservative. Materials limits are based either on laboratory or field experience. In both cases, factors playing a role in the corrosion/failure vary significantly, which modify the material choice between end-users. Additional care

Table 5.2 Typical vertical pump materials options. Reproduced with permission from [97]; Copyright 2007 @ Turbomachinery Laboratory, Texas A&M University

Component	Lowest cost (minimum useful life expected)	Medium cost (good service life)	Medium cost (good service life)	Higher cost (very good service life)	Highest cost (Excellent service life)	Highest cost (Excellent service life)
	316SS/C-steel construction	315SS construction	Al-BRZ/NAB construction	Duplex 2205/CD4MCuN	superduplex	superaustenitic
Bowls	CF-8 M 316SS	CF-8 M 316SS	Al-BRZ CA958	Duplex CD4MCuN	Alloy 2507	Alloy 6%Mo
Impellers	CF-8 M 316SS	CF-8 M 316SS	Al-BRZ CA958	Duplex CD4MCuN	2507	Alloy 6%Mo
Column/Disch. Head	Coated C-steel/316SS flanges	316SS	Al-BRZ CA958	Duplex 2205	2507	Alloy 6%Mo
Shafting	316SS	316SS Nitronic 50	Nitronic 50 Monel K500	Duplex 2205	2507	Duplex 2205

Table 5.3 Relative merits of four material options for an offshore seawater pump. (Source: Pump Engineer) [93]. NAB: Nickel Aluminum Bronze.

Property	Option			
	Ti	superduplex	NAB	316SS
Castability	1	3	3	4
Strength	3	4	3	1
Pressure tightness	3	3	3	3
Corrosion resistance	4	4	2	1
Weldability	2	3	2	4
Machinability	1	3	3	4
Availability	1	3	4	4
Cost	1	2	3	4
Total	16	25	23	25

should be paid to the pump type and associated piping and valve systems when selecting materials.

5.5.3   Monitoring

Corrosion monitoring within thermal desalination plants can be discussed from two perspectives. This includes corrosive agents monitoring and corrosion impact monitoring. The former is related to residual chlorine, oxygen, temperature, and flow. However, the latter consist of the periodic inspection, chemical analysis of product water and circulating brine and vibration for rotating equipment such as pumps.

Periodic inspection is based on non-destructive tests (NDT) including visual examination, surface crack detection and thickness measurements. Visual examination is supported with photography and crack detection is conducted by dye penetrant test (DPT). Thickness measurement is performed by ultrasound probes or eddy current depending on the inspected part i.e. evaporator structure or heat exchanger tubes. Thickness measurement data is used to calculate the general corrosion rate during life assessment exercise.

In addition, in-service monitoring is conducted by laboratory and inspection department including physical and chemical analysis that is related to the corrosion either directly or indirectly as summarized in Table 5.4.

5.6   Conclusions and Outlook

Corrosion in thermal desalination plants represents a serious concern affecting the lifetime of desalination assets, which merits multiperpsective mitigation practices. From a problem-understanding point of view, most of the forms are well-studied up to a certain extent in the literature. However, corrosion failures will continue due to

Table 5.4 Physical and chemical analysis of waters in thermal desalination process

Parameter/Medium	Feed seawater	Recycle brine	Blow down brine	Product water
Residual chlorine (Cl_2)	√			
pH	√	√	√	√
Conductivity	√	√	√	√
TDS	√	√		
Turbidity	√			
Chlorides (Cl^-)	√	√	√	
Oxygen	√			
M. Alkalinity[a]	√		√	
Copper		•	•	√
Iron		•	•	√

•: on demand
[a]Total alkalinity

different reasons that are human-based such as improper operation and maintenance practices, depraved supply chain quality, deceiving specifications, design faults, etc. and that are more concern for the asset integrity engineers rather than corrosion specialists.

However, some perspectives can be anticipated in terms of the material adopted for thermal desalination plants. For instance, the unification of evaporator and heat exchanger materials is an idea, which merits consideration. Indeed, a total solid stainless steel system or polymer composite material could be an interesting option for both MSF and MED plant designers offering a higher lifetime of the plant with low life cycle cost (LCC). However, the end-user asks often long-term performance data before switching to such choices.

Stainless steel heat exchangers are employed in some parts of MED process (e.g. condensate heater) and in steam/power plant utilities (e.g. steam condenser). In condensate heater, the austenitic 347 grade is already established with good performance. For the steam condenser using either brackish or seawater cooling medium, superferritic S44660 or S44735 grades are the most performing [100, 101].

Stainless steel manufacturers recently support the use of hyperduplex grades for heat exchangers combining excellent thermal conductivity, good corrosion resistance and better mechanical properties. However, the sensitivity of crevice corrosion in MSF and MED plant needs further studies with the challenge of crevice geometry control for tubes [102]. While this risk can be limited to the seawater/brine side due to the formation of calcareous deposit minimizing the cathodic process, it can be a concern in both MSF vapor side with acidic distillate and MED desalination as studied by Snis et al. [103]. On the other hand, duplex stainless steel as evaporator material can suffer from corrosion risks due to operation and maintenance practices or supply chain quality issues. This would need some research in different areas such as duplex welding corrosion, stress corrosion cracking in hot contaminated vapors.

For polymer composite materials, many efforts are conducted to enhance the thermal conductivity of these materials by the incorporation of thermally conductive fillers. These materials are already used in air conditioning but still facing some challenges for thermal desalination with limited real testing data. For evaporator made of Fiber Reinforcing Polymer (FRP), the situation is different where chemical and physical stability are different. Indeed, internal degradation concerns for this kind of material when in contact with neutral and acidified saline water and that did not receive enough research efforts to determine the expected lifetime. Also, the external degradation risks related to the saline humidity and sun radiations exposure merits more research efforts before drawing premature conclusions.

References

1. Cost estimator, Desaldata https://www.desaldata.com/cost_estimator
2. S. Arazzini, R. Borsani, G. Migliorini, U. Zuboli, Economics of MSF plants: Influence of variations in the design parameters and use of different construction materials. Desalination **52**, 97–103 (1985)
3. C. Sommariva, H. Hogg, K. Callister, Maximum economic design life for desalination plant: The role of auxiliary equipment materials selection and specification in plant reliability. Desalination **153**, 199–205 (2002)
4. R. Borsani, S. Rebagliati, Fundamentals and costing of MSF desalination plants and comparison with other technologies. Desalination **182**, 29–37 (2005)
5. D. Rozette, The road to reliability (2014), https://www.meridium.com/author/don-rozette
6. B. Todd, Materials selection for high reliability copper alloy seawater system (2000), http://innovations.copper.org/2000/0009/marine_supreme.html
7. B. Todd, The corrosion of materials in desalination plants. Desalination **3**, 106–117 (1967)
8. L. Giuliani, R. Cigna, Corrosion problems of copper alloys in desalination plants. Desalination **22**, 379–384 (1977)
9. U. Rohlfs, H.D. Schulze, Possibilities of corrosion prevention and removal in seawater exposed structures and desalination plants. Desalination **55**, 283–296 (1985)
10. A.U. Malik, P.C. Mayan-Kutty, I. Andijani, S. Al-Fozan, Materials performance and failure evaluation in SWCC MSF plants. Desalination **97**, 171–187 (1994)
11. R. Francis, Galvanic corrosion of highly resistant stainless steels in seawater. Br. Corros. J **29**, 53–57 (1994)
12. J.W. Oldfield, B. Todd, Technical and economic aspect of stainless steels in MSF desalination plants. Desalination **124**, 75–84 (1999)
13. S. Jacques, J. Peultier, J.C. Gagnepain, P. Soulignac, Corrosion resistance of duplex stainless steels in thermal desalination plants. Corrosion, NACE – 08261, NACE International, Houston,(2008)
14. B. Todd, Selection of materials for high reliability seawater systems, Supplement to Chemistry and Industry, (1977), p. 22
15. D.C. Silverman, Aqueous Corrosion, in *Corrosion: Fundamentals, Testing, and Protection*, ed. by S. D. Cramer, B. S. Covino, vol. 13A, (ASM Handbook, ASM International, 2003), pp. 190–195
16. F.L. LaQue, Theoretical studies and laboratory techniques in seawater corrosion testing evaluation. Corrosion **13**, 33–44 (1957)
17. I. Matsushima, Carbon steel-corrosion by seawater, in *Uhlig's Corrosion Handbook*, ed. by R. W. Revie, 3rd edn., (The electrochemical Society, INC, and John Wiley & Sons, INC., ISBN: 978-0-470-08032-0, 2011), p. 601

18. F.W. Fink, *Corrosion of Metals in Sea Water, PB 171344* (Battelle Memorial Institute, Columbus, 1960), p. 59. https://digital.library.unt.edu/ark:/67531/metadc11613/
19. K. Ichikawa, K. Nagano, S. Kobayashi, N. Kitajima, Ebara Eng. Rev **85**, 2 (1973)
20. R. Covert, Properties and applications of Ni-resist and ductile Ni-resist alloys. Nickel Institute Publication (1998), https://www.nickelinstitute.org/media/1770/propertiesand applicationsofni_resistandductileni_resistalloys_11018_pdf
21. A. Marshall, Corros. Prev. Contr., 177–181 (1983) Une nouvelle nuance de fonte Ni-resist a graphite spheroidal. INCO NICKEL 33 (1971). p. 4
22. A.U. Malik, S. Basu, I. Andijani, N.A. Siddiqui, S. Ahmad, Corrosion of Ni-resist cast irons in seawater. Br. Corros. J **28**, 209–216 (1993)
23. K. Lorentz, G. Medawar, Über das Korrosionsverhalten austenitischer chrom-nickel-(Molybdän) stähle mit und ohne stickstoffzusatz unter besonderer berücksichtigung ihrer beanspruchbarkeit in chloridhaltigen Lösungen. Thyssenforschung **1**, 97–108 (1969)
24. Bauernfeind D, Mori G Corrosion of superaustenitic stainless steels in chloride and sulfate containing media – Influence of alloying elements Cr, Mo, N, N and Cu. Corrosion (2003), NACE – 03257, NACE International, Houston
25. A.U. Malik, M. Mobin, I.N. Andijani, S. Al-Fozan, Investigation on the corrosion of flash chamber floor plates in a multistage flash desalination plant. J. Fail. Anal. Prev **6**, 17–23 (2006)
26. F. Elshawesh, A. Elhoud, O. Ragha, Role of surface finish and post-fabrication cleaning on localized corrosion of 316L austenitic stainless steel flash chambers. Paper No. 053, Eurocorr, (2001)
27. L. Giuliani, Pre-service corrosion control in multiflash desalination plants. Desalination **38**, 295 (1981)
28. Z. Szklarska-Smialowska, Influence of sulfide inclusions on the pitting corrosion of steels. Corrosion **28**, 388–396 (1972)
29. M. Bernas, I. Westermann, C. Lauritsen, R. Johnsen, M. Lannuzzi, Effect of microstructure on the corrosion resistance of duplex stainless steels: Materials performance maps. Corrosion 2017, NACE-2017–8923, NACE International, Houston, (2017)
30. R.A. Mohammed, K. Shahzad, N.S. Sabah, Control valves failures in seawater desalination plants-some case studies. 16th International Corros. Congress, Beijing, (2005)
31. H. Yakuwa, K. Sugiyama, M. Miyasaka, A.U. Malik, I. Andijani, M. Al-Hajri, K. Mitsuhashi, K. Matsui, NACE Middle East Corros. Conf., Paper No. 134, Bahrain, (2007)
32. N. Larché, D. Thierry, V. Debout, T. Cassagne, J. Peultier, E. Johansson, C. Tavel-Condat, Crevice corrosion of duplex stainless steels in natural and chlorinated seawater. Duplex World, 11–13 (2010)
33. T.S. Lee, R.M. Kain, Factors influencing the crevice corrosion behaviour of stainless steels in seawater. Corrosion 1983, NACE-69, NACE International, Houston, (1983)
34. J.R. Saithala, H.S. Ubhi, J.D. Atkinson, A.K.P. Patil, Pitting corrosion mechanisms of lean duplex and superduplex stainless steels in chloride solutions. Corrosion 2011, NACE-11255, NACE International, Houston, 2011
35. J. Olsson, H.L. Groth, Evaporators made of solid duplex stainless steel: A new approach to reduce cost. Desalination **97**, 67–76 (1994)
36. M. Streitcher, Austenitic and ferritic stainless steels, in *Uhlig's corrosion handbook*, ed. by R. W. Revie, 3rd edn., (The Electrochemical Society, INC, and John Wiley & Sons., INC, ISBN: 978–0–470-08032-0, 2011), p. 657
37. M.L. Falkland, M. Glaes, M. Liljas, Duplex stainless steels, in *Uhlig's Corrosion Handbook*, ed. by R. W. Revie, 3rd edn., (The Electrochemical Society, INC, and John Wiley & Sons, INC., ISBN: 978–0–470-08032-0, 2011), p. 695
38. A.M. Shams El-Din, M.E. El-Dahshan, H.H. Haggag, Carbon induced corrosion of MSF condenser tubes in Arabian Gulf seawater. Desalination **172**, 215–226 (2005)
39. N.M. Valota, Copper alloy tube for MSF desalination plant: Some practical considerations. Desalination **66**, 245–256 (1987)
40. R. Francis, *The Corrosion of Copper and its Alloys: A Practical Guide for Engineers*, NACE International, ISBN: 9781575902257, (2010)

41. CDA, Aluminium Bronze Alloys Corrosion Resistance Guide, Publication No, 80, Copper Development Association. (1981), https://copperalliance.org.uk/resources/aluminium-bronze-corrosion-resistance-guide/?download=start
42. Circuits Eau de Mer, Traitements et Matériaux, Edition Technip, ISBN: 2710806479 9782710806479. (1993)
43. A.U. Malik, S. Al-Fozan, F. Al-Muaili, Corrosion of heat exchanger in thermal desalination plants and current trends in material selection. Desalin. Water Treat **55**, 2515–2525 (2014)
44. S. Kido, T. Shinohara, Corrosion under heat flux encountered in desalination plant. Desalination **22**, 369–378 (1977)
45. J.A. Jr Mountford, Titanium – Properties, Advantages and Applications Solving the Corrosion Problems in Marine Service, Corrosion 2002, NACE – 02170, NACE International, Houston, (2002)
46. J. Been, J.S. Grauman, Titanium and titanium alloys, *In: The electrochemical Society*, INC, and John Wiley & Sons, INC, ISBN: 978-0-470-08032-0, (2011), p. 861
47. J.W. Oldfield, B. Todd, Vapor side corrosion in MSF plants. Desalination **66**, 171–184 (1987)
48. E.A. Al-Sum, A. Al-Radif, S. Aziz, Performance of materials used for the venting system in SED's desalination plants (MSF). Desalination **93**, 517–527 (1993)
49. A.M. Shams-El Din, R.A. Mohammed, Contribution to the problem of vapour side corrosion of copper nickel tubes in MSF distillers. Desalination **115**, 135–144 (1998)
50. N. Asrar, A.U. Malik, S. Ahmed, M. Al-Khalidi, K. Al-Moaili, Vapor side corrosion in thermal desalination plant. Mater. Perform **4**, 66–71 (1999)
51. Y.A. Alzafin, Condenser tube failures: MSF desalination plant Jebel Ali, G- PH1. Paper No. IDAWC/DB09, IDA World Congress – Atlantis, The Palm – Dubai, (2009)
52. D.H. Pope, Microbial corrosion in fossil-fired power plants – A study of microbiologically influenced corrosion and a practical guide for its treatment and prevention. EPRI Final Report, CS-5495, Project 2300-12 Electric Power Research Institute, Palo Alto, (1987)
53. K. Al-Nabulsi, T.Y. Rizk, F. Al-Abbas, O.C. Dias, *Seawater Cooler Tubes Corrosion and Leaks Due to Microbiologically Induced Corrosion, Corrosion 2016, NACE-7034* (NACE International, Houston, 2016)
54. T. Hodgkiess, D. Mantzavinos, Corrosion of copper-nickel alloys in pure water. Desalination **126**, 129–137 (1999)
55. M. Finšgar, Galvanic series of different stainless steels and copper and aluminum based materials in acid solutions. Corros. Sci **68**, 51–56 (2013)
56. A.A. Al-Mobayed, P. Bukovinszky, R. Harris, Evaporator roof corrosion in MSF desalination – A case study. IDA Paper No. IDAWC/026, IDA World Congress, (2002)
57. A. Meroufel, T. Green, A.U. Malik, MIC and desalination industry: What is the progress?. Paper No.1409, NACE Middle East Corros. Conference, (2014)
58. A. Mollica, A. Trevis, E. Traverso, G. Ventura, G. De Carolis, R. Dellepiane, Cathodic performance of stainless steels in natural seawater as function of microorganism settlement and temperature. Corrosion **45**, 48–56 (1989)
59. Shams El-Din AM, T.M.H. Saber, A.A. Hammoud, Biofilm formation on stainless steels in Arabian Gulf water. Desalination **107**, 251–264 (1996)
60. A. Meroufel, A. Al-Sahari, M. Ayashi, A. Alenazi, MIC of cupronickel tubes from MED desalination condenser. Mater. Perform **57**, 15–17 (2018)
61. V.K. Gouda, I.M. Banat, W.T. Riad, S. Mansour, Microbiologically induced corrosion of UNS N04400 in seawater. Corrosion **49**, 63–73 (1993)
62. K.R. Sreekumari, K. Nandakumar, Y. Kikuchi, Bacterial adhesion to weld metals: Significance of substratum microstructure. Biofouling **17**, 303–316 (2001)
63. M. James, D. Hattingh, Case studies in marine concentrated corrosion. Eng. Fail. Anal **47**, 1–15 (2015)
64. S. Al-Fozan, A.U. Malik, Effect of seawater level on corrosion behavior of different alloys. Desalination **228**, 61–67 (2008)
65. A. Ul-Hamid, H. Saricimen, A. Quddusa, A.I. Mohammed, L.M. Al-Hems, Corrosion study of SS304 and SS316 alloys in atmospheric, underground and seawater splash zone in the Arabian Gulf. Corros. Eng. Sci. Technol **52**, 134–140 (2017)

66. J. Chen, J.Y. Huang, X. Dong, Study on the splash zone corrosion protection of carbon steel by sacrificial anode. Int. J. Electrochem. Sci **7**, 4114–4120 (2012)
67. T. Hodgkiess, N. Eid, W.T. Hanbury, Corrosion of welds in seawater. Desalination **27**, 129–136 (1978)
68. N.M.A. Eid, Localized corrosion at welds in structural steel under desalination plant conditions Part II: Effect of heat treatment, test temperature and test media. Desalination **73**, 407–415 (1989)
69. H. Saricimen, M. Shamim, I.M. Allam, M. Maslehuddin, Performance of austenitic stainless steels in MSF desalination plant flash chambers in the Arabian Gulf. Desalination **78**, 327–341 (1990)
70. R. Pettersson, M. Johansson, E.M. Westin, Corrosion performance of welds in duplex, superduplex and lean duplex stainless steels. Corrosion 2013, NACE-2697, NACE International, Houston, (2013)
71. M. Eashwar, G. Subramanian, P. Chandrasekaran, K. Balakrishnan, Mechanism for Barnacle-induced crevice corrosion in stainless steel. Corrosion **48**, 608–612 (1992)
72. T. Rogne, U. Steinsmo, Practical consequences of the biofilm in natural sea water and of chlorination on the corrosion behaviour of stainless steels. Sea water corrosion, Stainless Steels – Mech. Exp. EFC Publ, 19, (1996)
73. T.G. Temperley, Material specification and the availability and life of desalination equipment in both Saudi Arabia and the Arabian Gulf. Desalination **33**, 99–107 (1980)
74. J.D. Birkett, E.H. Newton, Present trends in materials selection and performance in multistage flash desalination units. Desalination **38**, 207–222 (1981)
75. T. Hodgkiess, Current status of materials selection for MSF distillation plants. Desalination **93**, 445–460 (1993)
76. C. Sommariva, D. Pincirolli, E. Tolle, R. Adinolfi, Optimization of materials selection for evaporative desalination plants in order to achieve the highest cost-benefit ratio. Desalination **124**, 99–103 (1999)
77. A.U. Malik, S. Al-Fozan, Corrosion and materials selection in MSF desalination plants. Corros. Rev **29**, 153–175 (2011)
78. Y. Tazawa, Y. Nojiri, K. Fujii, Behaviour of concrete evaporator shells during operation of the module plant at the Chigazaki test facilities. Desalination **22**, 111–120 (1977)
79. O.J. Morin, R.A. Johnson, MSF evaporator fabricated from reinforced thermosetting materials. Desalination **31**, 355–363 (1979)
80. J.W. Oldfield, Survey of material usage in MSF plants over the past 25 years. IDA Paper No. IDAWC/024, IDA World Congress, Bahamas, (2003)
81. J.O. Olsson, H.L. Groth, Evaporator made of solid duplex stainless steel a new approach to reduce costs. Desalination **97**, 67–76 (1994)
82. J. Peultier, V. Baudu, P. Boillot, J.C. Gagnepain, News trends in selection of metallic material for desalination industry. IDA Paper No. IDAWC/DB09, IDA World Congress, Dubai, (2009)
83. M. Bernås, I. Westermann, C. Lauritsen, R. Johnson, M. Lannuzzi, Effect of microstructure on the corrosion resistance of duplex stainless steels: Materials performance maps. Corrosion 2017, NACE-8923, NACE International, Houston, (2017)
84. S.H. Mameng, A. Bergquist, E. Johansson, Corrosion of stainless steel in sodium chloride brine solutions. Corrosion 2014, NACE-4077, NACE International, Houston, (2014)
85. R.C. Barik, J.A. Wharton, R.J.K. Wood, K.R. Stokes, Galvanic corrosion of nickel aluminum bronze coupled to titanium or Cu15Ni alloy in brackish seawater, Paper No. 330, Eurocorr 2004, Nice, (2003)
86. H. Bos, Fiberglass for sea water boxes and MED modules – Long term solutions. IDA Paper No. IDAWC/156, IDA World Congress, (2002)
87. C.P. Tallman, Method for extending heat exchanger and condenser life. Desalination **73**, 231–246 (1989)
88. B. Todd, Copper-Nickel Alloys in Desalination Systems, Seminar Technical Report 7044–1919, Copper Development Association, (1999)

89. W. Fairhust, Some recent developments in stainless steels and their application to seawater cooled condenser. Mater. Des., **5**, 235–243 (1984)
90. R. Rautenbach, S. Schäfer, Development and testing of ultrathin superaustenitic tubes for heat exchangers in desalination plants. Desalination **127**, 13–17 (2000)
91. T. Orth, D.-C. Alarcón-Padilla, G. Zaragoza, P. Palenzuela, P. Mueller, Shaping the future of thermal desalination with high performance heat conducting polymer tubes. IDA Paper No. IDAWC/58054, IDA World Congress, São Paulo, (2017)
92. D.C. Deisenroth, R. Moradi, A.H. Shooshtari, F. Singer, A. Bar-Cohen, M. Ohadi, Review of heat exchangers enabled by polymer and polymer composite additive manufacturing. Heat Transfer Eng **39**, 1652–1668 (2018)
93. R. Francis, L. Phillips, Cost Effective Materials Selection for Pumps, Pump Engineer, (2003) p. 44–49 https://www.neonickel.com/wp-content/uploads/2016/12/66.-Cost-Effective-Materials-Selection-for-Pumps.pdf
94. NORSOK, Materials selection, Standard M-001 (2014), https://www.standard.no/pagefiles/1176/m-001.pdf
95. ISO 21457, Petroleum, petrochemical and natural gas industries – Materials selection and corrosion control for oil and gas production systems (2010), https://www.iso.org/standard/45938.html
96. B. Todd, (2010) Pump materials for desalination plants. Materials Selection and Corrosion, Vol. II, In: Encyclopedia of Desalination and Water resources, p. 261 http://www.desware.net/Sample-Chapters/D07/D14-020.pdf
97. S.J. Morrow, (2007) Vertical turbine pumps in seawater service materials of construction, ITT-Pumplines: 5–10 http://turbolab.tamu.edu/wp-content/uploads/sites/2/2018/08/Tutorial-05.pdf
98. NDI, (1995) Materials for Saline Water, Desalination and Oilfield Brine Pumps, 2nd Edition, Nickel Institute Reference Book, Series No. 11 004, Toronto,
99. J.I. Skar, S. Olsen, A review of materials application limits in NORSOK M-001 and ISO 21457. Corrosion **73**, 655–665 (2017)
100. E.R. Blessman, J.J. Dunn, J.F. Grubb, Stainless steels for desalination projects. IDA Paper No. IDAWC/157, IDA World Congress, (2003)
101. H. Richaud-Minier, H. Marchebois, P. Gerard, Titanium and super stainless steel welded tubing for solutions for seawater cooled heat exchangers, Paper No.1214, Eurocorr 2008, Freiburg, (2008)
102. C. Leballeur, N. Larché, D. Thierry, *Definition and Qualification of Crevice Assembly Adapted for Tube Geometry* (Stainless Steel World Conference, Maastricht, 2017)
103. M. Snis, J. Olsson, M. Willfor, D. Bjorklund, G. Antonopoulos, Stainless steels for LT-MED plants. IDA Paper No. IDAWC/MP07–052, IDA World Congress, Gran Canaria, (2007)

Chapter 6
Corrosion in Reverse Osmosis Desalination Processes: Forms and Mitigation Practices

Abdelkader A. Meroufel

6.1 Introduction

Seawater reverse osmosis (SWRO) is a widely used as desalination process due to its lower energy consumption compared to other processes like thermal desalination. The main components of an SWRO plant are the intake, pre-treatment, reverse osmosis (RO) process itself, energy recovery system and post-treatment.

When compared to the thermal desalination processes, many research efforts are going on in RO desalination technology to improve the operational performance, energy efficiency, produced water quality and membranes reliability. Within this context, corrosion and material selection in RO desalination attracted a lot of interest during the first 20 years (1975–1995) of large-scale RO plants implementation. Two major reasons that pushed RO professionals to consider corrosion and material selection are: the maintenance/capital cost reduction (need to be competitive with the thermal multi-stage flash, MSF process), and the membranes reliability issues (mainly affected by fouling). Membrane reliability can also be affected by the corrosion of equipment, which induces the blockage of membrane pores and declines the water product quality and quantity. All these issues are reflected in the water production cost where an important competition is undergoing globally.

Since RO desalination deals with aerated ambient seawater, common corrosion forms that are common with thermal desalination are observed in the seawater intake area of RO systems. The specific pretreatments of natural seawater followed and the diverse operating conditions used in the plants can dictate some typical corrosion forms similar to that observed in other industrial installations such as oil and gas offshore. Corrosion scientists and engineers are continuing working on different

A. A. Meroufel (✉)
Desalination Technologies Research Institute, Saline Water Conversion Corporation,
Jubail, Saudi Arabia
e-mail: ameroufel@swcc.gov.sa

© Springer Nature Switzerland AG 2020   131
V. S. Saji et al. (eds.), *Corrosion and Fouling Control in Desalination Industry*,
https://doi.org/10.1007/978-3-030-34284-5_6

aspects of corrosion forms, their mechanism, prevention methods, modelling, corrosion monitoring/detection and testing protocols specific to RO plants.

Seawater is a complex environment including physical parameters and living organisms that can affect corrosion phenomenon in different ways justifying the continuous research efforts and discussions. All desalination professionals agree on the fact that RO desalination performance and reliability are site-based. While there is a general agreement on the plant design, the variation of the seawater quality represents a challenge for designers and owners. The scarcity of field tests of the different suggested alloys and the scattering of the laboratory-performed tests raise the challenge of standardization as an additional barrier to overcome before concluding on material performance.

In the present chapter, an attempt to summarize the key points of this topic are made based on research outcomes, material development perspective and RO plant database. The interrelated corrosion forms considered for discussion are the crevice, pitting and microbiological corrosion.

6.2   Economics of Equipment Integrity

Several published information is available on economic analysis of RO desalination systems [1–3]. The major factors influencing the analysis are: plant configuration and capacity, membrane/alloy development, and accumulated operation/maintenance experience around the world.

The economics of corrosion in RO desalination is a challenging question especially in terms of operational cost. This can be explained by the plant design and performance affected by the site location and project procurement model. However, the CAPEX scheme indicates the significant contribution of alloys, pumps and piping in the total CAPEX of RO desalination plant (Fig. 6.1).

For operational expenditure (OPEX) cost, the corrosion costs represented through indirect factors such as parts, chemicals and labors are shown in Fig. 6.2. These contributions interfere with other functionalities such as mechanically failed parts, disinfection and anti-scaling activities. The two chemicals related to the corrosion cost analysis are biocides and acids for scaling control.

Since the total water cost is the sum of CAPEX and OPEX, Ghaffour et al. [2] suggested a typical water production cost breakdown as shown in Fig. 6.3. It appears clearly that corrosion cost is embedded in different parts such as the CAPEX, chemicals, spares, and O&M (operation and maintenance).

6.3   Reliability Philosophy

One of the measured performance indicators for RO desalination plant is the availability. Since RO desalination units should be in continuous operation, it is easier and appropriate to assess their availability. Improving this parameter for the whole plant is a challenging task to reduce the production cost. The improvement of availability induces technical solutions such as the duplication of some critical sys-

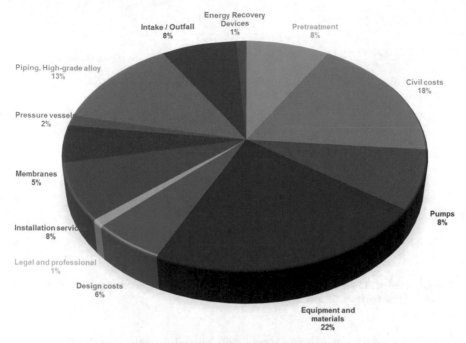

Fig. 6.1 CAPEX breakdown of RO desalination plant with capacity of 240,000 m³/d [4]

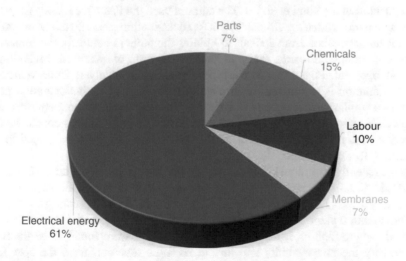

Fig. 6.2 OPEX breakdown of RO desalination plant with production capacity of 240,000 m³/d [4]

tems like the high-pressure pumps. In the same time, the intake system is very important to the whole plant operation. Any problem can induce a low seawater flow and will cause significant production decrease. Pretreatment section is also a crucial part that cannot be out of service for a long duration. Therefore, the RO desalination process is more sensitive to equipment reliability issues.

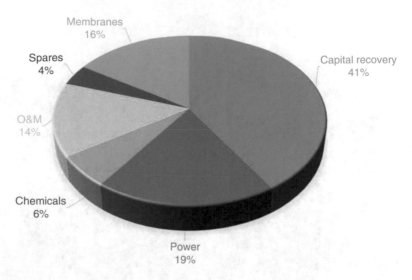

Fig. 6.3 Total water production cost for a RO desalination plant. Reproduced with permission from Ref. [2]; Copyright 2013 @ Elsevier

Unfortunately, there is limited work on the reliability analysis of RO plant except the ones conducted by some authors as done in early days of large-scale RO plant implementation by Kutbi et al. [5]. The authors used the Fault Tree Analysis (FTA) technique to determine the critical area in RO desalination plant. Based on operation and maintenance data from Jeddah RO plant, the authors could fix the impact of failures including corrosion. Pump failures in the form of corrosion, cavitation or erosion represented the major failure types. Based on the analysis, improvement in RO desalination is recommended in materials choice, and design/operation practices. In a similar way, Hajeeh et al. [6] assessed a RO desalination plant operating in the Gulf region where the redundancy and over design of the system showed an impact on the plant availability. More discussions on this topic are added in the material selection section.

More recently, Bourouni [7] discussed two methods of reliability analysis applied to RO desalination plants for an economic optimization objective. This includes Reliability Block Diagram (RBD) and FTA methods. Based on data from Kuwait RO desalination plant, it has been demonstrated a good agreement between the two methods where high-pressure pump and RO modules were found to be the most affected by the unavailability and that merit more attention from the specification phase.

6.4 Corrosion Forms in RO Desalination Plants

Since RO is dealing only with aerated seawater as environment, and corrosion resistant alloys (CRAs), the discussion will be articulated on the typical localized corrosion forms such as pitting, crevice and microbial corrosion. While the morphological appearance of these forms can be different, the electrochemical mechanism is tightly similar. The carbon and low alloy steels were never considered for this desalination process due to their low corrosion resistance making them a non-economical solution. Thus, the most used materials for RO desalination plants since the beginning are CRAs and in limited cases nickel-copper alloys such as Monel. In the below sections, the important corrosion forms in RO plants are briefly discussed. Further details on these corrosion forms can be found in the literature [8–11].

6.4.1 Pitting Corrosion

Pitting corrosion of stainless steels particularly in seawater has been subject to several studies. It is important to distinguish pitting corrosion in neutral to alkaline medium which has an electrochemical nature from other types of pitting (chemical and etch) as mentioned by Alvarez et al. [12]. It is well accepted that halides (chlorides and bromides), flowing condition, susceptible material (composition, surface heterogeneity), temperature, oxidants concentration (oxygen, hypochlorite) and microbial activity are the main factors that can affect pitting phenomenon [10, 13, 14]. The synergy between these factors determines the performance of any CRA depending on exposure duration.

Many studies are still undergoing to understand the passivity phenomenon and the role of alloy elements such as molybdenum to explain the pitting itself [15, 16]. The variation of passive layer characteristics between different CRAs implies the necessity of proper understanding the material's metallurgy. The passive layer formation depends on alloy composition, microstructure, surface treatment and the surrounding environment.

Localized electrochemical and surface analytical techniques continue to provide precious pieces of information in understanding the pitting mechanism. From the passive film breakdown to the pit growth giving macro damage (in the order of 100 μm), various competitive processes occur; under the effect of environmental parameters. That is why operators of RO desalination plant are guided for instance to the flushing when it comes to the shut-down of the plant as well as to keep the pH in a certain range. Indeed, as revealed by Marcus et al. [17], there is a competition between Cl^- and OH^- adsorption on the surface of the oxide layer and also on the underneath metal/alloy which hinders the repassivation inducing the enlargement of the formed pit.

Another aspect is the role of the alloying elements in the pitting corrosion resistance of different grades. Baroux made a distinction between elements entering the passive film and others that does not. While the first group reduce the pit initiation and growth, the second group of elements can reduce the ionic current through the metal/passive film interface which may slow down the dissolution rate [18]. Under the first group, it has been demonstrated that only chromium and molybdenum contribute to the protectiveness and chemistry of the passive layer. We recall that both chromium and molybdenum compounds are considered in the field of inorganic corrosion inhibitors. However, some other alloying elements belongs to the second group and still undergoing research studies to understand their role in the pitting corrosion resistance [19–21]. These elements include nitrogen, copper, and tungsten that reduce pitting corrosion through a different proposed mechanism. Since the role of each element is dependent on the material metallurgy and the environment parameters, no conclusion can be drawn here. However, one of the direct consequences of this alloying role concerns the corrosion engineer responsible for material selection. The classic material selection practice followed by corrosion engineers is based on the Pitting Resistance Equivalent Number (PREN) as discussed in Chap. 5. The different proposed formulas are directly related to the role of each element in either pitting or crevice corrosion resistance. Materials selection for RO desalination plant will be detailed later in this chapter.

Pitting corrosion failures in RO desalination was not subject to a high number of studies due to the absence of critical factors and the randomness (stochasticity) of its occurrence. While the chlorides concentration is high and the oxygen is at saturation level, the flowing condition makes the occurrence of pitting corrosion very rare as mentioned by Oldfield et al. [22]. In this regard, different critical velocities are suggested by corrosion specialists ranging from 1 to 1.5 m/s above which pitting corrosion is not expected [22, 23].

Electrochemically speaking, pitting corrosion of stainless steels can occur if the corrosion potential increase in an excessive way toward the pitting potential region. This kind of scenario can be expected in two situations i.e. at high biocide concentrations or in the presence of a biofilm.

Francis et al. [24] discussed the corrosion risk function of residual chlorine and seawater temperature. For instance, for a seawater temperature of 40 °C, the residual chlorine safe limit should not exceed 0.7 ppm. The usual practice in RO desalination plants is either a continuous chlorination followed by a dechlorination of the feed seawater to preserve RO membranes or the chock dozing where a high level of residual chlorine will be generated for short-term period. For both chlorination practices, the flow and the distance between the biocide injection point and CRA affect the pitting corrosion risk.

While the pitting corrosion due to the ennoblement of the corrosion potential in the presence of biofilm still discussed, this situation cannot be expected in RO desalination service due to the extensive seawater pretreatment and flowing conditions. Pitting corrosion due to the biofilm presence was observed by Neville et al. [25] on stainless steel alloys after long-term exposure in the north sea. The attacked samples were under anodic polarization which stimulates the activity of sulfate reducing bacteria (SRB) producing a black sulphide corrosion product.

Some rare cases of pitting during operation of RO desalination plant were reported during early days of large-scale RO plants implementation. The first case reported was in Malta RO plant within 2 years of operation [26]. However, the authors did not discuss or illustrate clearly the case. This resulted in some mix-up with crevice corrosion form since it happens on RO permeator manifolds and connections. Similarly the confusion between cavitation and pitting corrosion was noticed in the case reported on Energy Recovery Device (ERD) wheel buckets parts in Ras Abu Jarjur RO (Bahrain) desalination [27]. Darton et al. [28] reported a case study on pitting corrosion within 3 years of operation of RO desalination plant based at Gibraltar. Pitting occurred on high-pressure pump valves and seats due to the stagnant seawater. Pitting corrosion of welds were mentioned by Bullock et al. [29] but without enough details on their occurrence. The susceptibility of welds to the pitting corrosion still attracting significant research interest. The role of welding quality/parameters in pitting corrosion resistance is well established [30, 31]. Malik et al. [32] studied the pitting corrosion of 317 L brine reject pipe at the weld joints location (Fig. 6.4). Authors attributed the root cause to the preservation medium (saline water containing formalin) before plant commissioning. The preservation took 3 years and this can play a role in the pitting corrosion initiation. However, the welding quality and surface condition in these cases are the first issues that should be considered.

Similarly pitting corrosion of Monel alloy grades was studied in stagnant seawater by some authors [33]. Usually, this alloy is used for (seawater, flushing) pumps or ERD sleeves and bolts. So far, there were no pitting failures reported in the open literature for this material except the case discussed in thermal desalination corrosion forms in Chap. 5.

Therefore, pitting corrosion represents a risk to RO desalination plant owners during the shut-down period. There are several methods to rank the pitting corrosion resistance of the different alloys proposed to RO desalination. The quickest one in practical point of view is the use of PREN number. Scientists use various electrochemical laboratory parameters (breakdown potential, repassivation potential etc.),

Fig. 6.4 Pitting corrosion on the weld joint of brine rejection pipe in SWRO plant

critical pitting temperature (CPT), and pit generation rate. A detailed discussion of these pitting susceptibility indicators can be found in reference [8].

6.4.2   Crevice Corrosion

Many corrosion professionals qualify the crevice corrosion as a special case of pitting corrosion due to the similarities between these forms. Indeed, both of the two types belong to the localized corrosion where passive film breakdown occurs in a specific location and lead to the perforation or crack initiation on the metal. Thus, passive alloy, which resists to the crevice corrosion should resist to the pitting in stagnant conditions. Crevice corrosion happens in restricted flow areas where there is a contact between metal and non-metal or metal and metal. These contact areas are difficult to avoid in the design of equipment, and they can also form under deposits, in poor welds, pipe flanges, and joints.

Crevice corrosion is the major corrosion risk in all RO desalination plants. It occurs into two main phases i.e. initiation and propagation. From a corrosion engineering perspective, the lifetime of equipment susceptible to this kind of corrosion is limited to the initiation phase. This agrees well with the safety requirement for operators especially in high-pressure section where it is important to avoid the initiation since the progress of this crevice corrosion cannot be predicted.

This type of corrosion still undergoing many research efforts from scientists about the mechanism, the factors, the modelling, the testing procedure and the prevention methods [11, 34–41]. After describing the crevice corrosion as a differential aeration corrosion by Evans and Mears, four theories were developed to cover various aspects of crevice corrosion mechanisms: critical crevice solution, critical IR drop, metastable pitting and coupled environment [42, 43]. The two former theories correspond to the incubation and initiation steps while the last two theories govern the propagation stage. The major factors affecting crevice corrosion are: the geometry and type of crevice, the passivation and metallurgy of metals, environment parameters (temperature, oxidant, salinity, pH and biofouling) and electrochemical considerations (anode to cathode ratio and repassivation) [44]. All these factors are inter-related and different scenarios can occur. For instance, biofouling is known to accelerate the cathodic process affecting the propagation rate of crevice corrosion. However, this is dependent on the crevice geometry and seawater temperature. The later can accelerate or limit the biofilm development and at the same time can accelerate the kinetics of the cathodic reaction.

Scientists tried to model crevice corrosion in three ways i.e. mechanistic, empirical and probabilistic. For instance, Oldfield et al. [22] used his proposed model to predict the initiation of crevice corrosion on different stainless steels proposed to RO desalination. The authors could plot the crevice corrosion resistance for different RO alloys as a function of the crevice gap at ambient seawater. While the authors consider that practical crevice gaps are in the range 0.2–0.5 µm, it is still difficult to determine the geometry of crevice in real conditions. In a more developed way,

Chang et al. [45] developed a mathematical model including several processes (mass transport, chemical reactions, ion migration, etc.). The authors could determine the critical crevice depth/with ratio that have two impacts: controlling the initiation phase and defining two components of the IR drop (chemical and physical). This model was validated with experimental data on 316 L SS in NaCl solution. Some more complicated models aim to cover the phenomenon in an integrated and multiscale approach which increases the required inputs, expertise and challenge levels such as accuracy and validation [43, 46].

Crevice corrosion testing is usually conducted to compare or qualify alloys for certain service conditions. Sridhar et al. [47] conducted a review of the different crevice corrosion tests (electrochemical and non-electrochemical) where they discussed the limits of each test. Particularly, the non-uniformity of the compressive pressure is one of the major drawbacks of ASTM crevice corrosion testing standard (ASTM G48). This pushed the European Federation of Corrosion (EFC) to conduct a research program (named CREVCORR) during the 1990s to improve the crevice test assembly set-up [39]. Unfortunately, despite the excellent research findings of this project considered under ISO 18070, many corrosion professionals are not considering these findings when studying this type of corrosion. This with other influencing test parameters continues to scatter more the results and make their comparison quite impossible. For both the exposure and the electrochemical tests, the factors mentioned above continue to affect the accuracy and reproducibility of the results creating an additional gap with the field performance of the different alloys.

Contrary to the case of pitting corrosion, crevice corrosion failures in RO desalination plant was communicated since the early days of this industry [48–52]. These failures troubled austenitic low grades such as 316 L and 317 L, superaustenitic 904 L and conventional duplex 2205. Figures 6.5, 6.6 and 6.7 show crevice corrosion cases of 316 L, 317 L and 904 L on typical RO desalination parts in contact with pre-treated feed seawater. Due to the condition on connections to the RO cells to be demountable, this provides severe crevice sites and explains the important number of failures in these locations.

Fig. 6.5 Crevice corrosion of 316 L RO nozzles coupling high-pressure header to the RO membranes. Reproduced with permission from Ref. [49]; Copyright 1994 @ Elsevier

Fig. 6.6 Crevice corrosion of 317 L RO nozzles coupling high-pressure header to the RO membranes [52]

Fig. 6.7 Crevice corrosion
of 904 L micro-cartridge
filter top flange basket [52]

For high grades of stainless steels such as the superaustenitic high moly 254SMO or superduplex, rare cases of crevice corrosion are reported in RO desalination plants. However, both the alloys were mentioned in some cases of crevice corrosion in oil and gas industry [53]. This issue will be discussed in more details under material selection section of this chapter.

For instance, Byrne et al. [54] discussed one of the rare cases of crevice corrosion of high-pressure pipe spool made of superduplex Zeron 100® (Fig. 6.8). The failure occurred within 1 year of service showed a relation with stray currents from an undefined source. In the literature, stray-current was demonstrated to affect the crevice corrosion of low grade of stainless steels such as 316 L [55]. It is recommended to conduct further research to investigate how and when stray-currents can induce or accelerate crevice corrosion on stainless steels.

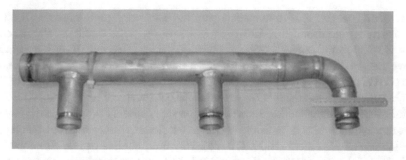

Fig. 6.8 Crevice corrosion of HP pipe spool made of Zeron 100® (Source: Rolled Alloys) [54]

6.4.3 Microbiological Influence Corrosion (MIC)

Microorganisms present in the seawater does not limit their impact to the membrane fouling which reduces the productivity of a RO desalination plant. They can also affect metallic parts in different ways. According to Dexter [56], microorganisms can either increase or decrease the corrosion of metals.

Many scientists studied microbial corrosion in seawater for different metals/ alloys with more focus on active alloys such as carbon/mild steels [57–59]. However, its impact on CRAs was explored more from the perspective of biofilm contribution to enhance the typical localized corrosion forms discussed above (pitting and crevice) [60–63]. For this case, the mechanism by which biofilm affect the corrosion of stainless steels remain unclear and debated among the corrosion community. Nevertheless, the creation of heterogeneous surface, participation to the cathodic process through direct electron uptake, secretion of biogenic products (hydrogen sulfide, acids) are among the mechanisms demonstrated in some cases [64–67].

Curiously, there are no MIC cases in the desalination industry reported in the open literature. This could explain the absence of this industry from the table proposed by Dexter summarizing the industries most affected by MIC [68]. From his side, Scott limited the impact of MIC in desalination to the biofilm formed on RO membranes [69]. However, these data are dependent on the desalination community understanding of MIC and the failure analysis, which can reveal the presence or absence of MIC cases. Within the context of this limited information, it is important to address one of the challenges to assess MIC in RO desalination plant: the ability to distinguish between MIC and non-MIC corrosion mechanism. While some authors claim certain corrosion geometry typical of MIC, this practice did not receive full agreement from corrosion scientists since it is case-based [70]. For instance, the weld joint pitting corrosion discussed above (Fig. 6.4) could be the result of certain contribution of microorganisms in a similar way to the case discussed by Hilbert et al. [71] in low chlorides waters (max. 130 ppm). However, most MIC reported cases on stainless steels occurred in stagnant or near stagnant conditions. In addition, within a RO desalination plant context, more efforts are needed to assess MIC where the continuous operation and very high pressure limit the possibility to monitor MIC. However, if corrosion cases occur as suggested by NACE TG304 and Javaherdahsti in their work, the first step in MIC assessment is to prove that the failure case is a non-MIC corrosion. If it is not the case, the prerequisite of MIC corrosion should be gathered to conclude, as illustrated in Fig. 6.9 [72].

One of the demonstrated effect of biofilm is enhancement of crevice corrosion of stainless steels. Many studies showed an enhanced cathodic process on the exposed area due to the biofilm establishment, which accelerates the crevice corrosion propagation of stainless steel grades [73–75]. However, studies in real RO conditions including high velocity and pH around 6.5–7 should be conducted to explore the level of biofilm contribution.

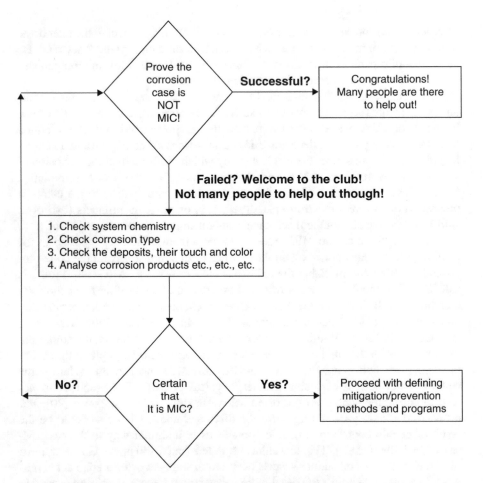

Fig. 6.9 MIC assessment guide proposed by Javaherdashti Reproduced with permission from Ref. [72]; Copyright 2008 @ Springer

6.5   Mitigation Practices

Corrosion mitigation practices in reverse osmosis plants can vary depending on plant conditions. In similar way to other industries, three main corrosion control practices are adopted, viz. environment control, material selection and corrosion monitoring.

6.5.1   Environment Control

The environment control for RO based desalination plant focuses on the aspect of desalting system performance. To avoid membranes pores blockage and fouling, intensive pre-treatments are conducted in desalination plants through injection of biocides, coagulants and anti-scalants. The impact of these chemicals on the localized corrosion initiation and propagation was limited to the biocides due to their well-established oxidizing power effect. In some limited cases, RO desalination plant uses copper sulphate as a disinfection chemical. The effect of this chemical on the localized corrosion of RO alloys was not explored. On the other hand, some RO processes operate under slightly acid conditions, i .e. at a pH value of 5–6. Although this will tend to increase the corrosivity of the water, the effect on stainless steels can be negligible.

Another aspect concerns the corrosivity of nano-filtered seawater in nanofiltration (NF)-RO desalination process. The NF of seawater will produce a very low sulfate content seawater making the water more corrosive as reported in the literature [23]. Curiously, this subject did not attract the attention of scientists.

Also, fully aerated waters with high chloride contents will be more severe than deaerated waters of the same chloride content. Except for the limited information from RO Tanajib desalination plant, there is no data in the open literature on the corrosion resistance in deaerated RO desalination plants. Material selection for Tanajib RO desalination plant considered 317LN and 904 L for high-pressure section where after 2 years of service, the experience was judged positive without details [76].

Recently, Francis et al. [77] outlined the importance of oxidation-reduction potential (ORP) as an indicator of the corrosivity of the feed seawater in RO desalination plant. The authors support this suggestion based on critical crevice temperature measurements correlated with real cases from SWRO plants. The usefulness of ORP parameter can be extended to the optimization of the chemical pretreatment and void corrosive seawater especially when using high grades of stainless steels.

6.5.2   Material Selection

Material selection for RO desalination plants was more dynamic compared to the thermal desalination plants for many reasons and particularly for the high-pressure section. Non-metallic materials (reinforced concrete and fiber reinforcing polymers, FRPs) dominate the low-pressure RO section. However, for the high-pressure section, there is an agreement on the necessarily required properties as follows [78]:

1. Adequate strength, fabricability and weldability
2. Availability
3. Good corrosion resistance

Some authors tried to include additional parameters such as successful case histories and cost factor [79]. Indeed, an economic solution in terms of material selection is always based on the long-term experience and cost parameters. However, such conditions are not easy to fulfill due to the variation in practices as well as the alloy market fluctuations. This induces decisions that are based on subjective conclusion and limited knowledge of alternatives. In general, stainless steels and nickel-base alloys meet the requirements but different grades with varying properties and costs are proposed in the market. Then, it is necessary to optimize the cost/performance equation and select the right grade.

When RO desalination plants started, austenitic 316 L was the most suggested stainless steel grade. However, with an accumulation of crevice corrosion failures, most of the plant designers started to recommend superaustenitic 254SMO which was made commercially available by the late 1970s. In parallel, some corrosion professionals stayed on the suggestion of other stainless steel grades such as 317 L and 904 L. Thus, the debate between stainless steel manufacturers and RO desalination plant owners started on the fact that although the capital cost of high stainless steel grade like 254SMO is about 10% more than 316 L, this could be justified by far better life cycle cost compared to low grades. Then, by the mid-2000s, superduplex availability and experience increased in the global market and start to penetrate RO desalination specifications offering an interesting option gathering performance and cost. This option allowed filling the great gap existing between conventional and high stainless steel grades. Shulz et al. [80] discussed the availability of superduplex among other duplex grades for different industries. While superduplex Zeron 100® was used by oil and gas offshore in the North Sea since the 1980s, it was necessary to wait a long time before seeing this grade present in RO desalination plant.

In terms of corrosion resistance ranking, RO desalination plant designers rely on the PREN number to specify the right stainless steel grade. The practice in this regard was slightly various where some authors suggested in the beginning, a minimum of PREN equal to 38 [81]. After some years, RO professionals started to be convinced by the NORSOK-M001 and ISO 21456 standards that require a minimum PREN equal to 40 for materials in contact with ambient aerated seawater [82, 83]. It is important to mention that this value is based on North Sea field experience and the situation can vary in tropical seawaters. This aspect fits well with temperature limits discussed by many authors in terms of chloride stress corrosion cracking risk [84].

On the other hand, Larché et al. [85] considered certain alloys with PREN equal or close to 40 having a borderline behavior in terms of crevice corrosion risk after their partial failure in seawater at a temperature of 30 °C. The results of these authors are in agreement with the previous study conducted in different seawater world locations where seawater temperature can reach up to 38 °C and chloride concentration upto 23,000 ppm [86].

In addition, some authors suggested another ranking of materials in terms of crevice corrosion resistance [87]. Crevice corrosion index (CCI) can be calculated by multiplying the maximum attack depth with the number of sides attacked. This would provide a ranking that accounts for both initiation and growth of crevice corrosion attack where lower values imply improved resistance.

Some corrosion engineering recommendations were suggested in the literature for RO desalination plant in high-pressure section [88]. For instance, it is always advised to adopt the same alloy for high-pressure pipe, pumps, valves and connectors. On the other hand, welding quality is an important aspect for all stainless steels and superduplex grades in a more particular way. These later are very sensitive to the precipitation on undesirable phases creating weak points highly susceptible to localized corrosion. The use of backing gas (argon) in the pipe, the purging and the pickling are strictly recommended to avoid internal oxidation of the pipe.

6.5.3 Electrochemical and Surface Treatment

Under this category of mitigation practices, we consider especially the crevice corrosion protection. This includes cathodic protection (CP) and crevice surface coating using different options. Some authors proposed CP to protect conventional low-grade stainless steel such as 316 L and 17-4PH [89, 90]. They used steel and aluminum anodes to prevent the crevice corrosion to occur in marine environment. More recently, Larché et al. [91] studied this possibility for different RO components (pump and valves) made of different stainless steel grades including 316 L, (rolled and cast) duplex 2205, superduplex 2507 and superaustenitic 254SMO. Laboratory and field tests were conducted considering the two CP approaches i.e. sacrificial anodes and impressed current. These authors noticed a competition between biofilm and calcareous deposit in terms of impact on the cathodic current demand. As expected also, the efficiency of the CP system was affected by the geometry of the component to be protected. For all these studies, good performance was obtained but there is no field application as per the open literature. This can be explained by two reasons; the necessity of proper CP design efforts and/or the end-user choice to go for maintenance-free choice.

The second mitigation practice, which is not well spread, is the filling or surface coating of the crevice area by a material than can play a role of physical barrier (ceramic or sacrificial material). This option needs strict control on the filler quality, application and performance monitoring.

6.5.4   Non-metallic Solutions

With their development, non-metallic materials and especially FRPs are attracting a lot of research interest to replace the metallic ones. The initial application was limited to the low-pressure section and RO membrane housing. Then, efforts were initiated for high-pressure pipes and pumps or ERDs applications. Indeed, with their excellent record in handling seawater and brine, these materials showed both excellent mechanical strength and almost corrosion-free performance. The recent efforts of Japanese scientists to develop high-pressure pipe made of FRP are very interesting trying to overcome the challenge of large diameter to be able to handle large seawater and brine quantities [92, 93]. These authors could develop 6 meter length FRP headers with 0.35 m diameter and capable to sustain a maximum working pressure of 82 bar. The second challenge remains with the joints where still high-grade stainless steels should be used.

In terms of seawater pumps, some manufacturers started to think about FRP parts that can be considered in seawater pumps to reduce corrosion problems. Hicks et al. [94] developed positive displacement high-pressure seawater pump made entirely from FRP. However, the model was limited to a pumping capacity of 158 m^3/d. FRP seawater pumps were first adopted by big aquarium and shipboard applications in US with a capacity, which does not exceed 100,000 m^3/d [95]. The recently available information from RO desalination plants reports the use of FRP in pumps with capacities, which does not exceed 25,000 m^3/d [96].

6.6   Conclusions and Outlook

In this chapter, it has been shown that if operation and maintenance practices respect corrosion-engineering recommendations, the dominant remaining corrosion form in RO plants is the crevice corrosion. The remaining research questions around this problem concern the materials qualification (testing) and the cost-effectiveness of CP as a mitigation method in field conditions. Whereas end-users are satisfied with high-grades of stainless steel in most of the time, if CP protection can be well-designed it can compete and break down significantly the CAPEX of RO desalination plants reducing further the water production cost.

The second interesting option is the FRP based materials for high-pressure components. This area merits more collaborative efforts to overcome the challenges of size and joints. Two research areas interconnect with this challenge i.e. nanotechnology and materials informatics. The exceptional properties of nano-materials can be considered as input in multiphysics modelling tools to design the optimal material for RO high-pressure conditions. Such approach can achieve an important saving in the time of material development if research entities collaborate efficiently.

Among the persisting research interest in seawater corrosion is the borderline behavior of certain high grade of stainless steels with regard to the crevice corrosion risk. More contribution from corrosion modelling community should clarify more the stochasticity aspect of this localized corrosion type. Taking benefit from the development of localized sophisticated techniques, in situ characterization of surface phenomena should bring more information to explain and justify certain macroscale test results.

References

1. R. Borsani, S. Rabagliati, Fundamentals and costing of MSF desalination plants and comparison with other technologies. Desalination **182**, 29–37 (2005)
2. N. Ghaffour, T.M. Missimer, G.L. Amy, Technical review and evaluation of the economics of water desalination: Current and future challenges for better water supply sustainability. Desalination **309**, 197–207 (2013)
3. A.F. Ismail, K.C. Khulbe, K. Matsuura, RO economics. In: Reverse osmosis, (Elsevier, 2019), ISBN: 9780128115398, pp.163–187
4. Cost estimator, Desaldata https://www.desaldata.com/cost_estimator
5. I.I. Kutbi, Z.A. Sabri, A.A. Husseiny, Reliability analysis of reverse osmosis plant. Desalination **42**, 291–313 (1982)
6. M. Hajeeh, D. Chaudhuri, Reliability and availability assessment of reverse osmosis. Desalination **130**, 185–192 (2000)
7. K. Bourouni, Availability assessment of a reverse osmosis plant: Comparison between reliability block diagram and fault tree analysis methods. Desalination **313**, 66–76 (2013)
8. G.S. Frankel, Pitting corrosion, in *Corrosion: Fundamentals, Testing, and Protection*, ed. by S. D. Cramer, B. S. Covino, vol. 13A, (ASM Handbook, ASM International, 2003), pp. 236–241
9. R.G. Kelly, Crevice Corrosion, in *Corrosion: Fundamentals, Testing, and Protection*, ed. by S. D. Cramer, B. S. Covino, vol. 13A, (ASM Handbook, ASM International, 2003), pp. 242–247
10. H.H. Strehblow, P. Marcus, Mechanisms of pitting corrosion. In: *Corrosion Mechanisms in Theory and Practice*, ed. by P. Marcus, (CRC Press, Taylor & Francis, 2012), ISBN: 9781138073630, p.349
11. P. Combrade, Crevice corrosion of metallic materials, In: *Corrosion Mechanisms in Theory and Practice*, ed. by P. Marcus, (CRC Press, Taylor & Francis, 2012), ISBN: 9781138073630, p. 449
12. M.G. Alvarez, J.R. Galvele, Pitting corrosion, in *Shreir's Corrosion*, (Elsevier, 2010), pp. 772–800
13. G.T. Burstein, C. Liu, R.M. Souto, S.P. Vines, Origin of pitting corrosion. Corros. Eng. Sci. Technol **39**, 25–30 (2004)
14. B.N. Popov, Pitting and crevice corrosion. *In: Corrosion Engineering Principles and Solved Problems*, (Elsevier, 2015), ISBN: 978-0444627223, pp. 289–325
15. I. Olefjord, The passive state of stainless steel. Mater. Sci. Eng **42**, 161–171 (1980)
16. C. Örnek, M. Långberg, J. Evertsson, G. Harlow, W. Linpé, L. Rullik, F. Carlà, R. Felici, E. Bettin, U. Kivisäkk, E. Lundgren, J. Pan, In-situ synchrotron GIXRD study of passive film evolution on duplex stainless steel in corrosive environment. Corros. Sci **141**, 18–21 (2018)

17. P. Marcus, V. Mauric, H.H. Strehblow, Localized corrosion (pitting): A model of passivity breakdown including the role of the oxide layer nanostructure. Corros. Sci. **50**, 2698–2704 (2008)
18. B. Baroux, Further insights on the pitting corrosion of stainless steels, in *Corrosion Mechanisms in Theory and Practice*, ed. by P. Marcus (Ed.), (CRC Press, Taylor & Francis, 2012), ISBN: 9781138073630, p. 419
19. M.K. Lei, X.M. Zhu, Role of nitrogen in pitting corrosion resistance of a high nitrogen face centered cubic phase formed on austenitic stainless steel. J. Electrochem. Soc **152**, 281–285 (2006)
20. L.F. Garfias-Mesias, J.M. Sykes, Effect of copper on active dissolution and pitting corrosion of 25%Cr duplex stainless steel. Corrosion **54**, 40–47 (1998)
21. E.B. Haugan, M. Næss, C.T. Rodriguez, R. Johnsen, M. Iannuzzi, Effect of tungsten on the pitting and crevice corrosion resistance of type 25Cr superduplex stainless steels. Corrosion **73**, 53–67 (2017)
22. J.W. Oldfield, B. Todd, The use of stainless steels and related alloys in reverse osmosis desalination plants. Desalination **55**, 261–280 (1985)
23. F.L. LaQue, *Marine Corrosion*, (Wiley, New York, 1975), ISBN 978-0471517450, p. 150
24. R. Francis, G. Byrne, Experiences with superduplex stainless steel in seawater. Corrosion 2003, NACE – 03255, NACE International, Houston, 2003
25. A. Neville, T. Hodgkiess, Localized effects of macrofouling species on electrochemical corrosion of corrosion resistant alloys. Br. Corros. J. **35**, 54–59 (2000)
26. W.T. Andrews, The Malta seawater RO facility- update on the first two years operation. Desalination **60**, 145–150 (1986)
27. M. Al-Arrayedh, B. Ericsson, M.A. Saad, H. Yoshioka, Reverse osmosis desalination plant Ras Abu Jarjur, state of Bahrain - two years operational experience for the 46,000 m³/d RO plant. Desalination **65**, 197–230 (1987)
28. E.G. Darton, A.G. Turner, Operating experiences in a seawater reverse osmosis plant in Gibraltar. Desalination **82**, 51–69 (1991)
29. R.J. Bullock, G. Byrne, R. Francis, C. Kuzler, G. Warburton, The selection design, fabrication and performance of Zeron®100 in SWRO applications. IDA Paper No. IDAWC/102, IDA World Congress, 2003
30. A.I. Grekula, V.P. Kujanpaa, L.P. Karjalainen, Effect of solidification mode and impurities on pitting corrosion in AISI 316 GTA welds. Corrosion **40**, 569–572 (1984)
31. T.P.S. Gill, U.K. Mudali, V. Seetharaman, J.B. Gnanamoorthy, Effect of heat input and microstructure on pitting corrosion in AISI 316 submerged arc welds. Corrosion **44**, 511–516 (1988)
32. A.U. Malik, I. Andijani, S. Al-Fozan, F. Al-Muaili, M. Al-Hajri, An Overview of the localized corrosion problems in seawater desalination plants: Some recent case studies. IDA Paper No. IDAWC/MP07–162, IDA World Congress, Gran Canaria, 2007
33. V.K. Gouda, I.Z. Selim, A.A. Khedr, A.M. Fathi, Corrosion behavior of monel-400 in chloride solutions. J. Mater. Sci. Technol. **15**, 208–212 (1999)
34. T.S. Lee, R.M. Kain, J.W. Oldfield, The effect of environmental variables on crevice corrosion of stainless steels in seawater. Mater. Perform. **23**, 9–15 (1984)
35. J.W. Oldfield, W.H. Sutton, Crevice corrosion of stainless steels I.A. mathematical model. Br. Corros. J **13**, 13–22 (1977)
36. S.M. Sharland, C.P. Jackson, A.J. Diver, A finite-element model of the propagation of corrosion crevices and pits. Corros. Sci **29**, 1149–1166 (1989)
37. T. Hakkarainen, A model for prediction of possibility of localized corrosion attack of stainless steels. Corrosion 96, NACE – 96363, NACE International, Houston, 1996
38. D. Gunasegaram, M. Venkatraman, I. Cole, Toward multiscale modeling of localized corrosion. Int. Mater. Rev **59**, 84–114 (2014)
39. J.P. Audouard, D. Thierry, C. Féron, C. Compère, V. Scotto, B. Wallen, A.H. Stigenberg, T. Rogne, Crevice corrosion resistance of stainless steels in natural sea water. Results of a paneuropean test programme. *Stainless Steels,* Düsseldorf 96, 3–5 June, 1996, pp. 83–88

40. A. Turnbull, Prevention of crevice corrosion by coupling to more noble materials? Corros. Sci **40**, 843–845 (1998)
41. B.G. Allen, R.H. Heiderbach, S.F. Mealy, Cathodic protection of stainless steels against crevice corrosion. Offshore Technology Conference, Paper No. 3856, Houston, 1980
42. I.L. Rosenfeld, I.K. Marshakov, Mechanism of crevice corrosion. Corrosion **20**, 115–125 (1964)
43. M. Raunio, Basic approaches and goals for crevice corrosion modelling. Project Research Report VTT-R-02078-15 (2015), https://www.vtt.fi/inf/julkaisut/muut/2015/VTT-R-02078-15.pdf
44. A. Meroufel, Crevice corrosion of stainless steels – The last 20 years research efforts focus on duplex. Keynote, Duplex Summit & seminar, Düsseldorf, 2018
45. H.Y. Chang, Y.S. Park, W.S. Hwang, Initiation modeling of crevice corrosion in 316L stainless steel. J. Mater. Proces. Technol **103**, 206–217 (2000)
46. C.D. Taylo, Corrosion informatics: An integrated approach to modelling corrosion. Corros. Eng. Sci. Technol **50**, 490–508 (2015)
47. N. Sridhar, D.S. Dunn, C.S. Brossia, G.A. Cragnolino, J.R. Kearns, Crevice corrosion, in *Corrosion Tests and Standards Application and Interpretation*, ed. by R. Baboian, 2nd edn., (ASTM International, West Conshohocken, 2005)
48. A. Muirhead, S. Beardsley, J. Aboudiwan, Performance of the 12,000 m³/d seawater reverse osmosis desalination plant at Jeddah, Saudi Arabia between January 1979 through January 1981. Desalination **42**, 115–128 (1982)
49. J. Nordström, J. Olsson, Stainless steel for high pressure piping in SWRO plants. Are they any options? Desalination **97**, 213–220 (1994)
50. K. Cosic, J. Olsson, Stainless steels for SWRO plants high-pressure piping, properties and experience. IDA Paper No. IDAWC/041, IDA World Congress, 2003
51. N. Larché, P. Dézerville, Review of material selection and corrosion in seawater reverse osmosis desalination plants. Desalin. Water Treat **31**, 121–133 (2010)
52. A. Meroufel, *Crevice Corrosion within RO Desalination Plants – Case Studies and Recommendations* (Jubail Corrosion & Materials Engineering Forum, NACE Jubail Saudi Arabia Section, Jubail, 2017)
53. Ø. Strandmyr, O. Hagerup, Field experience with stainless steel material sin seawater systems. Corrosion 98, NACE – 98707, NACE International, Houston, 1998
54. G. Byrne, R. Francis, A. Guamhusein, Stray current corrosion in SWRO systems. IDA Paper No. IDAWC/51462, IDA World Congress, San Diego, 2015
55. D.R. Lenard, J.G. Moores, Initiation of crevice corrosion by stray current on stainless steel propeller shafts. Corrosion **49**, 769–775 (1993)
56. S.C. Dexter, *Microbiological Effects, Corrosion Tests and Standards*, 2nd edn, ed. by R. Baboian, ASTM Manual 20, ASTM International
57. H. Liu, L. Xu, J. Zeng, Role of corrosion products in biofilms in microbiologically induced corrosion of carbon steel. Br. Corros. J **35**, 131–135 (2000)
58. R.E. Melchers, Influence of seawater nutrient content on the early immersion corrosion of mild steel-part 2: The role of biofilm and sulfate reducing bacteria. Corrosion **63**, 405–415 (2007)
59. Stipanicev M, Turcu F, Esnault L, Rosas O, Basseguy R, Sztyler M, B. Beech I (2014) Corrosion of carbon steel by bacteria from North Sea offshore seawater injection systems: Laboratory investigation. Bioelectrochemisry, 97, 76–88
60. J.F.D. Stott, B.S. Skerry, R.A. King, Laboratory evaluation of materials for resistance to anaerobic corrosion caused by sulphate reducing bacteria: Philosophy and practical design, the use of synthetic environments for corrosion testing. ASTM STP 970, ed. by P.E. Francis, T.S. Lee, ASTM International, 1998
61. C.W. Kovach, J.D. Redmond, A review of microbiological corrosion in high performance stainless steels, stainless steels, Dusseldorf, 3–5 June 1996
62. I.G. Chamritski, G.R. Burns, B.J. Webster, N.J. Laycock, Effect of iron-oxidizing bacteria on pitting of stainless steels. Corrosion **60**, 658–669 (2004)

63. A. Neville, T. Hodgkiess, Comparative study of stainless steel and related alloy corrosion in natural seawater. Br. Corros. J **33**, 111–119 (1998)
64. H.A. Videla, Biofilms and corrosion interactions on stainless steel in seawater. Int. Biodeter. Biodegr. **34**, 245–257 (1995)
65. S. Da Silva, R. Basséguy, A. Bergel, Electron transfer between hydrogenase and 316L stainless steel: Identification of a hydrogenase-catalyzed cathodic reaction in anaerobic mic. J. Electroanal. Chem **561**, 93–102 (2004)
66. H. Venzlaff, D. Enning, J. Srinivasan, K.J.J. Mayrhofer, A.W. Hassel, F. Widdel, M. Stratmann, Accelerated cathodic reaction in microbial corrosion of iron due to direct electron uptake by sulfate-reducing bacteria. Corros. Sci **66**, 88–96 (2013)
67. T. Gu, B. Galicia, Can Acid Producing Bacteria Be Responsible for Very Fast MIC Pitting?. Corrosion 2012, NACE – 2012–1214, NACE International, Houston, 2012
68. S.C. Dexter, Microbiologically influenced corrosion. *In*: Corrosion: Fundamentals, Testing, and Protection, Vol 13A, ed. by S.D. Cramer, B.S. Covino, (ASM Handbook, ASM International, 2003), pp. 398–416
69. P.J.B. Scott, Expert consensus on MIC-prevention and monitoring, part 1. The analyst xii: 1–16. (2005), https://apps.dtic.mil/dtic/tr/fulltext/u2/a442935.pdf
70. B.J. Little, J.S. Lee, R.I. Ray, Diagnosing microbiologically influenced corrosion: A state of the art review. Corrosion **62**, 1006–1017 (2006)
71. L.R. Hilbert, L. Carpén, P. Møller, F. Fontenay, T. Mathiesen, *Unexpected Corrosion of Stainless Steel in Low Chloride Waters –Microbial Aspect*. Proc. Eurocorr, Nice. (2009), https://cris.vtt. fi/en/publications/unexpected-corrosion-of-stainless-steel-in-low-chloride-waters-mi
72. R. Javaherdashti, (2008) Microbiologically Influenced Corrosion: An Engineering Insight. Springer, London, ISBN: 978-1-84800-074-2
73. V. Scotto, R. Di Cintio, G. Marcenaro, The influence of marine aerobic microbial film on stainless steel corrosion behavior. Corros. Sci **25**, 185–194 (1985)
74. N. Larché, P. Boillot, P. Dézerville, E. Johansson, J.M. Lardon, D. Thierry, Crevice corrosion performance of high alloy stainless steels and Ni-based alloy in desalination industry. IDA Paper No. IDAWC/151, IDA World Congress, Tianjin, 2013
75. A. Meroufel, N. Larché, S. Al-Fozan, D. Thierry, Crevice corrosion behavior of stainless steels and nickel based alloy in the natural seawater-effect of crevice geometry, temperature and seawater world location. Desalin. Water Treat **69**, 202–209 (2017)
76. W. Heyden, Seawater desalination by reverse osmosis plant design, performance data, operation and maintenance (Tanajib, Arabian Gulf coast). Desalination **52**, 187–199 (1985)
77. R. Francis, D.R. McIntyre, Case histories: The importance of oxidation-reduction potential on corrosion control. Mater. Perform **55**, 48–54 (2016)
78. J.W. Oldfield, R.M. Kain, Economic material selection for reverse osmosis desalination plants. Desalination **84**, 227–250 (1991)
79. A. Hassan, A.U. Malik, Corrosion resistant materials for seawater RO plants. Desalination **74**, 157–170 (1989)
80. Z. Schulz, D. Wachowiak, P. Whitcraft, Availability and economics of using duplex stainless steels. Corrosion 2014, NACE-2014–4345, NACE International, Houston, 2014
81. A. Martinho, The high pressure pump train on reverse osmosis plants. Experience and current trend. Desalination **138**, 219–222 (2001)
82. M. Norsok-001 Materials selection. (2014), https://www.standard.no/pagefiles/1176/m-001.pdf
83. ISO 21457, Petroleum, Petrochemical and Natural Gas Industries – Materials Selection and Corrosion Control for Oil and Gas Production Systems (2010), https://www.iso.org/obp/ui/#iso:std:iso:21457:ed-1:v1:en
84. J.I. Skar, S. Olsen, A review of materials application limits in NORSOK M-001 and ISO 21457. Corrosion **73**, 655–665 (2017)

85. N. Larché, P. Boillot, P. Dézerville, E. Johansson, J.M. Lardon, D. Thierry, Crevice corrosion performance of high alloy stainless steels and Ni-based alloy in desalination industry. Desalin. Water Treat **55**, 2491–2501 (2014)
86. H. Yakuwa, M. Miyasaka, K. Sugiyama, K. Mitsuhashi, Evaluation of Crevice Corrosion Resistance of Duplex and Superduplex Stainess Steels for Seawater Pumps. Corrosion 2009, NACE - 09194, NACE International, Houston, 2009
87. M.A. Streicher, Analysis of crevice corrosion data from two sea water exposure tests on stainless alloys. Mater. Perform. **22**, 37–50 (1983)
88. S. Nordin, J. Olsson, Corrosion engineering of high pressure piping in RO plants. Desalination **66**, 235–244 (1987)
89. B.G. Allen, R.H. Heidersbach, S.F. Mealy, Cathodic protection of stainless steels against crevice corrosion. Paper No. 3856, Offshore Technology Conference, Houston, 1980
90. T. Shahrabi, M.G. Hosseini, F. Evazzadeh, Sacrificial cathodic protection against crevice corrosion of austenitic stainless steels in seawater. Paper No 265, Eurocorr, Lisbon, 2005
91. N. Larché, D. Le Flour, P. Dezerville, P. Vinzio, K.H. Köfler, Étude de la protection cathodique interne d'un système en acier inoxydable pour l'industrie de dessalement par osmose inverse. Protection cathodique et revêtements associés théorie et applications terre, mer, Béton et surfaces internes, EFC Event No. 363, Juan les pins, 2014
92. H. Tanaka, Y. Kawai, K. Kumaki, T. Yamada, K. Andou, K. Tokunaga, Development high pressure piping and header with non-corrosive material. IDA Paper No. IDAWC/373, IDA World Congress, Tianjin, 2013
93. H. Tanaka, Y. Kawai, K. Kumaki, M. Fuse, M. Kuroda, K. Adachi, T. Ike, K. Tokunaga, Development high pressure piping and header with non-corrosive material. IDA Paper No. IDAWC/51817, IDA World Congress, San Diego, 2015
94. D.C. Hicks, C.M. Pleasss, W.A. Fearn, D. Staples, Development, testing and the economics of a composite/plastic seawater reverse osmosis pump. Desalination **73**, 95–109 (1989)
95. T. Beck, Beat the sea with FRP (pump, that is). Sea Technol **43**, 53–54 (2002)
96. G.A. D'Alterio, A thirst-quenching role. *Pumps Systems* **6**, June (2006), http://www.mp-gps.com/pdf/TechSpecs/MPGPS/Pumps_Systems_article.pdf

Chapter 7
Environmentally Assisted Cracking of Stainless Steels in Desalination

Abdelkader A. Meroufel

7.1 Introduction

Seawater desalination and corrosion resistant alloys (CRAs) is a long story that continues with the development of various alloys and processing techniques to overcome in-service failures. Stainless steel corrosion failures in desalination plants such as pitting and crevice corrosion were discussed in Chaps. 5 and 6. In the present chapter, the discussion will be limited to Environmental Assisted Cracking (EAC). Indeed, the combination of mechanical (intrinsic or extrinsic) stresses (static or dynamic) and corrosive environment leading to EAC represents one of the main challenges to the reliability of desalination equipment.

Stainless steels are present in the market with various grades offering a large spectrum of performance depending on the application requirements. The main advantage of these alloys is their very low general corrosion rate due to the stable passive film formed spontaneously. However, their susceptibility to localized corrosion induced great development efforts that continue in terms of alloying, microstructure control, and surface modifications. In parallel to this development, scientists continue to study the mechanisms of degradation of the different grades for diverse applications. The aim is to provide the industry with a complete figure on the performance map for the different stainless steels grades.

While a lot of progress has been made toward the understanding and mitigation of EAC for different alloys-environment combinations, still failures occur mostly for multiple reasons. Susceptible steel grades in old desalination plants, favorable operation and maintenance practices, quality control and quality assurance issues, and unpredicted environmental conditions represent the main reasons for these continuous failures. In terms of statistics, EAC failures and learned lessons are more

A. A. Meroufel (✉)
Desalination Technologies Research Institute, Saline Water Conversion Corporation,
Jubail, Saudi Arabia
e-mail: ameroufel@swcc.gov.sa

© Springer Nature Switzerland AG 2020
V. S. Saji et al. (eds.), *Corrosion and Fouling Control in Desalination Industry*,
https://doi.org/10.1007/978-3-030-34284-5_7

documented with other industries such as petrochemicals, oil and gas and nuclear power plants compared to desalination. Although the stress level (less than 5% of yield stress) and EAC rate are low in regard to the yield stress and asset life-time respectively, the catastrophic consequences emphasize the importance of the subject and justify the high level of research interest from scientific and engineering communities.

The scope of the present chapter will be limited to Stress Corrosion Cracking (SCC) for static equipment and Corrosion Fatigue (CF) for dynamic equipment. These two mechanisms and related research efforts will be discussed in the specific conditions relative to desalination plants including thermal and reverse osmosis (RO) desalination processes. For both cases, the fundamentals and proposed theories are presented. Typical EAC failure cases are also presented followed by mitigation practices.

7.2   Environmental Assisted Cracking Mechanisms

EAC include many failure types such as SCC, CF, and hydrogen damages (hydrogen embritllement (HE), hydrogen-induced cracking (HIC) etc.).

7.2.1   Stress Corrosion Cracking

SCC phenomenon is a combination of three interrelated elements as illustrated in Fig. 7.1. The three elements are in a complete synergy where susceptibility of a material to SCC has two perspectives; corrosion and stress. Thus, any material which can suffer from localized corrosion (pitting or crevice) can develop stress raisers where SCC will initiate leading to catastrophic failure. The transition from pits (and other localized corrosion sites) to SCC depends on the pit depth and shape, and local stresses/strains/stress-intensity factor.

In the same manner, any material which develops micro or macro cracks will provide localized corrosion site which initiates a SCC damage mechanism.

Usually, SCC risk is considered for certain combinations of material and environment as shown in Table 7.1. However, Anderson et al. [1] presented an emerging issue where service conditions that are not conducive to SCC may produce over time a SCC situation. This aspect can be anticipated from asset integrity perspective. Indeed, the combination, material/environment conducing to SCC can be either from normal service environment conditions or unanticipated environmental conditions.

In terms of environment, for each factor, there is a critical threshold value initiating a SCC phenomenon. Due to the synergy between SCC elements, this critical threshold value is specific for each material parameters (composition, microstructure) and stress level. In parallel, some factors impact the propagation rate of SCC. Then, the number of environments promoting SCC has increased significantly

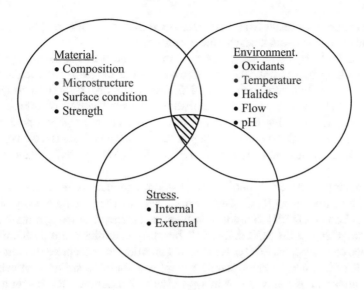

Fig. 7.1 Required elements for stress corrosion cracking mechanism

Table 7.1 Alloy-environment systems exhibiting SCC. Reproduced with permission from Ref. [2]; Copyright 2018 @ Springer

Metal	Environment (solutions)	Temperature (°C)
Carbon steel	Caustic	> 80
	Carbonate-bicarbonate	≥ 50/60
	Nitrate	≥ 50/60
	Phosphate	≥ 50/60
	Liquid ammonia with traces of water CO/ CO_2/H_2O	T_{amb}
Austenitic stainless steel	Neutral aerated containing chloride	> 80/100
	Acid containing chloride	≥ T_{amb}
	H_2S containing chloride	≥ T_{amb}
	Caustic	≥ 80/120
	Oxygen containing pure water	≥ 100
Nickel alloys	Caustic	> 100/200
Nickel alloys (Cr < 30%)	Water with dissolved H_2	> 250/280
Sensitized stainless steel and nickel alloy	Polythionates and thio-sulphates	≥ T_{amb}
Copper alloy	Ammonia containing	T_{amb}
Aluminum alloy	Chloride containing	T_{amb}
Titanium alloy	Alcohol containing chloride	T_{amb}

in recent times as a result from new developed SCC tests allowing better electro-chemical control and crack development monitoring [2].

As mentioned by Parkins [3], SCC is also influenced by the electrode potential where cracking occurs only within certain potential ranges for particular metal–environment combination. In this regard, Saithalla et al. [4] could plot critical SCC potentials versus pitting resistance equivalent number (PREN) of stainless steels and corresponding pitting potentials. In the same manner as pitting potential and other specific electrode potentials are dependent on the material surface-environment system, SCC critical potential can provide SCC susceptibility indication for any alloy.

Few authors studied the contribution of microorganisms or microbial biofilm on the SCC on carbon steel [5, 6]. These works were focused on the activity of Sulfur Reducing Bacteria (SRB) to produce H_2S as the precursor of the hydrogen absorption participating in the SCC failure in laboratory conditions. For austenitic stainless steels, one suspected SCC case in biotic conditions was reported by Duret-Thual [7]. On the other hand, duplex stainless steel was found to suffer from very rapid SCC in presence of a mixture of marine biofilm containing SRB bacteria during slow strain rate testing compared to the abiotic condition [8]. The contribution of biofilm in the electrochemical process during both SCC initiation and propagation phases is a fact which merits more investigation.

Among surface condition factors, roughness has a significant effect on SCC initiation. High roughness would correspond to deeper grooves that provide both corrosion initiation sites and stress raiser for crack initiation thus reducing the incubation time to cause SCC in the same manner as surface machining which induces residual stresses as studied by Turnbull et al. [9].

The complexity and synergy between SCC factors made the task for SCC risk assessment difficult where it was necessary to wait for the development of computing techniques to treat the high number of SCC data. Indeed, with Neural network processing techniques, corrosion community started to take benefit and address SCC failures and testing data to generate categorized output data for SCC initiation prediction [10, 11]. More focus was on SCC initiation risk prediction based on safe-life design approach philosophy where no significant damage is allowed and minimum inspection of the equipment is considered. However, SCC propagation prediction fit to the damage tolerance approach where fracture mechanics is the most adopted methodology. These two philosophies were discussed by Wanhill [12] for SCC testing methodologies.

Regarding stress causing SCC, it is believed that SCC occurrence is greater from internal residual than external service operation stress. Residual stresses that can be generated by welding or cold work operations could induce SCC if they could generate a dislocation movement at material defect regions. This residual stresses threshold for SCC of stainless steels still debated in the literature. From another perspective, it has been found that tensile residual stresses are more favorable to induce a SCC compared to compressive residual stresses. This lead to the use of laser peening as surface engineering to reduce SCC occurrence [13]. Usually, SCC is discussed with a focus on static stress. However, slow monotonic straining or low amplitude cyclic stress was also found to accelerate SCC in many alloys.

SCC mechanisms still discussed by scientists with different approaches within challenging testing issues context. From atomistic to macro levels, it is important to mention that there is a difficulty to unify SCC mechanism for different alloy-environment systems. Even for one material-environment combination, little consensus exists between scientists. This is due mainly to the difficulty to find techniques to obtain data on the atomistic level. In this regard, Persaud et al. [14] reviewed the recent efforts toward the use of microscopic techniques combined with computational science to develop SCC atomistic models from quantitative micro-nano (QMN) approach initiated by Roger Staehle in 2011. This promising approach is based on segmentation of SCC phenomenon into six segments where the critical segment is the precursor which need to be fixed to mitigate SCC initiation risk. One more challenge to identify the mechanism, is the presence of corrosion product on the fracture surface which hinds fine details. In this direction, Turnbull et al. [15] adopted the pulsed direct current potential drop (DCPD) to monitor the crack growth rate with reduced noise from corrosion products. This practice would provide some indirect mechanistic information.

At the end, most of the microscopic observations suggest the involvement of three environmental interactions in SCC damage i.e. adsorption at crack tips, hydrogen (generation, diffusion and segregation at grain boundaries), and dissolution at crack tips [13]. Table 7.2 summarizes the most discussed SCC mechanisms in the literature with relative references. According to Lynch [16], only three mechanisms have wide applicability based on model-system observations. This includes

Table 7.2 SCC proposed mechanisms

Mechanism	Brief description	Authors, Year
Dissolution-based (slip dissolution)	Atoms removal at crack tips preferentially in normal direction to the applied stress promoted by anodic precipitates, segregants along grain boundaries, or active path generated by stress/strain concentrated.	Scully, 1977 Parkins, 1992
Adsorption-based (adsorption induced dislocation-emission)	Adsorption reduces surface energy and weaken surface atoms bonding.	Uhlig, 1960 Lynch, 1976
Hydrogen-based	Involve either adsorbed H, dissolved H, or hydrides at crack tips that weaken interatomic bonds (easy dislocation emission/decohesion @ crack tip).	Lynch, 2003, Beachem, 1972 Jones, 1985
Vacancy-based	Vacancy generation and diffusion leading to deformation.	Aaltonen, 1997, Jones, 1996
Surface-mobility	Surface diffusion of atoms from elastically stressed crack tip to adjacent vacant lattice site.	Galvele, 1987
Film-induced-cleavage	Brittle film at crack tip (formation and fracture) proposed for trans and intra-granular cracking.	Sieradzki, 1985
Corrosion enhanced localized plasticity	Depassivation/localized dissolution along 111 plane/ localized shearing/pile-up of dislocations/propagation of crack/crack arrest.	Magnin, 1980

adsorption-induced dislocation emission (AIDE) or decohesion at crack tips, hydrogen-enhanced cohesion (HEDE) ahead the crack tip and localized strain assisted dissolution. These mechanisms can coexist in certain material-environment systems.

The validation of any mechanism should be conducted by the combination of detailed metallographic and fractographic studies. Then, mechanistic insights are useful information for material designers, industrial operator and failure investigators to control SCC failures. Practically, SCC failures are investigated at two levels i.e. macro and micro. For the former, SCC fractures are characterized by lack of macroscopic deformation, absence of general corrosion and difference in corrosion products between fracture and external surfaces. For instance, the external surface of stainless steel is shiny but the fracture surface is rusty colored. On the microscopic level, two types of cracks are observed usually i.e. intergranular and transgranular associated sometimes with plastic deformation. Depending on the crack type, the fractured surface can be either smooth or featureless (intergranular) or serrated steps, which form fan-like or herringbone patterns (transgranular) [14]. The fractography of SCC failed samples were used by some authors to determine the initiation and propagation rates of SCC [17].

The quantification of SCC growth rate has been tried through the use of electrical resistance measurement techniques which offer better accuracy of crack growth velocity [18]. However, this is conducted in laboratory testing conditions that are not usually correlating field service parameters. This pushes some authors to attempt the quantification based on SCC mechanisms but without high confidence due to the pending mechanisms questions. One typical example of this, is the work conducted by Galvele et al. [19] considering surface mobility mechanism which was discussed by Turnbull et al. [20] who pointed the unsatisfactory assumptions of this model.

The correlation of accelerated SCC laboratory tests with field applications is a common fact with other materials damages. However, efforts continue through the review and up-date testing international standards with various configurations and objectives. Specimen type (smooth or pre-cracked), load method (constant or slow strain) and corrosive environment ($MgCl_2$, $CaCl_2$, $NaCl$) are the main SCC testing parameters that vary with application and test purpose (qualitative or quantitative). Slow strain rate test (SSRT) seems the most adopted by industrials and scientists due to its convenience and aggressiveness which shorten the experiment duration. Henthorne reviewed the literature related to this test covering different industry applications (oil and gas, nuclear power plant, buried pipes, etc.) [21]. The author mentioned the weakness of discussion on the critical parameters (possible critical strain rate, potential susceptibility zones, and importance of aeration/deaeration) affecting SSRT test in international standards. However, the development of tools for electrochemical measurements during the test and its combination with fracture mechanics techniques seems a promising avenue to elucidate more mechanistic SCC features.

Due to the similarities between EAC mechanisms, the mitigation practices will be discussed at the end of the chapter.

7.2.2   Corrosion Fatigue

CF also belongs to EAC damage types where it is distinguished by the involvement of cyclic/fluctuating load. Some authors consider that CF is a mechanically assisted degradation [22]. Indeed, many alloys can fail below their yield stress due to the fatigue phenomenon. There is maximum stress during cycling below which material can sustain several cycles without failure. The fatigue life increases as the maximum stress during cycling decreases until the fatigue limit (stress threshold) is reached. Fatigue data are usually represented by stress versus the number of cycles to failure (S-N curve or Wöhler diagram). Wang distinguished the true fatigue limit, which does exist for a limited number of alloys [23]. In fact, due to the relation between EAC types, some authors distinguished between what they call true corrosion fatigue (TCF) and stress corrosion fatigue (SCF) depending if the metal is suscep-tible to SCC [2]. This lead to the interference between CF and SCC making a diffi-culty to distinguish between them.

The presence of localized corrosion phenomenon, in particular, reduces signifi-cantly the fatigue life of materials either by affecting initiation or propagation (growth) phases. Some authors include the effect of the improper design of cathodic protection (CP) system. However, this factor would induce a hydrogen embrittle-ment which is not covered in the present chapter. It is now established that higher damage of CF is observed compared to the sum of individual processes (corrosion and fatigue). To illustrate the impact of environment, some authors did not stop to the fatigue properties in the air but go up to the absolute non-corrosive conditions such as vacuum or under inert gas atmosphere based on some experimental findings which show certain role of the air's oxygen in the fatigue mechanism [24].

In terms of crack types, environment and mechanical load define the cracks of CF that can be transgranular, intergranular or mixed all being perpendicular to the tensile stress direction. It is important to notice that some authors addressed the cracks terminology in EAC; the cracks in EAC are classified into three categories; short, small and long. Zhou et al. [25] considered that short cracks correspond to crack with a short length and cracks developing from surface flaws correspond to small cracks (small in length, width and opening). These considerations are well supported and based on experimental microscopic observations.

CF initiation sites can be either from localized corrosion sites (pits) or material intrinsic defects such as slip bands, twins, interphases and grain boundaries. It is well accepted that 95% of structure/equipment life is spent in fatigue crack initia-tion. Then, crack propagation governed by the so-called "Paris law" which link the fatigue crack growth rate per cycle to the difference between maximum and minimum stress intensity factor has to be studied through fracture mechanics. In terms of materials and CF crack growth, Tice [26] mentioned that it is relatively well understood for low alloy steels contrary to the stainless steels where the pend-ing knowledge gaps were summarized through EPRI review in 2011 [27].

The factors that may affect CF progress rate will not go beyond the three contribut-ing elements i.e. material, environment and stress (amplitude, frequency, and wave-form) [22, 24]. One of the interesting metallurgical factors is the grain orientation, which originates from material processing and induces anisotropy in the material. This concept was noticed in mechanical fatigue of steels during the 1950s and continue to attract research interest. It has been found that grain orientation could affect both dissolution and hydrogen absorption, depending on surface finishing [28, 29].

The synergy between electrochemical corrosion and mechanical crack growth stay complicated as discussed previously for SCC. Recently, Xu et al. [30] developed a multiphysics model based on mechanochemistry of solid surfaces initiated by Gutman during 1990s. The authors assumed one dimensional SCC crack propagation direction for X100 pipeline steel. The obtained experimental and numerical results showed that the mechanical-electrochemical interaction affects the corrosion process only when the corrosion defect is under plastic deformation. These results are in agreement with more recent findings from Fatoba et al. [31] on X65 steel. Therefore, it would be interesting to adapt this model for CF of stainless steels. However, probabilistic analysis has to be associated with mechanochemistry models to address the randomness character of pitting, the behaviour of small cracks and microstructures [32]. Despite these efforts, the impact of corrosion on crack initiation for different materials and conditions remain unclear [33].

Fatigue usually appears transgranular unless it is associated with the SCC process where fatigue would contribute only to the initiation mechanism. Other combinations between EAC mechanisms including HE, SCC and TCF are discussed elsewhere [24]. The fingerprint on the macro aspect of fatigue is the presence of striations with borders that are smoothed out by corrosion, which is not the case for mechanical fatigue. Microscopically, these striations can be either ductile or brittle depending on the involvement of HE. This later or SCC contribution to the failure produce brittle striations as reported by Torronen et al. [34] for stainless steel grade 304 L tested in boiling water at different oxygen levels with some hydrogen coming from hydrazine treatment. Microscopic observations also allow the determination of fatigue propagation direction.

CF mechanisms are categorized into two processes; anodic slip dissolution and hydrogen embritlement [23]. Its evolution has been divided into different phases i.e. surface film breakdown, pit growth, pit to crack transition and cracking (small and long crack growth) [35]. Most of CF studies were conducted in applications that are energy-based such as boiler water reactors (BWR) and pressurized water reactors (PWR) called collectively light water reactors (LWRs). However, dynamic equipment such as pumps in the desalination industry can suffer from corrosion fatigue since it is handling corrosive seawater and brine liquids. In the following, both SCC and CF of stainless steels in typical desalination environment will be discussed.

7.3   Environmental Assisted Cracking on Stainless Steels

Material selection in the desalination industry was always centred on corrosion-resistant alloys among which stainless steels are the first choice. These alloys provide excellent corrosion performance due to their thin, stable and self-healing

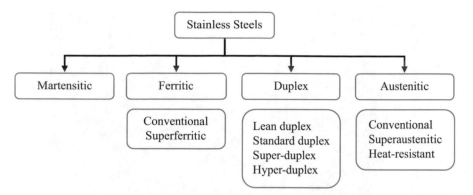

Fig. 7.2 Stainless steel grades

passive layer which is based on chromium oxide. The various challenges facing this industry pushed metallurgists and manufacturers to develop many alloys for different application purposes. Chemical composition and heat treatment are the most adopted approaches to tailor stainless steels. This lead to the formation of austenitic (FCC[1] structure), ferritic (BCC[2] structure), martensitic (BCT[3] structure) or duplex (ferritic and austenitic) microstructure based grades. Figure 7.2 shows the classification of stainless steel grades with typical sub-grades.

Due to the limitation in their use in the desalination industry, ferritic and martensitic stainless steels will not be covered. Austenitic stainless steels are the most important group adopted in aggressive environments since they do not undergo a ductile/brittle transition. However, the development of duplex grades competes successfully offering a lot of advantages of both mechanical and corrosion resistance properties breaking down the final cost. While they require careful processing and welding operations, they are gaining acceptance within the desalination industry from specifications stage.

Based on the interdependence between chromium and nickel equivalents, stainless steel metallurgy can be predicted as illustrated in the famous Schaeffler diagram as shown in Fig. 7.3.

The main mechanical properties including tensile and hardness of the different stainless steel grades are shown in Table 7.3 [36]. The superior mechanical properties of duplex SS compared to the austenitic family is clear.

The corrosion resistance of stainless steels is usually predicted using different parameters such as PREN, CPT, and CCT as already discussed in Chaps. 5 and 6. All these parameters provide a ranking which remains qualitative for the end-user. Then, testing including accelerated laboratory and field conditions remains the ultimate tool for decision-makers.

---

[1] Face Cubic Centered

[2] Body Cubic Centered

[3] Body Centered Tetragonal

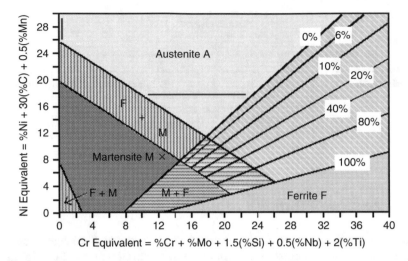

Fig. 7.3 Schaeffler diagram showing stainless steel grades

Table 7.3 Minimum room temperature tensile properties and maximum hardness of wrought solution annealed stainless steels. Reproduced with permission from Ref. [36]; Copyright © 1997 Woodhead Publishing and Elsevier

Family	Grade	$R_{p0.2}$ (MPa)	R_m (MPa)	As*(%)	Hardness	
					HB	HRC
Ferritic	S40900	205	380	20	179	–
	S44700	415	550	20	223	20
Austenitic	S31603	170	485	40	217	–
	S31254	300	650	35	223	–
Duplex	S31803	450	620	25	293	31
	S32304	400	600	25	290	32
	S32550	550	760	15	302	32
	S32750	550	795	15	310	32
	S32760	550	750	25	270	–

*Elongation

The resistance of the different stainless steel grades to EAC remains an attractive research topic where the publication of results continue. For instance, it is well established in the literature that austenitic stainless steels are susceptible to SCC in hot concentrated chloride solutions, chloride contaminated steam which represents typical desalination plant conditions [2]. Some laboratory testing conducted usually by stainless steel manufacturers for all their developed grade provides some indication on the susceptibility of certain grade to EAC depending on environmental parameters. Figure 7.4 shows the typical SCC risk prediction on different SS grades as a function of chlorides and temperature.

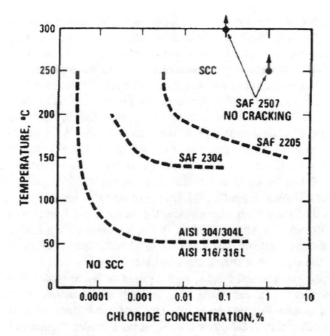

Fig. 7.4 SCC risk on different stainless steels in neutral chloride solution. Reproduced with permission from Ref. [37]; Copyright @ 2007 Taylor & Francis

However, it is important to consider other environmental parameters in the phenomenon, such as oxygen and pH. Besides, austenitic stainless steel grades exposure to seaside atmosphere environment could lead to SCC contrary to what was believed by Trumann et al. [38] where many research works are undertaken recently [39, 40]. It is commonly accepted that iron contamination on stainless steels surfaces results in transgranular SCC when stored in coastal environments at ambient temperature [41]. Mechanistically, stress intensity factor values in the second phase of SCC process for austenitic stainless steels in hot chloride solution indicates that a physical-chemical process (dissolution, diffusion or adsorption) is controlling the crack velocity rather than mechanical one [42].

Another source of chlorides in thermal desalination plant is the thermal insulation material. Indeed a contact between hot stainless steel and chloride containing thermal insulation can induce a SCC process if stress element is present. That is why regulation emphasizes certain limit of chlorides in the insulation in contact with hot stainless steel sensitized or non-sensitized.

Among stainless steel grades, duplex and ferritic structures are more resistant to SCC. SCC resistance of duplex was demonstrated by Tsai et al. [43] for the standard duplex 2205 in NaCl solution with concentration up to 26% and temperature up to 90 °C. The superior performance of duplex stainless steels in hot chloride environments is associated with 3–10 times threshold stress intensity values higher than for austenitic which means that duplex components must be highly stressed or defective to fail by SCC in hot chloride solutions.

However, it should be pointed out that not all duplex stainless steels are resistant to chloride SCC [44]. In all situations, the resistance of any SS grade is a function of localized corrosion resistance and mechanical strength. This led us to the chemical composition and metallurgical parameters (undesirable phases or precipitates, residual stresses, etc.) as the main factors affecting SCC resistance.

Regarding the impact of the chemical composition on SCC resistance, some authors suggested the concept of stacking fault energy (SFE), which should be high to offer a good SCC resistance [44]. SFE is related to the chemical composition of the SS grade where typical equations proposed can be found in the literature [45, 46]. It is important to mention that SFE decreases with temperature increase. Thus, SFE values should be compared at the temperature at which SCC resistance is aimed for the different SS grades [47]. On the other hand, lowering the carbon content reduces the SFE slightly where values for 304 and 304 L or 316 and 316 L are not quite different. For duplex stainless steels, intensive efforts are made to study the effect of certain alloying elements on EAC resistance of DSS. This includes manganese and nitrogen in certain environments [48, 49].

In terms of metallurgical factors, sigma phase presence in duplex SS has been shown to decrease SCC resistance [50]. Indeed, its formation creates depletion zones on its surrounding providing localized corrosion sites. This makes duplex welds more vulnerable to SCC risk than solution annealed duplex stainless steels. This is also the case of austenitic SS grades with sensitization where the remedial measures include lowering carbon content (typically below 0.03%) and introduction of carbide formers such as titanium and niobium. Some authors consider that it is necessary to control the combination carbon, chromium and nickel to avoid sensitization and intergranular corrosion (IGC) [51].

On the other hand, welding of duplex SS grades is known to coarsen the ferrite grain and reduce SCC resistance. However, coarser grain in austenitic SS was found to improve low cycle fatigue life above certain critical strain amplitude. This was observed for austenitic stainless steel 316 LN and attributed to the formation of martensite which absorbs energy during cyclic straining and introduces compressive stresses [52]. If welding is not controlled it can generate undesirable effects such as sensitization, depleted zones, and residual stresses. This later effect was explained by the high coefficient of thermal expansion and a low heat transfer rate of stainless steels creating a distortion or high level of residual stress in weldments.

At the microstructure level in modern duplex steels, SCC occurs in the ferrite and is arrested by the austenite. This has been explained through electrochemistry where ferrite plays the role of the anode and the austenite is the cathode. When the crack approaches a mixed potential, the ferrite is polarized above its normal cracking potential and cracks quickly while the austenite is below its cracking potential. However, in the presence of strong oxidant (oxygen, biocide), or if oxygen is supplied through a thin boiling layer (as in evaporating seawater, which leaves behind a layer of saturated $MgCl_2$ solution), then the austenite is no longer protected by the ferrite and threshold stress intensity drops significantly. The opposite situation can occur if the ferrite remains passive in the crack while the austenite corrodes which switch the protection role to the ferrite.

Selective dissolution in duplex stainless steels was found helping to remove the stress concentration by eliminating the slip bands. In the same time, it can modify the fracture surface morphology once created by fatigue [53].

7.4   Stress Corrosion Cracking of Static Equipment

In the present section, SCC of austenitic stainless steels will be discussed based on desalination failure history and laboratory studies. Very limited number of publications are available on SCC specific to desalination in the open literature. Prakash et al. [54] tried to assess through laboratory tests the SCC risk on different MSF engineered alloys in the presence of chlorides, CH_3COOH, H_2S and oxidants (NACE and Shell solutions). The list of materials included 316 L and 317 L tested at ambient temperature in a proof ring to define time to the rupture with parallel microscopic observations. The results showed a good resistance of these alloys in the tested conditions with high passive current densities relative to other alloys. More recently, Abuzeid et al. [55] studied the susceptibility of 316 L and superduplex 2750 to SCC in the proof ring as replacement of Ni-resist brine recycle pump casing. As expected, superduplex showed excellent SCC resistance in hot brine compared to the austenitic 316 L.

Practical cases of SCC on static equipment in desalination plant are occurring but not widely shared in publications. One of the most known cases is the SCC of austenitic 316 L venting pipe in MSF desalination plant. Indeed, the combination of hot vapors contaminated by less than 10 ppm of chloride and CO_2 non-condensable gas especially for the first high-temperature stages induced SCC attack with very slow progress. Figure 7.5 shows a typical SCC case from a MSF desalination venting line from stage 1, which was in service for 24 years. The cracks appear at and

Fig. 7.5   SCC attack on 316 L venting pipe from first stage of MSF evaporator

Fig. 7.6 Microstructure and crack path far from the weld joint location of 316 L venting pipe from 1st stage of MSF evaporator

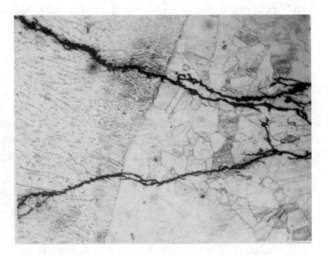

Fig. 7.7 Microstructure and crack path at the weld joint location of 316 L venting pipe from 1st stage of MSF evaporator

far from the weld joint and originate from the internal of the pipe where the corrosive environment is present.

Microscopic analysis on the failed sample at and far from weld joint are shown in Figs. 7.6 and 7.7 respectively. Based on a visual inspection and surface analysis, the SCC failure mechanism proposed included water droplets (entrained with noncondensable gases during suction) impingement on the internal surface of the venting pipe combined with a pitting corrosion process accelerated by chlorides, temperature and dissolved CO_2.

SCC of venting lines in MSF desalination is a chronic problem for pipes made of austenitic 316 L SS. Its growth is known to be slow where it has been observed that venting pipes can stay above 30 years before to decide to replace or material change. This lifetime of the venting pipe can be explained by the low aggressiveness of the environment and high thickness of the pipe. Then, material up-grade of venting pipes was suggested through the history of MSF plants design. In the mid-1990s, superaustenitic 904 L SS was proposed before the consideration of standard duplex 2205 grade which offers good SCC resistance and competitive life cycle cost. Non-metallic pipes for venting such as Fiber Reinforcing Polymer (FRP) are already adopted but up to certain temperature limit where improvements can be studied to extend their use up to high-temperature stages.

7.5 Corrosion Fatigue of Rotating Equipment

Corrosion fatigue of stainless steels in seawater environment was fully studied in the literature within the context of offshores structures. Many authors studied different grades considering environment parameters such as seawater temperature, salinity and depth [56–58]. Corrosion fatigue cases in desalination are not well shared in the open literature. One of the limited studied cases consists of the CF failure of make-up seawater pump shaft shown in Fig. 7.8 [59].

The cracked surface shows a rotating bending fatigue failure appearance with two zones i.e. fatigue and instantaneous. Large final fracture (instantaneous) zone ($\approx 45\%$ of the total surface) indicate a low cycle and high-stress load. The received shaft failed at threaded connection location as shown in Fig. 7.9. The interpenetration between fatigue and instantaneous zones indicating torsional load that could be due to the post-failure.

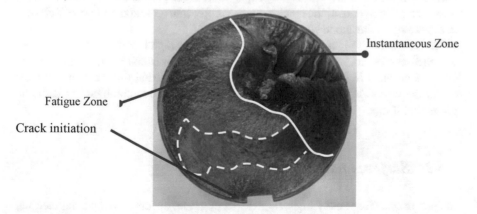

Fig. 7.8 Corrosion fatigue of seawater make-up pump shaft

Fig. 7.9 Complete view of the fractured shaft as received

Chloride ingress at the keyway area combined with stress concentration at threaded connections contributed to the initiation of a corrosion fatigue process. A ductile mode of fracture was observed associated with post-failure damage (torsion and rubbing).

7.6 Mitigation Practices

7.6.1 Material Selection

Depending on the aggressiveness of the environment and the equipment design, the material selection can dictate the use of high grades of stainless steel. Before jumping to the costly higher SS grades, some initial attempts to add some alloying elements was successful in certain applications. This is the case of 316 L which has been alloyed with nitrogen offering an enhanced CF life in sodium chloride solution. Based on microstructural observations, it was proposed that nitrogen addition induced slip reversibility. CF crack propagation proceeded in a transgranular mode prior to tensile overload; the cracking was wholly transgranular and no switch over to the intergranular mode was observed [60].

The selection of higher SS grades is based on their high localized corrosion resistance (discussed in Chaps. 5 and 6) combined with their excellent mechanical properties mentioned in Table 7.3. It is now well accepted that duplex stainless steel offers the best EAC resistance keeping strict control when combined to cathodic protection if any.

7.6.2 Surface and Materials Engineering

Instead of selecting costly high stainless steel grade, different surface engineering techniques has been studied with the aim to reduce both localized corrosion attack and crack development. This includes laser surface treatment where proper control of laser parameters (beam power, size and transverse speed) affect positively carbides, chromium depleted zones, impurities or unwanted precipitates [61].

Ion nitriding is another example of surface treatment which is widely industrialized allowing to harden steel surface via nitrogen diffusion which increases surface resistance to wear and fatigue initiation. However, its direct application to stainless steels, while enhancing their mechanical properties, also causes a marked degradation in their corrosion resistance. Similarly, by the adaptation of the plasma process with nitrogen and/or carbon elements, the compromise between tribological and corrosion resistance properties can be satisfied. It was observed that carbon surface treatment improve better surface tribology compared to nitrogen [62, 63]. Stainless steel surface nitriding can also be achieved by low-temperature liquid weld cladding process. The resultant higher surface hardness and corrosion resistance were demonstrated in the case of 316 L grade [64].

Chromization of 316 L grade was performed by fluid bed reactor chemical vapor deposition (CVD). The material was studied in terms of localized corrosion resistance in reverse osmosis desalination conditions. Excellent localized corrosion resistance was obtained including welded samples. Chromium-enriched surface layers are found to be retained after the welding process, which leads to the improvement in corrosion resistance of weld zones [65].

The elimination of residual stresses and undesirable phases within materials susceptible to EAC can be obtained through a proper heat treatment either during manufacturing or after installation when performing joining operations (welding). For instance, Post-Welding-Heat-Treatment (PWHT) on austenitic stainless steel grades is a debatable topic between engineers. While ASME code does not allow such as treatment due to dimensional variation and loss of mechanical properties risk, it is suggested for some purposes. This includes the anticipation of corrosive service, the homogenization of the structure and stabilization.

7.6.3 Operation and Maintenance Practices

The initiation of localized corrosion followed by EAC process can be avoided through proper operation practices. This is especially related to start-up and shutdown procedures. Flushing with distilled water and air drying are the main recommendations adopted in desalination plants. Stagnant chloride-based solution avoidance is the general rule followed not only in desalination but in all industrial processes.

From a maintenance point of view, SS surface remains a precious part to be protected from mechanical damage or contamination such as by iron and salts.

7.7 Conclusions and Outlook

In this chapter, an attempt was made to cover the main issues related to EAC of stainless steels with a focus to desalination conditions. Corrosion scientists still looking to the mechanism aspect within a context of testing standardization and capabilities challenges. Lynch summarized some pending mechanistic questions for SCC process [15]. This includes the crack growth mechanisms versus localized plasticity, crack continuity, fracture planes and direction. These questions should be addressed to predict the behavior of the different grades and anticipate the failure with proper inspection and maintenance timing. Also, the understanding of phenomenon will allow the material developer to design and manufacture suitable alloys with required properties.

From his side, Magnin discussed some research needs for CF such as the crack tip chemistry modelling, quantitative analysis of corrosion-deformation interaction at crack tip and interaction between CF and SCC near the fatigue threshold [66].

It is important to mention that due to the nature of EAC process, multiphysics modelling seems the best way to solve all modeling questions. However, the validation of these models will require intensive testing investments that need to be addressed by industrial professionals. For desalination industry, there is no doubt on the EAC resistance superiority of duplex stainless steel grades over austenitic grades. However, when the lifetime question is raised, EAC models become a necessity.

References

1. P.L. Anderson, T.M. Angeliu, L.M. Young, Immunity, thresholds, and other SCC fiction. Proceedings of the Staehle Symposium Chemistry and Electrochemistry of Corrosion and SCC, The Materials Society, 2001
2. Lazzari L, Pedeferri M, Corrosion science and engineering, (Springer, 2018), ISBN: 978–3–319-97625-9, p. 255
3. R.N. Parkins, Stress corrosion cracking, in *Uhlig's Corrosion Handbook*, 3rd edn., (Wiley, 2011)
4. J.R. Saithala, S. McCoy, A. Houghton, J.D. Atkinson, H.S. Ubhi, Critical stress corrosion cracking potentials of stainless steels in dilute chloride solutions. Corrosion 2010, NACE-10287, NACE International, Houston
5. R. Javaherdashti, R.K. Raman Singh, C. Panter, E.V. Pereloma, Microbiologically assisted stress corrosion cracking of carbon steel in mixed and pure cultures of sulfate reducing bacteria. Int. Biodeter. Biodegr **58**, 27–35 (2006)
6. F. Xie, X. Wang, D. Wang, M. Wu, C. Yu, D. Sun, Effect of strain rate and sulfate reducing bacteria on stress corrosion cracking behaviour of X70 pipeline steel in simulated sea mud solution. Eng. Fail. Anal **100**, 245–258 (2019)
7. C. Duret-Thual, The effect of H$_2$S on the corrosion of steels. *In*: *Understanding Biocorrosion*, ed. by T. Liengen, R. Basseguy, D. Feron, I. Beech, (Elsevier, 2014), p. 404, ISBN: 9781782421207
8. R. Javaherdashti, R.K. Raman Singh, C. Panter, E.V. Pereloma, Stress corrosion cracking of duplex stainless steel in mixed marine cultures containing sulphate reducing bacteria. Proceedings of Corrosion and Prevention 2004 (CAP04), 21–24 November 2004, Perth

9. A. Turnbull, K. Mingard, J.D. Lord, B. Roebuck, D.R. Tice, K.J. Mottershead, N.D. Fairweather, A.K. Bradbury, Sensitivity of stress corrosion cracking of stainless steel to surface machining and grinding procedure. Corros. Sci **53**, 3398–3415 (2011)
10. H.M.G. Smets, W.F.L. Bogaerts, SCC analysis of austenitic stainless steels in chloride-bearing water by neural network techniques. Corrosion **48**, 618–623 (1992)
11. S. Zhou, D. Coleman, A. Turnbull, Application of neural networks to predict sulphide stress corrosion cracking of duplex stainless steels. NPL Report MATC (A) 21 (2001), http://eprint-spublications.npl.co.uk/1953/
12. R.J.H. Wanhill, (1991) Fracture Control Guidelines for Stress Corrosion Cracking of High Strength Alloys, NLR Technical Publication TP 91006 L, Amsterdam
13. A.J. Sedriks, *Corrosion of Stainless Steels*, 2nd edn. (Wiley, New York, 1996)
14. S.Y. Persaud, J.M. Smith, R.C. Newman, Nanoscale precursor sites and their importance in the prediction of stress corrosion cracking failure. Corrosion **75**, 228–239 (2019)
15. A. Turnbull, S. Zhou, Electrochemical short crack effect in environmentally assisted cracking of a steam turbine blade steel. Corros. Sci **58**, 33–40 (2015)
16. S.P. Lynch, Mechanistic and fractographic aspects of stress corrosion cracking (SCC). in *Stress Corrosion Cracking Theory and Practice*, ed. by V.S. Raja, T. Shoji, (Woodhead publishing, 2011), ISBN: 9780857093769
17. T. Kobayashi, D.A. Shockey, R.L. Jones, Deriving SCC initiation times and growth rates from Posttest fractographic analysis. Corrosion **47**, 528–535 (1991)
18. M.F. Hurley, A. Cunningham, D. Lysne, S. Acharya, B.J. Jaques, D.P. Butt, A Condition Monitor for Atmospheric Induced Stress Corrosion Cracking, NACE-11634, Corrosion 2018, NACE International, Houston, 2018
19. J.R. Galvele, Stress corrosion cracking mechanism based on surface mobility. Corros. Sci **27**, 1–33 (1987)
20. A. Turnbull, Modeling of environment assisted cracking. Corros. Sci. **34**, 921–960 (1993)
21. M. Henthorne, The slow strain rate stress corrosion cracking test – a 50 years retrospective. Corrosion **72**, 1488–1518 (2016)
22. W. Glaeser, I.G. Wright, Forms of mechanically assisted degradation, in *Corrosion: Fundamentals, Testing, and Protection*, ed. by S. D. Cramer, B. S. Covino, vol. 13A, (ASM Handbook, ASM International, 2003), pp. 322–330
23. Y.Z. Wang, Corrosion fatigue, in *Uhlig's Corrosion Handbook*, 3rd edn., (Wiley, New York, 2011), p. 195
24. P.P. Milella, *Fatigue and Corrosion in Metals*. (Springer, 2013), ISBN: 978–88–470-2336-9 p.767
25. S. Zhou, M. Lukaszewicz, A. Turnbull, Small and short crack growth and the solution-conductivity dependent electrochemical crack size effect. Corros. Sci. **97**, 25–37 (2015)
26. D. Tice, Contribution of research to the understanding and mitigation of environmentally assisted cracking in structural components of light water reactors. Corros. Eng. Sci. Technol. **53**, 11–25 (2018)
27. D.R. Tice, D. Green, A. Toft, Environmentally assisted fatigue gap analysis: A roadmap for future research activities. Product ID 071128, Palo Alto (CA): EPRI. (2011), https://www.epri.com/#/pages/product/1026724/?lang=en-US
28. Z. Hua, B. An, T. Iijima, C. Gu, J. Zheng, The finding of crystallographic orientation dependence of hydrogen diffusion in austenitic stainless steel by scanning Kelvin probe force microscopy. Scr. Mater. **131**, 47–50 (2017)
29. A. Schreiber, C. Rosenkranz, M.M. Lohrengel, Grain-dependent anodic dissolution of iron. Electrochim. Acta **52**, 7738–7745 (2007)
30. L.Y. Xu, Y.F. Cheng, Development of a finite element model for simulation and prediction of mechanochemical effect of pipeline corrosion. Corros. Sci. **73**, 150–160 (2013)
31. O.O. Fatoba, R. Leiva-Garcia, S.V. Lishchuk, N.O. Larrosa, R. Akid, Simulation of stress-assisted localised corrosion using a cellular automaton finite element approach. Corros. Sci. **137**, 83–97 (2018)

32. D.G. Harlow, R.P. Wei, Probability approach for prediction of corrosion and corrosion fatigue life. AIAA J. **32**, 2073–2079 (1994)
33. N.O. Larrosa, R. Akid, R.A. Ainsworth, Corrosion-fatigue: A review of damage tolerance models. Int. Mater. Rev. **63**, 283–308 (2018)
34. K. Torronen, M. Kemppainen, H. Hanninen, Fractographic evaluation of specimens of A 533 B pressure vessel steels. Final Report of EPRI, Contract RP 1325–7, Report NP 3483, 1984
35. R. Akid, The role of stress-assisted localised corrosion in the development of short fatigue cracks, in *Effects of the environment on the initiation of crack growth*, ed. by W. A. Van Der Sluys, R. S. Piascik, R. Zawierucha, (ASTM STP 1298, American Society of Testing and Materials, 1997), pp. 3–17
36. G. Gunn, Duplex Stainless Steels: Microstructure, Properties and Applications, Woodhead Publishing Series in Metals and Surface Engineering, (Elsevier, 1997), ISBN: 9781845698775
37. M. Kaneko, Stress corrosion cracking of stainless steels. Weld. Int. **21**, 95–99 (2007)
38. J.E. Truman, The influence of chloride content, pH and temperature of test solution on the occurrence of stress corrosion cracking with austenitic stainless steel. Corros. Sci. **17**, 737–746 (1977)
39. T. Prosek, A. Le Gac, D. Thierry, S. Le Manchet, C. Lojewski, A. Fanica, E. Johansson, C. Canderyd, F. Dupoiron, T. Snauwaert, F. Maas, B. Droesbeke, Low-temperature stress corrosion cracking of austenitic and duplex stainless steels under chloride deposits. Corrosion **70**, 1052–1063 (2014). s
40. C. Ornek, D. Engelberg, Stress corrosion cracking and hydrogen embrittlment of type 316L austenitic stainless steel beneath $MgCl_2$ and $MgCl_2$:$FeCl_3$ droplets. Corrosion **75**, 167–182 (2019)
41. J.B. Gnanamoorthy, Stress corrosion cracking of un-sensitized stainless steels in ambient temperature coastal atmosphere. Mater. Perform. **29**, 63–65 (1990)
42. R.C. Newman, Stress corrosion cracking mechanisms. in *Corrosion Mechanisms in Theory and Practice*, ed. by P. Marcus, 3rd edn. (CRC Press, 2012), p. 499, ISBN 9781138073630
43. W.T. Tsai, M.S. Chen, Stress corrosion cracking behavior of 2205 duplex stainless steel in concentrated NaCl solution. Corros. Sci. **42**, 545–559 (2000)
44. V. Kain, Stress corrosion cracking (SCC) in stainless steels. in *Stress Corrosion Cracking: Theory and Practice*, ed. by V.S. Raja, T. Shoji, (Woodhead publishing, 2011), ISBN: 9780857093769
45. C.G. Rhodes, A.W. Thompson, The composition dependence of stacking fault energy in austenitic stainless steels. Metall. Trans. A. **8**, 1901–1906 (1977)
46. Q.X. Dai, A.D. Wang, X.N. Cheng, X.M. Luo, Stacking fault energy of cryogenic austenitic steels. Chin. Phys. **11**, 596–600 (2002)
47. M. Fujita, Y. Kaneko, A. Nohara, H. Saka, R. Zauter, H. Mughrabi, Temperature dependence of the dissociation width of dislocations in a commercial 304L stainless steel. ISIJ Int. **34**, 697–703 (1994)
48. W.T. Tsai, C.M. Tseng, H.Y. Liou, *Corrosion Fatigue Crack Growth Behaviour of Duplex Stainless Steels with Different Nitrogen Contents*. NACE-00214, Corrosion 2000, NACE International, Houston, 2000
49. Y. Jang, S. Kim, J. Lee, Effect of different Mn contents on tensile and corrosion behavior of CD4MCU cast duplex stainless steels. Metall. Mater. Trans. A **36**, 1229–1236 (2005)
50. H. Spaehn, in *Environmental-Induced Cracking of Materials*, ed. by R. P. Gangloff, M. B. Ives, (NACE, Houston, 1990), pp. 449–487
51. V. Kain, R.C. Prasad, P.K. De, H.S. Gadiyar, Corrosion assessment of AISI 304L stainless steel in nitric acid environments – An alternate approach. J. Test. Eval. **23**, 50–54 (1995)
52. K. Basu, M. Das, D. Bhattacharjee, P.C. Chakraborti, Effect of grain size on austenite stability and room temperature low cycle fatigue behaviour of solution annealed AISI 316LN austenitic stainless steel. Mater. Sci. Technol. **23**, 1278–1284 (2007)
53. W.T. Tsai, I.H. Lo, Effect of potential and loading frequency on corrosion fatigue behavior of 2205 duplex stainless steel. Corrosion **64**, 155–163 (2008)

54. T.L. Prakash, A.U. Malik, Studies on the stress corrosion cracking behaviour of some alloys used in desalination plants. Desalination **123**, 215–221 (1999)
55. O.A. Abuzeid, A. Al-Jouboury, M. Abou Zour, Effect of corrosion inhibition on the stress corrosion cracking of UNS S31603 austenitic stainless steel. Adv. Mater. Res. **476-478**, 256–262 (2012)
56. J. González-Sánchez, N. Acuña, Corrosion fatigue of stainless steel in tropical seawater, in *Environmental Degradation of Infrastructure and Cultural Heritage in Coastal Tropical Climate*, ed. by A. J. González-Sánchez, F. C. Pérez, (Transworld Research Network, 2009), pp. 87–114
57. P. Woollin, S.J. Maddox, D.J. Baxter, Corrosion fatigue of welded stainless steels for deepwater riser applications. Proceedings of the OMAE 2005: 24th International Conference Mechanics and Arctic Engineering, Halkidiki Greece, 12–16 June 2005
58. M. Ahsan, Z. Gassem, Corrosion fatigue crack growth inhibition of duplex stainless steel, in *Damage and Fracture Mechanics*, ed. by C. A. Brebbia, (WIT Press, 2004), p. 59
59. A. Meroufel, A.U. Malik, S. Al-Fozan, A. Al-Sahari, M. Ayashi, A. Ansari, Investigation on the failure of make-up seawater ump shaft failure, Technical report No. 3814/13002, SWCC, 2013
60. A. Poonguzhali, M.G. Pujar, C. Mallika, U.K. Mudali, Characterization of microstructural damage due to corrosion fatigue in AISI type 316 LN stainless steels with different nitrogen contents. Corros. Eng. Sci. Technol. **51**, 408–415 (2016)
61. R.K. Dayal, Laser surface modification for improving localized corrosion resistance of austenitic stainless steels. Surf. Eng. **13**, 299–302 (1997)
62. L. Poirier, Y. Corre, J.P. Lebrun, Solutions to improve surface hardness of stainless steels without loss of corrosion resistance. Surf. Eng. **18**, 439–441 (2002)
63. U.K. Mudali, H.S. Khatak, B. Raj, M. Uhlemann, Surface alloying of nitrogen to improve corrosion resistance of steels and stainless steels. Mater. Manuf. Process. **19**, 61–73 (2004)
64. J.A.R. Jayachandran, N. Murugan, Development of eco-friendly surface modification process for 316L austenitic stainless steel weld cladding. Surf. Eng. **28**, 5–10 (2012)
65. J. Xiong, G. Manjaiah, D. Fabijanic, M. Forsyth, M.Y. Tan, Enhancing the localized corrosion resistance of 316L stainless steel via FBRCVD chromising treatment. Corros. Eng. Sci. Technol. **53**, 114–121 (2018)
66. T. Magnin, Corrosion fatigue mechanisms in metallic materials. in *Corrosion Mechanisms in Theory and Practice*, ed. by P. Marcus, 3rd edn. (CRC Press, 2012), p. 545, ISBN 9781138073630

Chapter 8
Corrosion Monitoring in Desalination Plants

Mahbuboor Rahman Choudhury, Wesley Meertens, Liuqing Yang, Khaled Touati, and Md. Saifur Rahaman

8.1 Introduction

Corrosion monitoring in desalination plants is the measurement of corrosion occurring under real operational situations. It refers to the determination of the occurrence (extent and rate) of degradation of material (e.g., evaporators, storage tanks, piping, pumps, etc.) exposed to the desalination process environments. The corrosion monitoring data represent key information for corrosion management strategies in the desalination plants. Depending on the corrosion monitoring techniques adopted, the monitoring data might provide a time-averaged or instantaneous corrosion rate. Modern corrosion monitoring technologies are directed towards real-time data acquisition, which provides a better understanding of instantaneous corrosion rates in an aqueous system and helps in the overall management of process parameters in a desalination plant.

A desalination plant includes a wide range of corrosive media like seawater, salt-air vapors, and non-condensable volatile gases at different temperatures, flow velocities, and particulate concentrations. There are different forms of corrosion that can take place in a seawater desalination plant. Some corrosion forms are identifiable by visual inspection (e.g., uniform corrosion, pitting corrosion, crevice corrosion, galvanic corrosion, etc. taking over a substantial time period to make them appear visible to naked eyes); some are identifiable with special inspection tool (e.g., erosion, cavitation, fretting, exfoliation, inter-granular,

M. R. Choudhury
Department of Building, Civil and Environmental Engineering Department,
Concordia University, Montreal, QC, Canada

Civil and Environmental Engineering Department, Manhattan College, Bronx, NY, USA

W. Meertens · L. Yang · K. Touati · M. S. Rahaman (✉)
Building, Civil and Environmental Engineering Department, Concordia University,
Montreal, QC, Canada
e-mail: saifur.rahaman@concordia.ca

© Springer Nature Switzerland AG 2020
V. S. Saji et al. (eds.), *Corrosion and Fouling Control in Desalination Industry*,
https://doi.org/10.1007/978-3-030-34284-5_8

etc.); and others require microscopic examination (e.g. stress corrosion crack, corrosion fatigue, etc.) [1]. Corrosion monitoring is more complex than monitoring other process parameters mainly due to the different forms corrosions taking place in the system, which can be either uniform, localized, or even have a wide variability over relatively short distances. It is common to use more than one monitoring techniques in a system, as one single measurement technique cannot detect all types of corrosion.

A wide range of corrosion monitoring techniques has evolved for identifying, measuring, and predicting the occurrence of corrosion. The efficient use of corrosion measurement techniques allows timely remedial actions and allows a cost-effective operation in desalination plants as it reduces the life-cycle costs of associated components and operations. This chapter aims to provide a concise account of commonly employed corrosion monitoring techniques in desalination industries.

8.2 Significance of Corrosion Monitoring in Desalination Plants

Desalination plants can implement four different approaches concerning its corrosion issues. Firstly, the corrosion rates can be accepted until a failure occurs in the system. This approach is based on the excessive thickness of different components. In this "run to fail" technique, the equipment replacement is undertaken after the failure occurrence. This practice is inefficient from a maintenance cost perspective, particularly for extending the life of old engineering systems. Secondly, corrosion-related issues could be inspected at scheduled intervals, where required repairs and maintenance can be provided. This process can provide a somewhat reasonable safeguard against corrosion failure and can keep the system operational for a longer duration when compared to no maintenance. The inspection, if unaided by corrosion monitoring, can lead to maintenance schedules set too conservatively and having excessive system downtimes. Reversely, if the inspection intervals are set too far apart, the system may experience excessive corrosion and remain vulnerable to the associated cost and safety consequences. Thirdly, corrosion rates can be kept within desired levels by adopting different corrosion prevention schemes (e.g., dosing of corrosion inhibitors, protective coatings, durable corrosion-resistant materials, etc.). The excess cost of corrosion prevention schemes can be optimized using information from real-time or time-averaged corrosion monitoring data. Finally, corrosion control can be applied selectively in locations where immediate action is required, which is facilitated by continuous corrosion monitoring tools. Although instantaneous corrosion rate monitoring is clearly helpful in assessing real-time corrosion inspection and maintenance needs, the cost-saving aspects in the overall operation of the plant often guide the deployment of highly sophisticated corrosion-monitoring schemes.

The importance of corrosion monitoring techniques for desalination plants from an investment perspective can be related to achieving process safety, reduced downtime, reduced maintenance costs, longer intervals between maintenance, reduced operating costs, and life extension of the plant. These techniques can provide early warning to impending corrosion-induced failures for preemptive action. They can also relate corrosion scenarios to different process parameters and guide corrosion prevention schemes for optimal system management.

8.3 Corrosion Monitoring Techniques

Corrosion monitoring in desalination plants is complicated due to the wide variety in process setup and conditions in different desalination units. A report by NACE International organized different corrosion monitoring techniques in field applications into different categories [2]. By following the NACE report [2] and various desalination industry approaches, the most commonly used corrosion monitoring techniques in desalination plants can be tabulated as, shown in Table 8.1 [2]. Direct techniques measure parameters that are directly influenced by the corrosion process. An indirect technique provides data on parameters that either influence or have been influenced by the corrosiveness, or by corrosion end products in that environment. A monitoring tool is defined as intrusive if it requires access through a vessel or conduit wall to make a measurement, such as scraping a sample off a wall. This is usually done using a probe or obtaining a test specimen. Among the techniques listed in Table 8.1, gravimetric mass-loss coupons, ER, and LPR techniques are the prominent and the widely used methods in desalination.

In addition to the methods mentioned above, there are several methods that are suitable for corrosion monitoring in desalination plants; that include various electrochemical methods (impedance spectroscopy, potentiodynamic polarization, and noise analysis), harmonic distortion analysis method and non-intrusive techniques (field signature, and acoustic emission). More details can be found elsewhere [2].

Table 8.1 Different corrosion monitoring techniques that are currently used in desalination plants

Direct corrosion monitoring	Physical techniques (intrusive) – Gravimetric mass-loss – Electrical resistance (ER) – Visual inspection Electrochemical (DC) techniques (intrusive) – Linear polarization resistance (LPR) – Zero-resistance ammeter (ZRA)
Indirect corrosion monitoring	– Corrosion potential – Water quality analyses – Residual inhibitor
Microbiologically induced corrosion monitoring	– Planktonic microorganism – Sessile organism – Biological assessment

8.3.1   Direct Intrusive Techniques

8.3.1.1   Physical Techniques

Physical techniques involve gauging the change in the geometry or other physical properties of exposed coupons, or structural components of the desalination plant, in the process environment to establish a corrosion rate over the exposure time. The properties that are influenced by corrosion include mass, electrical resistance, magnetic flux, reflectivity, stiffness, porosity, topography, etc. In some instances, the properties can be read frequently by electrochemical means while the test specimen remains in-situ. In other cases, the test specimen requires to be removed physically from the process to get a measurement of property change due to corrosion. Frequent readings become difficult in the latter case.

a. Gravimetric mass-loss method

The gravimetric mass-loss method uses the weight difference of small metal, or metal alloy specimens/coupons before and after exposure to a process stream to evaluate the corrosiveness of the environment. It is an inexpensive and facile technique with well-documented standard operational procedures [3–5]. This method gives an average corrosion rate over the entire exposure period of the test coupon [6]. The gravimetric method is employed to observe corrosion rates occurring on existing equipment and to evaluate the potential of alternative materials for construction/maintenance. A coupon holder allows mounting of metal or metal alloy specimens on a rod, which is then inserted in the process stream. The coupon holder is placed in a way that it does not blowout of the system under pressure nor affect flow patterns within the process stream. Many coupons can be exposed simultaneously in the process streams to evaluate corrosion rates in this process, which enables duplicate or triplicate testing. This process also enables testing of corrosion under various conditions (e.g., welding, residual stresses, crevice corrosion, pitting, de-alloying corrosion, etc.) by fabricating different coupon geometries to meet the specifications of the conditions [2]. These variations can enhance the confidence of the engineer in selecting materials for new units, maintenance, or repair [7]. The coupons can be placed either in the main process stream or in a side stream that is isolated from the main process stream, where operational conditions are duplicated for representative corrosion analysis.

The test coupons are cleaned as soon as possible after removal from the process stream following ASTM standard G1–03 [3]. The gravimetric process only provides a time-averaged corrosion rate. However, the associated visual inspection helps in isolating critical localized corrosion effects, which may not be reflected by the corrosion rate data. The coupon needs to be cleaned thoroughly through the repetitive cleaning process, and the corrosion rate can be estimated from a mass plot curve, as indicated in Fig. 8.1 [3]. The average corrosion rate is then estimated by the following equation (Eq. 8.1):

$$R = \frac{K\ (m_1 - m_2)}{A\ (t_1 - t_2)\rho} \tag{8.1}$$

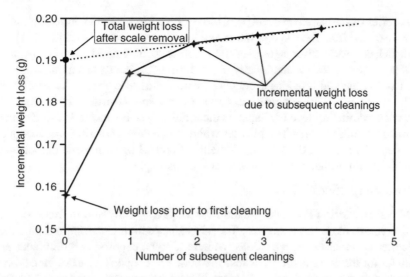

Fig. 8.1 Mass loss of corrosion coupons from repetitive cleaning cycles carried out for weight loss determination. Reproduced with permission from [8]; Copyright 2007; John Wiley & Sons Inc

Where R is the penetration (corrosion) rate (mm/year); A is the exposed area (cm²); m_1 and m_2 are the initial and final masses (g), with m_2 being the intercept made by extrapolating line BC to the y axis in Fig. 8.1; t_1 and t_2 are the starting and ending times (h); ρ is the density (g/cm³); and K is a constant for unit conversion. The value of K in Eq. 8.1 is 1.142×10^{-5}.

b. Electrical resistance (ER) method

The ER method quickly became popular for industrial corrosion monitoring in the 1950s [9, 10]. The principle of the widely used ER technique is very simple. It measures the change in electrical resistance of a corroding metal probe specimen, which is exposed to the process stream environment. An increase in electrical resistance is related directly to the metal loss: as corrosion reduces the cross-sectional area of the probe, its electrical resistance increases. The electrical resistance of a metal alloy is given by the following equation (Eq. 8.2),

$$R = r\frac{L}{A}$$
(8.2)

Where L is the probe element length (cm); A is the cross-sectional area (cm²); and r is the specific resistance of the probe metal (Ω cm).

Temperature influences the electrical resistance of the probe element. Hence, ER sensors usually measure the resistance of a corroding sensor element relative to that of an identical shielded element. Commercial ER sensors are available in the forms of plates, tubes, plates, or wires. While reduced sensor thickness can increase their sensitivity, it will consequently provide a shorter life span for the sensor.

A series of measurements can be made over a period, and the corrosion rate can be determined from the slope of the resulting metal loss versus time plot [11]. This method has several advantages over the gravimetric method; the sensors or probes are small, they can be installed easily, and can connect directly to a monitoring facility. The ER method provides average corrosion rates over periods of several days [12]. Presence of pitting corrosion, crevice corrosion, corrosion byproducts, and mineral precipitation severely hampers the efficiency of ER method [13]. Erroneous corrosion results will manifest in cases where conductive corrosion products or surface deposits form on the sensing element. Microbial corrosion and carbonaceous deposits in atmospheric corrosion are relevant examples.

c. Visual inspection method

Visual inspection is a basic method for analyzing corrosion in metal or metal alloys exposed in process streams. The visual inspection is carried out on coupon specimens after cleaning operations, which are to be exposed to a particular environment for the gravimetric-mass loss method. The coupons are examined with the unaided eye, and then at increased magnifications using a microscope or scanning electron microscope. These examinations help in identifying different forms of superficial localized defects and corrosion damages. Coupons can be further bent or cut to reveal underlying types of corrosion damage. The visual inspection helps in identifying some localized corrosion effects that may otherwise jeopardize the estimation of realistic corrosion rates using other physical methods.

8.3.1.2 Electrochemical Techniques

Measurements of electrical properties on the metal solution interface are carried out to evaluate corrosion scenarios over a wide range of circumstances. Electrochemical monitoring techniques involve the determination of specific interface properties like corrosion potential, current density, surface impedance, etc. Some methods study the interface using a direct current (DC), while other methods employ alternating current (AC) to provide further characterization of the corrosion interface between the metal alloy and process fluids.

a. Linear polarization resistance (LPR) method

The electrochemical LPR method can respond to the small variations in the corrosiveness of the aqueous environment, in which the metal specimen remains exposed, and can measure the corrosion rate in short time intervals [6, 14, 15]. It can be employed for long-term corrosion monitoring in field conditions as well [16].

In this method, a small potential perturbation (usually in the range of ±10 mV to ±30 mV) around the free corrosion potential is applied to the sensor electrode (also referred to as the working electrode), and the corresponding current values with the potential scan recorded. The polarization resistance (R_P) is defined as the differential of the overpotential (E) over the withdrawn current (i) when ΔE approaches zero, i.e., the slope of the potential-current curve at the free corrosion potential

(Eq. 8.3). The R_P is inversely proportional to the uniform corrosion rate, and it can be related to the corrosion current (i_{corr}) using the Stern and Geary equation (Eq. 8.3):

$$R_P = \left(\frac{\Delta E}{\Delta I}\right)_{\Delta E \to 0} = \frac{B}{i_{corr}} \tag{8.3}$$

Where B is an empirical polarization resistance constant (also known as Stern and Geary constant) that can be related to the anodic (b_a) and cathodic (b_c) Tafel slopes with the following equation (Eq. 8.4):

$$B = \frac{b_a \cdot b_c}{2.3 \left(b_a + b_c\right)} \tag{8.4}$$

The Tafel slopes can be either determined empirically from polarization plots or obtained from the literature. They can also be determined by other techniques, such as curve fitting of polarization resistance curves by potentiodynamic polarization scans or by harmonic distortion evaluation.

A standard test method is available for the LPR method [17], however, the coefficient values used to convert polarization resistance to instantaneous corrosion rate are not constant and vary with system parameters (like water quality, metal alloys, etc.) [15, 16]. Typically, a two or three-electrode probe configuration is employed in the LPR measurement. With a two-electrode system, the corrosion measurement is an average of the rate for both electrodes. Both electrodes would then be made of the alloy being monitored. The three-electrode system comprises a working electrode (corrosion rate of the working electrode is measured in the process), a reference electrode, and a counter electrode. A stainless steel reference electrode is suitable for field application, which offers more robustness than the capillary salt bridge arrangement used for laboratory measurement [16]. The distance between the working and reference electrode are kept minimum (usually within 2–3 mm) to reduce the interference of solution resistance [16]. The solution resistance can offer significant error to LPR measurements. However, for seawater, the solution in the desalination plant will offer low resistivity. High resistivity solution will render erroneously low corrosion rates. The probes in LPR measurement will require access to the vessel or pipe walls, hence, there is always an added chance of leakage in the system. One limitation of LPR measurement is that it only provides an average corrosion rate, hence, this method does not provide any indications of localized corrosion (like pitting corrosion).

b. Zero resistance ammeter (ZRA) method

A zero resistance ammeter is a current to voltage converter that produces a voltage output proportional to the current flowing between its two input terminals while imposing no voltage drop to the external circuit. In the ZRA method, galvanic currents between dissimilar electrode materials are measured with a zero resistance ammeter. This method is used to study the electrochemical reactions that take place when two different metals or metal alloys, immersed in the same stream, are electrically coupled to one another. The differences in the electrochemical property of two electrodes when exposed to a process stream give rise to differences in the redox potential at these electrodes. Once the two electrodes are externally connected, the more stable electrode acts predominantly as a cathode, while the more active electrode acts predominantly as an anode, becoming sacrificial. When the anodic or cathodic reaction is relatively stable, the galvanic current monitors the response of the cathodic or anodic reaction, respectively, to the variations in process stream conditions [2].

The ZRA method can provide a quantitative measure of corrosion rate when the number of influencing factors is limited and easily verifiable by other processes. The corrosion rate results may not always reflect the actual corrosion rates as the galvanic corrosion depends on the relative areas and geometries of the electrode components, which can vary between the probe and the actual plant components. Moreover, this method cannot distinguish between the activation of the anodic or the cathodic reaction. The incremental measurement of current can result from either a cathodic activation by increased dissolved oxygen, an anodic activation by increased bacterial biofilm activity, or due to a combination of the two. Hence, a separate analysis is warranted to differentiate between these electrochemical processes.

8.3.2 Indirect Corrosion Monitoring Techniques

Several on-line and off-line indirect corrosion-monitoring methods have been developed to assess the corrosiveness in a process. These indirect techniques measure the indirect changes, either in the solution environment or in the metallic component of interest, that take place during the corrosion process or the change in corrosion rates. Different chemical analysis methods have been used over the years that aid the process plant operators gain insight on the corrosive environment existing in the metal-solution interface or on the efficacy of different corrosion inhibitors dosed in the process stream. The following section describes different popular indirect corrosion monitoring methods that are used in desalination plants.

8.3.2.1 Corrosion Potential Method

This method only indicates a corrosion risk and does not measure corrosion rates. The open-circuit potential or freely corroding potential (referred to as E_{corr}) is measured with respect to a reference electrode, which is characterized by a stable half-cell potential. The E_{corr} measurement in the long term depends on the stability of the reference electrode. Environmental factors like temperature, pressure, pH, electrolyte composition, and other such variables can limit the scope of use of the reference electrode for corrosion-monitoring service. This method can be valuable in cases where a metal alloy can demonstrate both active and passive corrosion characteristics in a given exposure stream. For example, stainless steel can provide excellent service when they remain passive at the metal-liquid interface. However, the occurrence of chloride or reducing agents in the process stream can make stainless steel become active and exhibit excessive corrosion rates. E_{corr} measurements would indicate the development of active corrosion, and the corrosion rate could be measured by other electrochemical means [7].

8.3.2.2 Water Quality Analyses

Probing of process stream water quality provides important information to a corrosion monitoring system for a desalination plant. Some of the water qualities can be measured on-line using appropriate sensors, while others require further analysis in the laboratory to collect relevant information. Parameters like pH, conductivity, dissolved oxygen (DO) concentration, and oxidation-reduction (redox) potential of the process stream can be measured on-line. Alternatively, alkalinity, metal ion concentration, total dissolved solids (TDS), residual oxidant analysis, and scaling indices can be measured off-line after further laboratory analysis of collected water samples. The scaling indices (like Langelier, Ryzner, etc.) can be used to assess the scale formation tendency of a process stream, which in turn can infer on the corrosiveness of the metal-liquid interface [18, 19].

A higher pH indicates fewer hydrogen ions in solution. Process streams, which have a pH < 7 are acidic, whereas substances with pH > 7 are alkaline. In general, for steel and copper piping, low pH (or high acidity) of process streams could create extremely corrosive environments. The rate of corrosion would vary from one metal to the other. The hydrogen ion can get reduced to molecular hydrogen through participation in the cathodic reaction of the corrosion process. Also, pH influences the solubility of different chemicals in solution. The dissolution of different corrosion products (e.g., oxides, sulfides, or carbonates) and the rate of corrosion would depend on the pH of the solution.

Conductivity refers to the current-carrying capacity of water, which is due to the presence of ions from dissolved salts. The charged ions can move through the solution under the influence of an externally applied electric field. Since corrosion is an electrochemical process, an increase in conductivity corresponds with an increase in solution corrosiveness. Seawater, which contains many dissolved ions, has high conductivity and renders a highly corrosive environment to the pipes and vessels in a desalination plant. However, zero conductivity does not guarantee corrosion inhibition, as the use of pure water has demonstrated corrosion in other systems. Solution conductivity can also influence the electrochemical readings for corrosion assessment. Similarly, TDS concentration refers to the sum of minerals and salts dissolved in water. Increased TDS can indicate a corrosive environment in the process stream of a desalination plant.

DO refer to the amount of oxygen dissolved in a liquid (expressed as parts per million or milligrams per liter). DO concentration depends on temperature, pressure, and molarity of the solution. The attraction of oxygen for most structural metal is the cause of many corrosion phenomena. Oxygen is responsible for both corrosive attack and passivation. Monitoring the DO level in the process stream thus provides valuable information on the corrosive environment existing in the metal-liquid interface.

Redox potential is the potential of a reversible oxidation-reduction electrode measured with respect to a reference electrode in a given electrolyte. Oxidation potential measures the power of a substance to add oxygen or to remove hydrogen as well as to lose electrons. Reduction potential indicates the power to add hydrogen, lose oxygen, or gain electrons. When the redox potential increases in value and turns positive, its ability to oxidize is enhanced. When it decreases and turns negative, its reducing ability is increased. As corrosion involves both oxidation and reduction, the redox potential becomes an indicator of the possible electrochemical activity, which may either lead to corrosion or resistance to corrosion.

Alkalinity indicates the ability of a liquid to buffer acid. Along with pH and hardness, the alkalinity helps to identify whether a body of water may corrode or cause scaling. Increased alkalinity reduces calcium carbonate solubility, which results in carbonate scaling and reduced corrosion propensity. Presence of microorganisms can influence microbiologically induced corrosion (MIC) in the presence of such scaling. Therefore, the alkalinity of water should be monitored to help indicate the possible impact of water corrosiveness to a process plant.

Concentrations of metal ions in solution can provide some idea on the corrosiveness of the water. Metals like iron, copper, nickel, zinc, and manganese can be analyzed quickly and inexpensively in solution. An increase in metal ion concentration can indicate an increase in corrosion. However, low metal ion concentration does not confirm low corrosion rates, as localized corrosion and metal deposition can hamper such readings. Metal ion analysis provides more confident assessments in closed systems and if the corrosion products are soluble or relate to particular concentrations of soluble species. It is critical to obtain a representative sample for metal ion analysis from a process stream under varying fluid velocity, temperature, and pressure.

Residual oxidants (like ozone and halogens) are maintained in the process stream to control microbiological fouling. Residual halogen or halogenated compounds can directly oxidize the protective coating and enhance corrosion. Dissolved halogens or halogenated compounds in the process stream can be measured using redox potential or one of a variety of colorimetric techniques.

8.3.2.3 Residual Inhibitor Analysis

Corrosion inhibitor residual measurements provide an idea of the consumption of inhibitors in different systems of a process plant. A residual inhibitor concentration above the required inhibitor concentration ensures adequate corrosion protection in the system. Loss in residual inhibitor concentration can occur due to an aggressive environment or interactions with the process components (e.g., adsorption, neutralization, precipitation, adsorption on solids or corrosion products, etc.). Once such loss is detected, it can be compensated by an additional dosage of inhibitor in the system.

8.3.3   Microbiologically Induced Corrosion (MIC) Monitoring

The development of biofilm on membranes and other components of desalination plants can lead to significant MIC issues. A growing biofilm increasingly clogs the pores of the membrane, making it more difficult to purify saline water. As the pores continue to clog, there is either an increasing pressure, thus cost, requirement to ensure continuous flux, or a decreasing flux with constant pressure. The interaction of microbial metabolism and corrosion processes can also produce localized corrosion at a very high rate. The presence of a biofilm on a metallic surface can greatly alter the local corrosion processes such as electrochemical changes, pressure drop, and heat-transfer resistance [8]. MIC monitoring is necessary to ensure adequate pre-treatment and anti-bacterial chemical dosing, and to conduct an effective bio-corrosion mitigation program [20]. The following sections describe the common MIC monitoring techniques in desalination plants.

8.3.3.1   Planktonic Organisms Monitoring

Planktons are marine organisms that are incapable of making their way against a current [21], and many of these organisms may find themselves in reverse osmosis (RO) plants. Planktonic organisms include unattached algae, diatoms, fungi, and other microorganisms present in the system's bulk fluids [22]. In most cases, planktonic bacteria monitoring is preferred in MIC as sampling influent water is easier than sampling the channel walls or membranes within a purification facility. Water quality, redox potential measurements, the timing of added biocides/other chemical input, and the identification of microorganisms are the key information with planktonic monitoring.

While it is easier to sample the saline influent, the levels of planktonic bacteria are not necessarily indicative of if MIC has taken place or the severity of the corrosion. The detection of viable planktonic bacteria serves as an indicator that living microorganisms are present and are capable of participating in a microbial attack. It is generally agreed that additional monitoring methods are important to confirm corrosion due to microbial processes has occurred [22].

8.3.3.2   Sessile Organisms Monitoring

The sessility of a microorganism is attributed to the inability of self-mobilization, and sessile organisms are commonly categorized by the tendency to attach to surfaces and alter topography [21]. The accumulation of sessile and planktonic microorganisms on membrane surfaces is commonly referred to as biofilm. Scraping of accessible surfaces can facilitate sampling for the analysis of sessile microorganisms. Proper sample handling must be emphasized to ensure that no contamination occurs during the transport or sampling process. Biofilms are particularly susceptible to dehydration, air exposure, temperature change, and mechanical damage [8]. To ensure no growth or death of cells by the change of environment, laboratory processing should occur as quickly as possible.

8.3.3.3 Biological Assessment

Biological assays can be performed on liquid samples, suspensions of solids, or solid deposits (such as biofilm) to enumerate viable microorganisms, quantify metabolic and specific enzyme activity, or quantify amounts of key metabolites [23]. These can be used to quantify the biomass within the system and clarify what microorganisms are present.

a. Direct inspection

This method is best suited for relatively clean influent waters, or after the pre-treatment step of saline water. The purpose of the direct inspection is to identify and enumerate microorganisms and to provide insight into the location, growth rate, and activity of specific microorganisms within a biofilm. Quantities can be easily found using a counting chamber and phase-contrast microscopy on a thin water layer of known volume. This technique is performed in a laboratory and can be enhanced using fluorescent dyes to illuminate cells [2]. Another inspection method is to apply a stain to a water sample and to filter the cells through a 0.25 μm membrane to visualize and count the cells using the epifluorescent technique. The usage of dyes can discover the active metabolism within water samples. The formation of new fluorescent products when using new stains, such as fluorescein diacetate or p-iodonitrotetrazolium violet, is used as cell viability and active metabolism measurement [24].

b. Growth assays

The most common microbial assessment is through commercially available growth media to estimate the most probable number (MPN) of viable cells present in a sample [25]. The incubation period can be completed in a few days, but typically the test takes about 14–28 days [2]. Despite the common use of these assays, only a small fraction of organisms actually grow in the commonly available artificial media.

8.3.3.4 Deposition Monitors

A clear indicator of biological fouling in the membrane is the pressure drop across the membrane. By maintaining constant influent pressure with the continuous accumulation of foulants across the membrane, the permeate flux will drop proportionally to the amount of clogging of the membrane pores. The main disadvantage of monitoring the pressure drop is that it measures the effects of all scaling and membrane depositions as opposed to the biofouling specifically. Organic and inorganic deposits affect the heat transfer efficiency, but biofilms are especially effective at disrupting heat transfer. For example, A 165-pm thick biofilm shows a thermal conductivity 100 times less than carbon steel and 100 times the relative roughness of the calcite scale [24].

8.3.3.5   Detailed Coupon Examinations

A more advanced coupon technology can be used for sampling during the operations of a high-pressure RO plant. This coupon is placed to ensure flow is not inhibited, and the biofilm developing on this coupon is indicative of the behavior throughout the system. These coupons can be injected and released into the system as needed. Once the coupon is retrieved with the sessile organism deposition, a great deal of information can be learned by a careful and in-depth examination of the corrosion coupon surfaces through direct observations and biological assessments [22].

8.3.3.6   Electrochemical Monitoring

Electrochemical methods can be employed for constant monitoring of biofilm formation. Biofilm activity is an electrochemical process and can be tracked by the generated currents of their activity. In a typical set-up, two sets of electrodes are placed in the plant environment, with one of the sets being polarized. The electrochemical monitoring is completed by comparing the measured current when applying an external potential, and the current generated without external potential caused just by the biofilm activity. When the current deviates from the baseline amount, it is indicative of biofilm formation. The amount of variation is proportional to the amount of biofilm formation [22].

8.4   Conclusions and Outlook

Corrosion occurring in desalination plants challenges the long-term efficient operation of plants. A lot of work has been focused on minimizing corrosion through the application of corrosion-resistant alloys, chemical additives, protective coatings, and cathodic protection. However, the aggressive nature of seawater always causes corrosion-related issues. Hence corrosion monitoring techniques always play an important role in the safe operation of desalination plants.

The corrosion monitoring techniques provide timely warning of expensive corrosion damage and critical information (e.g., rate of deterioration) during the occurrence of damaging events. This information is essential to make real-time decisions on urgent preventive actions on site. Among the different types of corrosion monitoring techniques, gravimetric weight loss coupons, electrical resistance probes, visual inspection, linear polarization resistance, zero-resistance ammeter, corrosion potential, water quality analyses, and residual inhibitor concentration, are applied commonly in desalination plants. The advanced electrochemical approaches such as impedance spectroscopy, and electrochemical noise analysis, and methods such as harmonic distortion analysis, field signature method, which are widely researched as corrosion monitoring techniques need to be considered for its routine application in desalination systems.

The use of remote corrosion monitoring systems, especially with electrochemical and electronic methods, enables real-time assessment of corrosion in different parts of a desalination plant from a central monitoring location. Application of advanced computational programs like artificial neural networks and other predictive tools can enable forecast of impending corrosion failure in a system. Use of continuous corrosion monitoring data with suitable predictive programming can provide the desalination plant operators valuable insights to evaluate and design corrosion prevention programs.

References

1. Dillon CP (1982) Forms of Corrosion: Recognition and Prevention. Vol. 1, NACE International, Houston, TX, ISBN: 0915567873
2. NACE (1999) Techniques for monitoring corrosion and related parameters in field applications. Report no. 24203, NACE International publication 3T199, Houston, TX
3. ASTM (1999) standard practice for preparing, cleaning, and evaluating corrosion test specimens. ASTM standard G1-90, Vol. 03.02, Philadelphia, PA
4. ASTM (2005) Standard Test Method for Corrosivity of Water in the Absence of Heat Transfer (Weight Loss Method), ASTM Standard D2688–05, Vol. 11.01 and 11.02, Philadelphia, PA
5. ASTM (2014) Standard Guide for Conducting Corrosion Tests in Field Applications, ASTM Standard G4–01, Vol. 03.02, Philadelphia, PA
6. D.A. Jones, *Principles and Prevention of Corrosion*, 2nd edn. (Prentice-Hall, Prentice-Hall, Upper Saddle River, NJ, 1996)
7. S.W. Dean, Corrosion monitoring for industrial processes, in *ASM Handbook, Vol. 13A: Corrosion: Fundamentals, Testing, and Protection*, ed. by S. D. Cramer, B. S. Covino, vol. 13A, (ASM International, Metals Park, OH, 2003), pp. 533–541
8. Roberge PR (2007) Corrosion Inspection and Monitoring. John Wiley & Sons Inc., Hoboken, New Jersey, ISBN:9780470099766
9. A. Dravnieks, H.A. Cataldi, Industrial applications of a method for measuring small amounts of corrosion without removal of corrosion products. Corrosion **10**, 224–230 (1954)
10. A.J. Freedman, E.S. Troscinski, A. Dravnieks, An electrical resistance method of corrosion monitoring in refinery equipment. Corrosion **14**, 29–32 (1958)
11. ASTM (2001) Standard Guide for on-Line Monitoring of Corrosion in Plant Equipment (Electrical and Electro-Chemical Methods), ASTM Standard G96–90, Vol. 03.02, Philadelphia, PA
12. Denzine AF, Reading MS (1997) An improved, rapid corrosion rate measurement technique for all process environments. *In* Corrosion 97, NACE-97287, NACE International
13. Boffardi BP (1995) Water Treatments. *In* Corrosion Tests and Standards: Application and Interpretation, Baboaiain R (Ed.), ASTM Manual Series MNL 20, Philadelphia, PA, p. 704
14. S. Keysar, D. Hasson, R. Semiat, D. Bramson, Corrosion protection of mild steel by a calcite layer. Ind. Eng. Chem. Res. **36**, 2903–2909 (1997)

15. M. Hsieh, D.A. Dzombak, R.D. Vidic, Bridging gravimetric and electrochemical approaches to determine the corrosion rate of metals and metal alloys in cooling systems: Bench scale evaluation method. Ind. Eng. Chem. Res. **49**, 9117–9123 (2010)

16. M.R. Choudhury, M.-K. Hsieh, R.D. Vidic, D.A. Dzombak, Development of an instantaneous corrosion rate monitoring system for metal and metal alloys in recirculating cooling systems. Ind. Eng. Chem. Res. **51**, 4230–4239 (2012)

17. ASTM (2005) Standard Test Method for Conducting Potentiodynamic Polarization Resistance Measurements, ASTM G59–97, Vol. 03.02, Philadelphia, PA

18. Tchobanoglous G, Burton FL, Stensel HD (2003) Wastewater Engineering: Treatment and Reuse. 4th Ed., Metcalf & Eddy Inc., McGraw-Hill, New York, NY, ISBN: 0071122508

19. M.R. Choudhury, M.A.Z. Siddik, M.Z.E.I. Salam, Use of shitalakhya river water as makeup water in power plant cooling system. KSCE J. Civ. Eng. **20**, 571–580 (2015)

20. S.C. Dexter, Microbiologically influenced corrosion, in *ASM handbook, corrosion: Fundamentals, testing, and protection*, ed. by D. S. Cramer, B. S. Covino, vol. 13A, (ASM International, Metals Park, OH, 2003), pp. 398–416

21. Lalli CM, Parsons TR (1997) Biological Oceanography: An Introduction, 2nd Ed., Elsevier-Butterworth-Heinemann, Burlington, MA, ISBN: 9780750633840

22. T.P. Zintel, G.J. Licina, T.R. Jack, Techniques for MIC monitoring, in *A practical manual on microbiologically influenced corrosion*, ed. by J. G. Stoecker II, vol. 2, (NACE International, Houston, TX, 2001)

23. T.R. Jack, Biological corrosions failures, in *ASM handbook, failure analysis and prevention*, ed. by R. J. Shipley, W. T. Becker, vol. 11, (ASM International, Metals Park, OH, 2002), pp. 881–898

24. T.R. Jack, Monitoring microbial fouling and corrosion problems in industrial systems. Corr. Rev. **17**, 1–32 (2011)

25. Colwell RR (1977) Enumeration of specific population by the most-probable-number (MPN) method. *In* Native aquatic bacteria: Enumeration, activity, and ecology. Costertor JW, Colwell RR (Eds.) ASTM Special Technical Publication 695. ASTM, West Conshohocken, PA, pp. 56–64

Chapter 9
Chemical Additives for Corrosion Control in Desalination Plants

Saviour A. Umoren and Moses M. Solomon

9.1 Introduction

Water is life! The human body has as much as 60% of water and averagely, we drink four litres of water per day in one form or another [1]. Meanwhile, not all waters are suitable for drinking; hence, 'potable' is often used to describe drinking water. The World Health Organization (WHO) defined potable water as water without significant risk to health over a lifetime of consumption [2]. Unfortunately, about 40% of the global population live in water-scarce regions, with more than three-quarter of a billion people without access to potable water. According to the recent report by the WHO [3], by 2025, half of the world's population will be living in water-stressed areas.

The desalination of seawater and highly brackish water is one of the options for water supply enhancement [4]. It is a long-standing technology for the removal of salts and contaminants from water [5]. Globally, there are about 15,906 operational desalination plants producing around 95 million m³/day of fresh water for human consumption [6]. A desalination plant is a diversified assembly of equipment such as the storage tanks, pumps, heat exchangers, and pipes of varying sizes for operational processes. The equipment are made from metals such as carbon steel, stainless steel, copper and nickel based alloys, titanium and aluminium based alloys *etc*.

Desalination plants mostly operate at high temperatures [6]. The multistage flash and the multiple-effect distillation are the two common thermal distillation techniques. The reverse osmosis (RO) systems, on the other hand, uses high pressure pumps to force saltwater through a membrane that blocks solid particles but allows the penetration of water molecules [6]. The RO plants could operate at pressure as high as 8000 kPa [7]. The high temperature and pressure employed in desalination

S. A. Umoren (✉) · M. M. Solomon
Center of Research Excellence in Corrosion, King Fahd University of Petroleum & Minerals, Dhahran, Saudi Arabia
e-mail: umoren@kfupm.edu.sa

© Springer Nature Switzerland AG 2020
V. S. Saji et al. (eds.), *Corrosion and Fouling Control in Desalination Industry*,
https://doi.org/10.1007/978-3-030-34284-5_9

plants make the operating environments highly corrosive. Further details on desalination processes are found in Chaps. 1, 2, and 3.

One of the essential features of the operation of a thermal desalination plant is the necessity to undertake occasional acid cleaning of the brine side of heat-transfer tubes [8]. This is to avoid a reduction in performance due to scale precipitation [8]. This exercise encourages the corrosion of metallic components. Corrosion damage increases maintenance expenses and generates problems in desalination plant operation and in extreme cases even lead to equipment shut down. For instance, the corrosion of couplings of a State Government's $1.2 billion desalination plant at Tugun, Australia induced a forced shutdown for 5 weeks just 2 months after the taps were turned on [9]. El-Twaty and Karshman [10] reported that, the evaporator body of a Tripoli west desalination plant suffered severe corrosion after 6 years of operation that led to 60% reduction in the design capacity. According to the NACE-IMPACT report in 2016 [11], the estimated global cost of corrosion is US$2.5 trillion, which is equivalent to 3.4% of the global Gross Domestic Product (GDP).

Corrosion mitigation strategies include the use of corrosion resistant alloys (CRAs), protective coatings, cathodic protection, deaeration of the feed water and the use of corrosion inhibitors. Amongst these techniques, the use of corrosion inhibitors is the most effective and practical method of corrosion control in desalination plants, especially during the acid cleaning exercise. Anodic inhibitors (oxidising and non-oxidising such as chromates, nitrites, molybdates, and phosphates), cathodic inhibitors (such as polyphosphates and phosphonates), and organic corrosion inhibitors (film forming amines and azoles) are commonly employed to inhibit corrosion in desalination plants [12]. This chapter focuses on chemical additives such as corrosion inhibitors, biocides, and oxygen scavengers that are used to control corrosion in the desalination industry.

9.2 Inhibitors for Corrosion Control in Desalination Plants

Periodically, acid cleaning is carried out on thermal desalination plants to descale heat exchanger tubes. This exercise utilizes acid solutions, mainly hydrochloric acid, sulfuric acid, or sulfamic acid in the concentration range of 2–5% [13]. The operation temperature could be up to 50 °C and the process normally last for 72 h [13]. Because of the corrosive nature of the acids, there is always the concern of potential corrosion risk of metallic components.

Mostly, desalination plant component materials are carbon steel (flash chamber), stainless steel (flash chamber), copper-nickel alloys, titanium grade 2 (heat transfer tubes), and Ni-resist (brine recycle pump). Normally, an effective corrosion inhibitor is added to the acid solutions as a preventive measure against corrosion. The general corrosion rate of the metallic components during acid cleaning exercise should not exceed the maximum allowance as listed in Table 9.1. It is, however, a tedious task to find an inhibitor capable of protecting the multi-metallic components of desalination plants.

Table 9.1 Corrosion rate maximum allowance during acid cleaning [14, 15]

| Alloy | Maximum corrosion rate allowance (mpy) | | | |
| | HCl | | H_2SO_4 | |
	Atmospheric	Deaerated	Atmospheric	Deaerated
Carbon steel	20–50 fair	1–5 excellent	20–50 fair	1–5 excellent
316 L SS	< 1 outstanding	< 1 outstanding	1–5 excellent	1–5 excellent
Ni-resist	20–50 fair	< 1 excellent	20–50 fair	< 1 outstanding
90Cu-10Ni alloy	1–5 excellent	< 1 outstanding	1–5 excellent	< 1 outstanding
70Cu-30Ni alloy	1–5 excellent	< 1 outstanding	1 excellent	< 1 outstanding
Titanium	< 1 excellent	< 1 outstanding	< 1 excellent	< 1 outstanding

Table 9.2 Some commercial inhibitors for acid cleaning

S/N	Inhibitor code	Chemistry	Recommended acid	Ref.
1	Armohib CI-28	Proprietary surfactant blend	Hydrochloric	[18]
			Hydrochloric + hydrofluoric	
2	Armohib CI-31	Proprietary surfactant blend	Sulfuric	[18]
			Sulfamic	
			Citric	
			Phosphoric	
3	IBIT®		Hydrochloric	[19]
			Sulfuric	
4	TH-503	Organophosphine and polycarboxylic acid-based	Sulfuric	[20]
			Sulfamic	
			Hydrochloric	
5	Nevamine CN 356	Amine-based	Hydrochloric	[21]
			Sulfuric	
6	Nevamine WNF		Sulfuric	[21]
7	Rodine 213	Amine-based	Hydrochloric	[22]
8	Nevamin CP-20	Blend of ß (ethyl phenyl keto cyclohexyl) amino hydrochloride, formaldehyde, cinnamaldehyde, and methanol	Hydrochloric	[23]
			Sulfuric	

There are a handful of commercially available chemicals for use as inhibitors for acid cleaning processes (Table 9.2). These chemicals are formulations (also called cocktails or packages) consisting of active inhibitor substance(s), wetting agent(s), detergent(s), foaming agent(s), solvent, and co-solvent. The selection of a particular inhibitor for acid cleaning exercise is dependent on the range of alloys it can effectively protect, its cost effectiveness, availability, compatibility with other chemicals, as well as ecological friendliness. For instance, IBIT 570S and IBIT were reported to be excellent inhibitors for titanium and copper-nickel alloys [16] while ARMOHIB

28 was found to be outstanding for 316 L SS but corrosive towards copper-nickel alloys. Malik et al. [17] noted that, Nevamin CP-20 performance was inferior to that of IBIT but more cost-effective than IBIT and recommended Nevamin CP-20 for use in the cleaning of the Desal tubes of the Al-Jubail desalination plant.

9.2.1   Inhibitors for Corrosion Control of Carbon and Stainless Steels

The effective corrosion inhibitors for carbon and stainless steels are mostly nitrogen-based compounds [24, 25] and as listed by Schmitt [25] are mainly amines and its derivatives. This class of compounds has the ability to undergo chemical adsorption by denoting lone electron pair into the vacant orbital of iron [25]. They are components of most commercially available corrosion inhibitors.

Frequently used N-containing compounds in inhibitor cocktails are derivatives of acetylene. From the group of hydroxy acetylenes, propargyl alcohol, 1-hexyn-3-ol and l-iodo-3-methyl-1-butyn-3-ol are the most efficient derivatives [24, 25]. In the class of propargyl ethers and propargyl thioethers, 1-phenoxy- 2-butyn-4-ol and dipropargyl-sulphide exhibits outstanding inhibitive property as single compounds [24, 25]. The acetylenic compounds easily form Fe-complex protective films and temperature rise favors this type of bonding. Therefore, in commercial inhibitors designed for use in hydrochloric acid at elevated or high temperatures, acetylenic compounds are very likely to be present.

Sulphur-containing compounds like mercaptans, thioethers, sulphonium compounds, sulphoxides, thiocyanates, thioureas, thiazoles [25], *etc.* are also effective corrosion inhibitors for carbon steel and are component of several commercial inhibitors. However, the use of some of these substances, notably thiocyanates, thioureas and mercaptans [25, 26] can be hazardous in acid cleaning process because, some of them upon degradation produce H_2S gas, which is toxic and a promoter of hydrogen embrittlement [25]. Schmitt [25] recommended the inclusion of additives like the formaldehyde in inhibitor package containing the S-containing compounds to prevent hydrogen uptake by metals in the presence of H_2S gas.

Two classes of compounds appear to be the focus of corrosion scientists in recent times regarding the corrosion mitigation of carbon and stainless steels. There are ionic liquids and plant extracts. They are marked as green chemicals for tomorrow. Ionic liquids; have negligible vapour pressure, large liquidus range, high ionic conductivity and thermal stability, and large electrochemical window [27]. They exhibit two distinctive parts: the organic cationic part (mostly ammonium, pyridinium, imidazolium, pyrrolidinium, piperidinium, phosphonium, and sulfonium) and the inorganic anionic part (commonly bromide, chloride, cyanide, and borate). Most reports indicate that, they are capable of suppressing steels corrosion above 90% [28–30]. The plant extracts are naturally abundant and are cheap sources for inhibitor formulation. Some plant extracts with notable protection efficiency (above 90%) of carbon steel corrosion in acidic media are given in Table 9.3.

Table 9.3 Some plant extracts with potential to serve as active in corrosion inhibitor cocktails for the acid cleaning of carbon steel

Inhibitor	Highest IE (%)	Conc. with highest IE	Ref.	Inhibitor	Highest IE (%)	Conc. with highest IE	Ref.
Punica granatum Linne	95.0	1000 mg/L	[31]	*Ligularia fischeri*	92.0	500 ppm	[45]
Musa paradisica (Banana)	90.0	300 mg/L	[32]	*Ficus hispida*	90.0	250 ppm	[46]
Aloe Vera gel	92.6	200 ppm	[33]	*Gentiana olivieri*	93.7	800 mg/L	[47]
Longan	92.4	600 mg/L	[34]	*Diospyros kaki* (persimmon)	91.0	225 ppm	[48]
Lychee	98.0	600 mg/L	[35]	*Diospyros kaki* L.	94.3	1000 mg/L	[49]
Euphorbia falcata	93.2	3.0 g/L	[36]	*Pelargonium*	90.6	4 mL/L	[50]
Aniba rosaeodora	95.3	200 mg/L	[37]	*Chrysophyllum Albidum*	96.2	1000 mg/L	[51]
Orange (*Citrus sinensis*)	95.0	10%	[38]	*Tilia cordata*	96.0	300 mg/L	[52]
Capsella bursa-pastoris	97.0	60 mg/L	[39]	Egyptian *Schinus_ terebinthifolius*	93.3	900 ppm	[53]
Pisum sativum	91.0	400 mg/L	[40]	Olive	92.9	4.0 g/L	[54]
Eleusine aegyptiaca and Croton rottleri	91.3 and 94.5	2400 ppm	[41]	*Mentha rotundifolia*	92.9	35% (v/v)	[55]
Retama monosperma (L.) Boiss	94.4	400 mg/L	[42]	Roselle	91.0	500 ppm	[56]
Pimenta dioica	97.4	4% (v/v)	[43]	*Argemone Mexicana*	94.0	400 mg/L	[57]
Watermelon	90.2	1000 ppm	[44]	*Petersianthus macrocarpus*	93.5	1000 mg/L	[58]
Ligularia fischeri	92.0	500 ppm	[45]	*M. pulegium*	90.0	1 g/L	[59]

IE = Inhibition efficiency

9.2.2   *Inhibitors for Corrosion Control of Copper and its Alloys*

The corrosion of copper and its alloys during acid cleaning process can be effectively mitigated using corrosion inhibitors. Research studies [60–68] have shown that, the most effective corrosion inhibitors for copper and its alloys are organic compounds containing both sulphur and nitrogen heteroatoms in their moiety. It was demonstrated by Tan et al. [61] that, 5 mM 2,2′-dithiodipyridine and 5,5-dithiobis(1-phenyl-1H-tetrazole) respectively provided 98.7% and 99.4% protection to copper in acid medium. Rao and Kumar [62] reported an inhibition efficiency of 99.72% for 6.5 mM 5-(3-aminophenyl) tetrazole against Cu–Ni (90/10) alloy corrosion in synthetic seawater and synthetic seawater containing 10 ppm sulphide.

The sulphur-containing and nitrogen-containing organic compounds inhibit corrosion by coordinating with Cu^0, Cu^+, or Cu^{++} through lone pair electrons to form protective complexes on the metal surface. The complexes are polymeric in nature [62] as such, form adherent protective films that act as barriers against corrosive ions penetration. Some potent sulphur and nitrogen-containing organic inhibitors that are component or could be a component of copper corrosion inhibitor cocktails are listed in Table 9.4.

9.2.3   Inhibitors for Corrosion Control of Titanium

Titanium exhibits high corrosion resistance when exposed to air or any environment containing a trace of moisture or oxygen due to the formation of stable, protective, and strongly adherent passive films of TiO_2 [69]. However, in media such as, hydrofluoric acid, caustic solutions, and uninhibited concentrated hydrochloric or sulphuric acid solutions at medium temperatures (~ 50 °C), severe corrosion of titanium occur [70] due to the dissolution of the protective oxide film.

The most effective corrosion inhibitors for titanium are oxidizing inorganic compounds [71]. They are capable of inducing passivation of the metal in the corrosive medium. As seen in Table 9.5, oxidizing inorganic compounds can keep the corrosion rate of titanium in acid solution at less than 0.01 mm/yr up to 24 h. However, environment concern over some of these compounds has necessitated a search for replacement.

The nitroaromatic compounds (nitrobenzene, trinitrobenzoic acid, picric acid) and benzenearsonic acid are the most effective organic corrosion inhibitors for titanium in acid environment [25, 71]. They act as passivators, shifting the corrosion potential in the positive direction [25, 71]. In a typical study carried out by Deyab [70], 4-nitro-o-phenylenediamine (NI), 3-nitro-phydroxyethylaminophenol (NII), and N, N-bis(2-hydroxyethyl)-2-nitrop-phenylenediamine (NIII) inhibited the corrosion of titanium tubes when added to acid cleaning solution (1.0 M H_2SO_4). The aromatic nitro compounds exhibited a maximum inhibition efficiency of 73.0% (NI), 82.6% (NII) and 94.6% (NIII) at 600 ppm. Protection of titanium alloys in acid cleaning solutions is also possible using the condensation product of formaldehyde with aromatic amines [25, 71].

9.2.4   Inhibitors for Corrosion Control of Ductile Ni-Resist
          Cast Iron

Discharge columns, diffusers of brine recycle, and blow down pumps are usually made of Ni-resist, a highly alloyed cast iron with Ni content in the range of 18–22% [72]. Two grades of Ni-resist: ASTM A439 D2 (often designated as D-material) and

Table 9.4 Some potent corrosion inhibitors (single compounds) for copper corrosion in HCl and H_2SO_4 solutions

S/N	Inhibitor	Chemical structure	Medium studied	Maximum IE (%)	Conc. with max. IE	Ref.
1	Glycine		0.5 M H_2SO_4	92.0	50 mM	[69]
2	Tyrosine		0.5 M H_2SO_4	96.8	50 mM	[69]
3	Phenyl disulfide		0.5 M H_2SO_4	96.5	5 mM	[70]
4	2,2'-dithiodipyridine		0.5 M H_2SO_4	98.7	5 mM	[70]
5	5,5-dithiobis(1-phenyl-1H-tetrazole)		0.5 M H_2SO_4	99.4	5 mM	[70]
6	1,2-bis(4-chlorobenzylidene)azine		1 N HCl	92.93	1 mM	[61]
7	Montelukast sodium		0.5 M H_2SO_4	95.6	50 mg/L	[62]

(continued)

Table 9.4 (continued)

S/N	Inhibitor	Chemical structure	Medium studied	Maximum IE (%)	Conc. with max. IE	Ref.
8	N-(5,6-diphenyl-4,5-dihydro-[1, 2, 4] triazin-3-yl)-guanidine	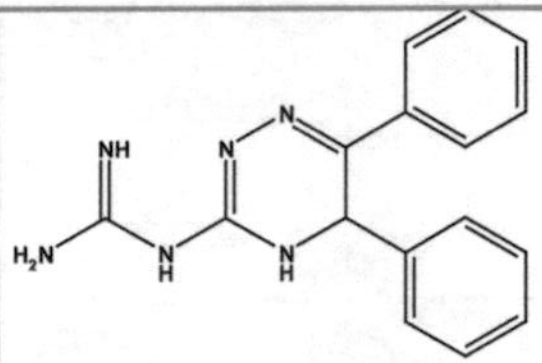	0.5 M H_2SO_4	95.03	0.1 mM	[63]
9	N-hydroxyacetamide (Acetohydroxamic acid)	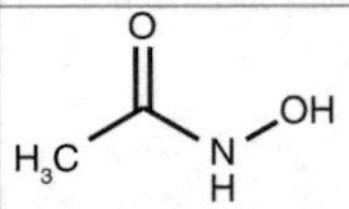	1 M HCl	92.91	150 ppm	[64]
10	N-hydroxybenzamide (Benzohydroxamic acid)	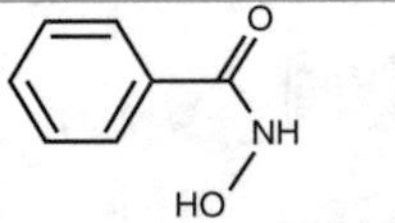	1 M HCl	92.91	150 ppm	[64]
11	5-(4-methoxyphenyl)-2-amino-1,3,4-thiadiazloe	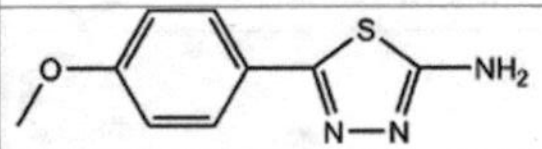	0.5 M H_2SO_4	90.0	6 mM	[65]
12	4-phenylthiazole-2-amine	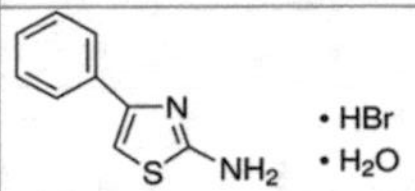	1 M HCl	90.5	0.1 mM	[66]
13	4-(2-aminothiazole-4-yl) phenol	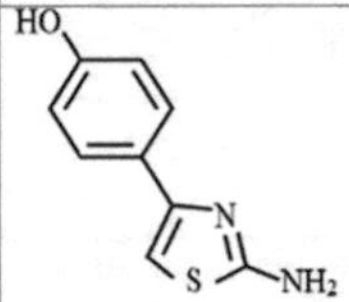	1 M HCl	90.3	1 mM	[67]

IE = Inhibition efficiency

Table 9.5 Oxidizing inorganic compounds as potent corrosion inhibitor for titanium in acid cleaning solution. Reproduced with permission from Ref. [71]. Copyright 1981 @ Elsevier

Inhibitor	Conc. (mol/L)	Corrosion rate ($g\ m^{-2}\ h^{-1}$)	
		1% H_2SO_4	3% HCl
Blank	–	4.710	2.800
$Fe_2(SO_4)_3$	0.01	0.000	–
$Ce(SO_4)_3$	0.01	0.000	0.025
$NaNO_3$	0.01	0.004	0.008
$KMnO_4$	0.01	0.000	0.004
$Na_2Cr_2O_7$	0.01	0.004	0.000
$NaMoO_4$	0.01	0.000	0.000
$NaWO_4$	0.01	0.000	0.000
$NaIO_3$	0.01	0.000	0.004
$NaBrO_4$	0.01	0.004	0.016
H_2O_2	0.10	0.000	0.050
$NaNO_3$	0.01	0.012	0.067

BS 3468 S2W (designated as G-material) are widely used for the fabrication of brine recycle pump. Generally, Ni-resist exhibits better corrosion resistance in dilute non-oxidizing acids than low carbon steel. Nevertheless, corrosion can be severe upon continuous exposure to acid solution and at elevated temperature. There have been cases of corrosion failure of pressure parts of brine pumps in desalination plants [73, 74]. In a typical report [73], a brand of pump made with G-material failed after 5 years. A brand of pump fabricated with D-materials was reported to live for 18 years before failure [73]. More so, a corrosion rate as high as 46 mpy was reported for Ni-resist during acid cleaning of recycle pump [13]. The use of effective corrosion inhibitors is therefore desirable during acid cleaning process to elongate the service life of brine recycle pumps. Surprisingly, information on the corrosion inhibition of Ni-resist is scanty in the corrosion literature. Nevertheless, a commercial inhibitor, Nevamin CP-20 had been recommended for use during acid cleaning of recycle pumps [13]. The best recommended condition is 0.5% of the inhibitor along with de-aeration of the acid solution (dissolved oxygen <20 ppb). In this case, corrosion rate of Ni-resist in the range of 4.5–6.8 mpy is achievable.

9.3   Inhibitors for Microbial Influenced Corrosion Control

Fouling is a common challenge in desalination plants, particularly the RO desalination plants. It is a term used to describe the accumulation, deposition, and/or adsorption of foulants onto the surface of a material. In RO, fouling causes membrane's performance to degrade [75]. It often results in a decrease in the flux and quality of the permeate, which consequently leads to an increase in the operating pressure with time [75, 76].

Table 9.6 Some known microorganisms that cause corrosion [76, 82]

Bacteria	• Iron oxidising bacteria (IOB) and metal depositing bacteria
	• Nitrate reducing bacteria, *e.g. Pseudomonas aeruginosa*
	• Sulphate reducing bacteria (SRB) *e.g. Desulfovibrio*
	• Sulphur oxidising and acid producing bacteria, *e.g. Acidithiobacillus*
	• Iron oxidizing and metal depositing bacteria, *e.g. Gallionella, Crenothrix, Leptothrix*
	• Metal reducing bacteria, *e.g. Pseudomonas, Shewanella*
Fungi	• *Aspergillus fumigatus*
	• *Cladosporium resinae*
	• *Paecilomyces varioti*
	• *Aspergillus niger*
	• *Penicillium cyclospium*
Algae	• Blue green algae
Microbial consortia	• Mutualism among different groups of microorganisms

There are four main types of fouling: (i) crystalline (deposition of inorganic material precipitating on a surface); (ii) organic (deposition of organic substances (e.g. oil, proteins, humic substances, *etc.*); (iii) particulate and colloidal (deposition of clay, silt, particulate humic substances, debris, silica); and microbiological (biofouling, adhesion and accumulation of microorganisms, forming biofilms). The crystalline, organic, and the particulate and colloidal fouling can be reduced by feed pretreatment (e.g. microfiltration or biocide application) [75] but biofouling cannot. Biofouling, apart from causing flux decline is also responsible for microbially induced corrosion (MIC) of materials [75].

MIC includes a variety of processes by which microbes contribute to corrosion directly or indirectly. Various microorganisms including bacteria, fungi, and algae induce corrosion (Table 9.6). In the category of bacteria, sulfate-reducing bacteria (SRB) have the most pronounced impact on corrosion because they are widely distributed in anoxic environments [76–78]. SRB were found to be responsible for the breakdown of passive film layers on Monel 400 alloy bolts of seawater intake pump, which led to intergranular corrosion [77]. *Pseudomonas aeruginosa*, a nitrate reducing bacteria was also found to induce the corrosion of 304 stainless steel [79]. The acid producing bacteria (APB), *e.g. Acidithiobacillus caldus* cause corrosion by producing weak organic acids that lowers the pH underneath their biofilms [82]. Iron oxidizing bacteria (IOB) contribute to MIC by encouraging precipitate formation and deposition on metallic component surfaces [80, 81]. The precipitates alter the acidity of the environment. Deposits can create anaerobic regions that promote the growth of SRB. The combined effects of IOB and SRB can cause severe corrosion [81].

MIC can be controlled by reducing the population of microorganisms in the water. This can be done by using biocides. There are various biocides, which can be divided into oxidizing and non-oxidizing. The oxidizing biocides include chlorine,

Table 9.7 Mechanism of action, advantages and disadvantages of oxidizing biocides [76, 82]

Biocide	Mechanism of action	Advantages	Disadvantages
Hypochlorite, chlorine	• Hydrolyze into hypochlorous and hydrochlorous acid • Hydrochlorous acid oxidizes the cytoplasm of organisms	• High inactivation efficiency. Hypochlorite is, however less effective than chlorine. • Organic matter removal • Relatively low cost	• Corrosive to steels • Removes bisulphite oxygen scavengers from water
Chloro isocyanurates	• Hydrolyse into hypochlorous acid and cyanuric acid	• The cyanuric acid reduces chlorine loss due to photochemical reactions with UV-light, hence more effective • Easily handled powdered compound	• Same as chloride and hypochlorite
Chlorine dioxide	• It does not form hydrochlorous acids in water. • It exists as dissolved chlorine dioxide, a compound that is a more reactive biocide at high pH ranges.	• No damage on membrane • Less damaging effects to the environment and human health than chlorine	• Explosive gas, need to be generated on site by reacting sodium chlorite ($NaClO_2$) and chlorine or hydrochloric acid.
Ozone	• Same as chlorine. Microorganisms die from loss of life-sustaining cytoplasm	• Effective inactivation, particularly at higher pH • High oxidation potential for organic matter	• Bromate formation • Very small half life • Damage by residual ozone
Mono chloramine	• Same as other oxidizing biocides • Needs to be used at high pH because it turns into di- and trichloramine at pH lower than 8.5.	• Lower oxidation potential than free chlorine • Less toxic than chlorine	• Degrades reverse osmosis membranes

chlorine dioxide, chloroisocyanurates (*e.g.* sodium dichloroisocyanuric acid), hypochlorite, ozone, and monochloramine. The oxidizing biocides enjoyed greater patronage compared to the non-oxidizing counterparts due to numerous advantages listed in Table 9.7. However, there are environmental concerns over the use of oxidizing biocides because of their high toxicity [82].

The non-oxidizing biocides are film formers and include quaternary ammonium compounds, formaldehyde, and glutaraldehyde [83–85]. In an investigation [83] involving the use of didecyldimethylammonium chloride (DDAC) as a biocide for carbon steel exposed to saline water containing SRB, it was found that, the present

of 1.3 mM DDAC prevented completely the growth of SRB (i.e 100% biocidal efficiency). The corrosion rate of the metal substrate was reduced from
5.40×10^{-4} g cm^{-2} h^{-1} in the absence of DDAC to 0.34×10^{-4} g cm^{-2} h^{-1} in the
presence of 1.3 mM DDAC corresponding to 93.7% inhibition efficiency. It was
also reported that, with 0.018 mM hexamethylene-1,6-bis(N,N dimethyl-N-
dodecyloammonium bromide), biocorrosion caused by SRB was effectively controlled [84]. Nevertheless, long-term application of this class of biocides may lead
to acclimation of microbes to be resistant [86]. This constitutes the major drawback
to the use of non-oxidizing biocides in water treatment processes.

9.4   Oxygen Scavengers

Oxygen-promoted corrosion is one of the common corrosion cases in seawater
desalination plants [87]. The genesis is the presence of dissolved oxygen in seawater. Due to the predominance of chloride in seawater, the presence of oxygen even
at low concentration can result in metal failures. To avoid corrosion in desalination
system, a near zero oxygen level has to be maintained. This can be achieved through
mechanical de-aeration of seawater followed by chemical treatment with an oxygen
scavenger. Common oxygen scavengers used in desalination plants are sulphite
based compounds *e.g.* sodium sulphite, sodium hydrogen sulphite, sodium metabi-
sulphite, and ammonium sulphite. Their efficiency is dependent on time, temperature, concentration and system pH. In general, as the pH, time, and temperature
increase, the oxygen depletion reaction becomes more complete [88].

Sulphites ($SO_3^{2?}$) are chemicals that deplete oxygen by reacting to form sulphates ($2SO_3^{2?} + O_2$? $2SO_4^{2?}$). According to Nada et al. [89], it is strongly recommended that sulphite be injected in the make-up after deaerator to reduce dissolved
oxygen content in the brine recycle to near zero level. This is to ensure good protection against corrosion in the flashing brine side as well as in the vapour side.
Dissolved oxygen levels lower than 20 ppb can be achieved by using sulphite [90].
However, addition of sulphite increases the solid contents of the water [90].

Several works on other oxygen scavengers such as hydrazine, carbohydrazide,
hydroquinone, and hydroxylamines are available in the literature that in general are
applicable to water boilers; but not typical of desalination.

9.5   Conclusions and Outlook

The use of chemical additives is an effective and practical method of controlling
various forms of corrosion in desalination systems. Film forming organic compounds are the most common corrosion inhibitors used during acid cleaning process. They are effective but some of them are challenged by their high toxicity and

high cost. Attention has been drawn to the enormous gain that could be derived in utilizing plant and ionic liquids-based materials as corrosion inhibitors. These materials are green, cheap, and readily available. Future researches should also focus on the development of inhibitors for Ni-resist. At present, information in this regard is insufficient.

Inhibitors for microbial influenced corrosion have been classified as oxidizing and film forming (non-oxidizing). The advantages and the disadvantages of the two classes have been highlighted. Also emphasized in the chapter is the use of sulphite-based compounds as oxygen scavengers. They remove dissolved oxygen in water by a reduction reaction, and by so doing inhibit corrosion caused by oxygen.

References

1. F.R. Spellman, *The Science of Environmental Pollution*. 2nd edn, (CRC press, Taylor and Francis, 2009), ISBN 9781138626607
2. World Health Organization., Guidelines for drinking-water quality, p. 631, ISBN 978-92-4-154995-0, 2017
3. World Health Organization., Drinking water (2019), https://www.who.int/news-room/fact-sheets/detail/drinking-water
4. Food and Agriculture Organization of the United Nations, Coping with Water Scarcity: An Action Framework for Agriculture and Food Security, FAO water reports 38, Rome, ISBN 978-92-5-107304-9, 2012
5. R.J. Forbes, A short history of the art of distillation: From the beginnings up to the death of Cellier Blumenthal, Ams Pr Inc, ISBN-13: 978–0404184704, 1948
6. E. Jones, M. Qadir, M.T.H. vanVliet, V. Smakhtin, S. Kang, The state of desalination and brine production: A global outlook. Sci. Total Environ. **657**, 1343–1356 (2019)
7. T. Outteridge, Preventing desalination plant corrosion: The role of molybdenum. The International Molybdenum Association (IMOA), http://www.sswnews.com/pdf/Preventing_Desalination_Plant_Corrosion.pdf
8. T. Hodgkiess, K.H. Al-Omari, N. Bontems, B. Lesiak, Acid cleaning of thermal desalination plant: Do we need to use corrosion inhibitors? Desalination **183**, 209–216 (2005)
9. The Courier Mail, Corrosion Hits Tugun Desalination Plant, April 18 2009. https://www.couriermail.com.au/news/special-features/corrosion-hits-desal-plant/news-story/0056a6868b238497868cc24cdb8127a9
10. A.I. El-Twaty, S.A. Karshman, Experience with desalination plants in Libya. Desalination **73**, 385–396 (1989)
11. NACE, NACE study estimates global cost of corrosion at $2.5 trillion annually, March 8, NACE International (2016), https://inspectioneering.com/news/2016-03-08/5202/nace-study-estimates-global-cost-of-corrosion-at-25-trillion-ann
12. M. Schorr, B. Valdez, J. OcaMpo, A. So, A. Eliezer, Materials and corrosion control in desalination plants. Mater. Perform. **51**, 56–60 (2012)
13. A.U. Malik, S. Ahmad, I. Andijani, N. Asrar, Acid cleaning of some Desal units at Al-Jubail plant. Technical Report No. TR3804/APP95007 (1997), https://www.scribd.com/document/258538217/Acid-Cleaning-of-Some-Desal-Units
14. M.G. Fontana, *Corrosion Engineering*. (Mc Graw Hill Intl., 1987), 3rd edn., p. 172
15. H. Uhlig, R. Revie, *Corrosion and Control*. (Wiley, 1985), 3rd cdn., p. 13
16. Japan Titanium Society, (1994). Counter measures against deposit of scale oceanic lives – light gauge titanium tubes for seawater desalination plants – 3, Q&A practical application, p. 14

17. A.U. Malik, I.N. Andijani, A. Nadeem Siddiqi, A. Shahreer, A.S. Al-Mobayaed, Studies on the role of sulfamic acid as a descalant in desalination plant. Proc. VI Middle East Corros. Conf., January 24–26 1994, pp. 65–78

18. ARMOHIB® corrosion inhibitors, https://surfacechemistry.nouryon.com

19. IBIT® – for acid cleaning, https://www.asahi-chem.co.jp/english/products/ibit/cleaning.html. Accessed 2 July 2019

20. Corrosion inhibitor for hydrochloric acidic cleaning, http://www.thwater.net/06-Hydrochloric-Acid-Cleaning.htm. Accessed 2 July 2019

21. Corrosion   inhibitors,   https://www.indiamart.com/proddetail/corrosion-inhibitors-10793739288.html

22. Rodine 213, HCl acid inhibitor, http://www.chemequal.com/supplier/rodine-213

23. S.D. Wadekar, V.N. Pandey, G.J. Hipparge, Environmentally friendly corrosion inhibitors for high temperature applications. US 0233872 A1, 2017

24. Y. Feng, S. Chen, J. You, W. Guo, Investigation of alkylamine self-assembled films on iron electrodes by SEM, FT-IR, EIS and molecular simulations. Electrochim. Acta 53, 1743–1753 (2007)

25. G. Schmitt, Application of inhibitors for acid media. Br. Corros. J. 19, 166–176 (1984)

26. R.H. Hausler, Corrosion inhibition and inhibitors, in Corrosion Chemistry, ed. by G.R. Brubaker, P.B.P. Phipps, ACS Symp. Ser., 89, (American Chemical Society, Washington, 1979), 263–320

27. S. Zhang, N. Sun, X. He, X. Lu, X. Zhang, Physical properties of ionic liquids: database and evaluation. J. Phys. Chem. 35, 1475–1517 (2006)

28. P. Kannan, J. Karthikeyan, P. Murugan, T.S. Rao, N. Rajendran, Corrosion inhibition effect of novel methyl benzimidazolium ionic liquid for carbon steel in HCl medium. J. Mol. Liq. 221, 368–380 (2016)

29. T. Tüken, F. Demir, N. Kıcır, G. Sığırcık, M. Erbil, Inhibition effect of 1-ethyl-3-methylimidazolium dicyanamide against steel corrosion. Corros. Sci. 59, 110–118 (2012)

30. Q.B. Zhang, Y.X. Hua, Corrosion inhibition of mild steel by alkylimidazolium ionic liquids in hydrochloric acid. Electrochim. Acta 54, 1881–1887 (2009)

31. G. Chen, M. Zhang, M. Pan, X. Hou, H. Su, J. Zhang, Extracts of Punicagranatum Linne husk as green and eco-friendly corrosion inhibitors for mild steel in oil fields. Res. Chem. Intermed. 39, 3545–3552 (2013)

32. G. Ji, S. Anjum, S. Sundaram, R. Prakash, Musa paradisica peel extract as green corrosion inhibitor for mild steel in HCl solution. Corros. Sci. 90, 107–117 (2015)

33. A.K. Singh, S. Mohapatra, B. Pani, Corrosion inhibition effect of Aloe Vera gel: gravimetric and electrochemical study. J. Ind. Eng. Chem. 33, 288–297 (2016)

34. L.L. Liao, S. Mo, H.Q. Luo, N.B. Li, Longan seed and peel as environmentally friendly corrosion inhibitor for mild steel in acid solution: Experimental and theoretical studies. J. Colloid Interface Sci. 499, 110–119 (2017)

35. L.L. Liao, S. Mo, H.Q. Luo, N.B. Li, Corrosion protection for mild steel by extract from the waste of lychee fruit in HCl solution: experimental and theoretical studies. J. Colloid Interface Sci. 520, 41–49 (2018)

36. M. Tabyaoui, B. Tabyaoui, H. El Attari, F. Bentiss, The use of Euphorbia falcata extract as eco-friendly corrosion inhibitor of carbon steel in hydrochloric acid solution. Mater. Chem. Phys. 141, 240–247 (2013)

37. M. Chevalier, F. Robert, N. Amusant, M. Traisnel, C. Roos, M. Lebrini, Enhanced corrosion resistance of mild steel in 1M hydrochloric acid solution by alkaloids extract from Anibarosaeodora plant: Electrochemical, phytochemical and XPS studies. Electrochim. Acta 131, 96–105 (2014)

38. N. M'hiri, D. Veys-Renaux, E. Rocca, I. Ioannou, N.M. Boudhrioua, M. Ghoul, Corrosion inhibition of carbon steel in acidic medium by orange peel extract and its main antioxidant compounds. Corros. Sci. 102, 55–62 (2016)

39. Q. Hu, Y. Qiu, G. Zhang, X. Guo, Capsella bursa-pastoris extract as an eco-friendly inhibitor on the corrosion of Q235 carbon steels in 1 mol·L^{-1}hydrochloric acid. Chin. J. Chem. Eng. **23**, 1408–1415 (2015)
40. M. Srivastava, P. Tiwari, S.K. Srivastava, A. Kumar, G. Ji, R. Prakash, Low cost aqueous extract of *Pisumsativum* peels for inhibition of mild steel corrosion. J. Mol. Liq. **254**, 357–368 (2018)
41. V. Rajeswari, D. Kesavan, M. Gopiraman, P. Viswanathamurthi, K. Poonkuzhali, T. Palvannan, Corrosion inhibition of *Eleusineaegyptiaca* and *Croton rottleri* leaf extracts on cast iron surface in 1M HCl medium. Appl. Surf. Sci. **314**, 537–545 (2014)
42. N. El Hamdani, R. Fdil, M. Tourabi, C. Jama, F. Bentiss, Alkaloids extract of *Retamamonosperma (L.) Boiss* seeds used as novel eco-friendly inhibitor for carbon steel corrosion in 1M HCl solution: Electrochemical and surface studies. Appl. Surf. Sci. **357**, 1294–1305 (2015)
43. K.K. Anupama, K. Ramya, K.M. Shainy, A. Joseph, Adsorption and electrochemical studies of *Pimentadioica* leaf extracts as corrosion inhibitor for mild steel in hydrochloric acid. Mater. Chem. Phys. **167**, 28–41 (2015)
44. N.A. Odewunmi, S.A. Umoren, Z.M. Gasem, S.A. Ganiyu, Q. Muhammad, L-Citrulline: An active corrosion inhibitor component of watermelon rind extract for mild steel in HCl medium. J. Taiwan Inst. Chem. Eng. **51**, 177–185 (2015)
45. M. Jokar, T. ShahrabiFarahani, B. Ramezanzadeh, Electrochemical and surface characterizations of *Morus alba pendula* leaves extract (MAPLE) as a green corrosion inhibitor for steel in 1M HCl. J. Taiwan Inst. Chem. Eng. **63**, 436–452 (2016)
46. P. Muthukrishnan, P. Prakash, B. Jeyaprabha, K. Shankar, Stigmasterol extracted from Ficushispida leaves as a green inhibitor for the mild steel corrosion in 1M HCl solution. Arab. J. Chem. (2015). https://doi.org/10.1016/j.arabjc.2015.09.005
47. E. Baran, A. Cakir, B. Yazici, Inhibitory effect of *Gentiana olivieri* extracts on the corrosion of mild steel in 0.5M HCl: electrochemical and phytochemical evaluation. Arab. J. Chem. (2016). https://doi.org/10.1016/j.arabjc.2016.06.008
48. H. Gerengi, I. Uygur, M. Solomon, M. Yildiz, H. Goksu, Evaluation of the inhibitive effect of *Diospyros kaki* (Persimmon) leaves extract on St37 steel corrosion in acid medium. Sustain. Chem. Pharm. **4**, 57–66 (2016)
49. G. Chen, X. Hou, Q. Gao, L. Zhang, J. Zhang, J. Zhao, Research on *Diospyros* Kaki L.f leaf extracts as green and eco-friendly corrosion and oil field microorganism inhibitors. Res. Chem. Intermed. **41**, 82–92 (2015)
50. Y. El Ouadi, A. Bouyanzer, L. Majidi, J. Paolini, J.M. Desjobert, J. Costa, A. Chetouani, B. Hammouti, S. Jodeh, I. Warad, Y. Mabkhot, T.B. Hadda, Evaluation of *Pelargonium* extract and oil as eco-friendly corrosion inhibitor for steel in acidic chloride solutions and pharmacological properties. Res. Chem. Intermed. **41**, 7125–7149 (2015)
51. C.O. Akalezi, E.E. Oguzie, Evaluation of anticorrosion properties of *Chrysophyllum albidum* leaves extract for mild steel protection in acidic media. Int. J. Ind. Chem. **7**, 81–92 (2016)
52. A.S. Fouda, A.S. Abousalem, G.Y. EL-Ewady, Mitigation of corrosion of carbon steel in acidic solutions using an aqueous extract of *Tiliacordata* as green corrosion inhibitor. Int. J. Ind. Chem. **8**, 61–73 (2017)
53. K. Shalabi, A.A. Nazeer, Adsorption and inhibitive effect of *Schinus terebinthifolius* extract as a green corrosion inhibitor for carbon steel in acidic solution. Prot. Met. Phys. Chem. Surf. **51**, 908–917 (2015)
54. M. Larif, A. Elmidaoui, A. Zarrouk, H. Zarrok, R. Salghi, B. Hammouti, H. Oudda, F. Bentiss, An investigation of carbon steel corrosion inhibition in hydrochloric acid medium by an environmentally friendly green inhibitor. Res. Chem. Intermed. **39**, 2663–2677 (2013)
55. A. Khadraoui, A. Khelifa, H. Hamitouche, R. Mehdaoui, Inhibitive effect by extract of *Mentharotundifolia* leaves on the corrosion of steel in 1 M HCl solution. Res. Chem. Intermed. **40**, 961–972 (2014)

56. A.A. Nazeer, K. Shalabi, A.S. Fouda, Corrosion inhibition of carbon steel by Roselle extract in hydrochloric acid solution: electrochemical and surface study. Res. Chem. Intermed. **41**, 4833–4850 (2015)

57. G. Ji, P. Dwivedi, S. Sundaram, R. Prakash, Aqueous extract of *Argemone mexicana* roots for effective protection of mild steel in an HCl environment. Res. Chem. Intermed. **42**, 439–459 (2016)

58. C.O. Akalezi, C.K. Enenebaku, E.E. Oguzie, Inhibition of acid corrosion of mild steel by biomass extract from the *Petersianthus macrocarpus* plant. J. Mater. Environ. Sci. **4**, 217–226 (2013)

59. M. Chraibi, K. Fikri Benbrahim, H. Elmsellem, A. Farah, I. Abdel-Rahman, B. El Mahi, Y. Filali Baba, Y. Kandri Rodi, F. Hlimi, Antibacterial activity and corrosion inhibition of mild steel in 1.0 M hydrochloric acid solution by *M. piperita* and *M. pulegium* essential oils. J. Mater. Environ. Sci. **8**, 972–981 (2017)

60. M.A. Amin, K.F. Khaled, Copper corrosion inhibition in O_2-saturated H_2SO_4 solutions. Corros. Sci. **52**, 1194–1204 (2010)

61. B. Tan, S. Zhang, W. Li, X. Zuo, Y. Qiang, L. Xu, J. Hao, S. Chen, Experimental and theoretical studies on inhibition performance of Cu corrosion in 0.5 M H_2SO_4 by three disulfide derivatives. J. Ind. Eng. Chem. **77**, 449–460 (2019)

62. B.V. Appa Rao, K.C. Kumar, 5-(3-Aminophenyl) tetrazole – A new corrosion inhibitor for Cu–Ni (90/10) alloy in seawater and sulphide containing seawater. Arab. J. Chem. **10**, S2245–S2259 (2017)

63. K. Abderrahim, I. Selatnia, A. Sid, P. Mosset, 1, 2-bis (4-chlorobenzylidene) azine as new and effective corrosion inhibitor for copper in 0.1 N HCl: A combined experimental and theoretical approach. Chem. Phys. Lett. **707**, 117–128 (2018)

64. B. Tan, S. Zhang, Y. Qiang, L. Feng, C. Liao, Y. Xu, S. Chen, Investigation of the inhibition effect of Montelukast Sodium on the copper corrosion in 0.5 mol/L H_2SO_4. J. Mol. Liq. **248**, 902–910 (2017)

65. K.F. Khaled, Adsorption and inhibitive properties of a new synthesized guanidine derivative on corrosion of copper in 0.5 M H_2SO_4. Appl. Surf. Sci **255**, 1811–1818 (2008)

66. D.K. Verma, E.E. Ebenso, M.A. Quraishi, C. Verma, Gravimetric, electrochemical surface and density functional theory study of acetohydroxamic and benzohydroxamic acids as corrosion inhibitors for copper in 1M HCl. Results Phys. **13**, 102194 (2019)

67. Y. Tang, W. Yang, X. Yin, Y. Liu, R. Wan, J. Wang, Phenyl-substituted amino thiadiazoles as corrosion inhibitors for copper in 0.5 M H_2SO_4. Mater. Chem. Phys. **116**, 479–483 (2009)

68. R. Farahati, A. Ghaffarinejad, S.M. Mousavi-Khoshdel, J. Rezania, H. Behzadi, A. Shockravi, Synthesis and potential applications of some thiazoles as corrosion inhibitor of copper in 1 M HCl: Experimental and theoretical studies. Prog. Org. Coat. **132**, 417–428 (2019)

69. E.M.M. Sutter, A. Cornet, J. Pagetti, The inhibition of the corrosion of titanium in 10 N sulphuric acid by cupferron (n-nitrosophenyl hydroxylamine). Corros. Sci. **27**, 229–238 (1987)

70. M.A. Deyab, Corrosion inhibition of heat exchanger tubing material (titanium) in MSF desalination plants in acid cleaning solution using aromatic nitro compounds. Desalination **439**, 73–79 (2018)

71. J.A. Petit, G. Chatainier, F. Dabos, Inhibitors for the corrosion of reactive metals: Titanium and zirconium and their alloys in acid media. Corros. Sci. **21**, 279–299 (1981)

72. A.U. Malik, S.A. Al-Fozan, M.A. Romiah, Relevance of corrosion research in the material selection for desalination plants. 2nd Science Symposium on Maintenance Planing and Operations, King Saud University, Riyadh, 24–26 April 1993

73. Y.A. Alzafin, A.-H.I. Mourad, M. Abou Zour, O.A. Abuzeid, A study on the failure of pump casings made of ductile Ni-resist cast irons used in desalination plants. Eng. Fail. Anal. **14**, 1294–1300 (2007)

74. Y.A. Alzafin, A.-H.I. Mourad, M. Abou Zour, O.A. Abuzeid, Stress corrosion cracking of Ni-resist ductile iron used in manufacturing brine circulating pumps of desalination plants. Eng. Fail. Anal. **16**, 733–739 (2009)

75. A. Matin, Z. Khan, S.M.J. Zaidi, M.C. Boyce, Biofouling in reverse osmosis membranes for seawater desalination: Phenomena and prevention. Desalination **281**, 1–16 (2011)
76. R. Jia, T. Unsal, D. Xu, Y. Lekbach, T. Gu, Microbiologically influenced corrosion and current mitigation strategies: A state of the art review. Int. Biodeter. Biodegr. **137**, 42–58 (2019)
77. F.A. Abd El Aleem, K.A. Al-Sugair, M.I. Alahmad, Biofouling problems in membrane processes for water desalination and reuse in Saudi Arabia. Int. Biodeter. Biodegr. **41**, 19–23 (1998)
78. A.U. Malik, T.L. Prakash, I. Andijani, Failure evaluation in desalination plants some case studies. Desalination **105**, 283–295 (1996)
79. R. Jia, D. Yang, D. Xu, T. Gu, Anaerobic corrosion of 304 stainless steel caused by the *Pseudomonas aeruginosa* biofilm. Front. Microbiol. **8**, 2335 (2017)
80. I.G. Chamritski, G.R. Burns, B.J. Webster, N.J. Laycock, Effect of iron-oxidizing bacteria on pitting of stainless steel. Corrosion **60**, 658–669 (2004)
81. K.M. Usher, A.H. Kaksonen, I. Cole, D. Marney, Critical review: Microbially influenced corrosion of buried carbon steel pipes. Int. Biodeter. Biodegr. **93**, 84–106 (2014)
82. S. Lattemann, T. Höpner, Environmental impact and impact assessment of seawater desalination. Desalination **220**, 1–15 (2008)
83. M.A. Deyab, Efficiency of cationic surfactant as microbial corrosion inhibitor for carbon steel in oilfield saline water. J. Mol. Liq. **255**, 550–555 (2018)
84. M. Pakiet, I. Kowalczyk, R.L. Garcia, R. Moorcroft, T. Nichol, T. Smith, R. Akid, B. Brycki, Gemini surfactant as multifunctional corrosion and biocorrosion inhibitors for mild steel. Bioelectrochemistry **128**, 252–262 (2019)
85. M. Lavania, P.M. Sarma, A.K. Mandal, S. Cheema, B. Lal, Efficacy of natural biocide on control of microbial induced corrosion in oil pipelines mediated by *Desulfovibrio vulgaris* and *Desulfovibrio gigas*. J. Environ. Sci. **23**, 1394–1402 (2011)
86. T. Nguyen, F.A. Roddick, L. Fan, Biofouling of water treatment membranes: A review of the underlying causes, monitoring techniques and control measures. Membranes **2**, 804–840 (2012)
87. A.U. Malik, P.C. Mayan Kutty, N.A. Siddiqi, I.N. Andijani, T.S. Thankachan, Effect of deaeration and sodium sulfite addition to MSF make-up water on corrosion of evaporator and heat exchanger materials. J. King Saud Univ. **8**, 21–36 (1996)
88. J.D. Zupanovich, Oxidation and Degradation Products of Common Oxygen Scavengers. The Analyst: *The Voice of the Water Treatment Industry,* Fall, 1–8 (2002), https://www.awt.org/pub/0149322F-0C20-5CEC-AE62-1E826AF61A4C
89. N.A. Nada, A. Zahrani, B. Ericsson, Experience on pre- and post-treatment from sea water desalination plants in Saudi Arabia. Desalination **66**, 303–318 (1987)
90. O. Kattan, K. Ebbers, A. Koolaard, H. Vos, G. Bargeman, Membrane contactors: An alternative for de-aeration of salt solutions? Sep. Purif. Technol. **205**, 231–240 (2018)

Chapter 10
Corrosion Control during Acid Cleaning of Heat Exchangers

Abdelkader A. Meroufel

10.1 Introduction

Heat exchangers represent important equipment in the thermal desalination process. Their application spectrum include different purposes such as distilled water production, heat recovery, cooling, etc. In all these applications, seawater is the dominantly used fluid due to its excellent heat transfer properties, availability and low cost. However, due to the presence of organic and inorganic fouling constituents, heat exchangers start to loose performance after a certain operation time. This induces a decline in the overall process performance and especially on water production. The restoration of the initial performance is necessary for sustainable water production.

Different chemical and physical methods were suggested to clean heat exchangers with various performance and limitations. The choice of the method depends on the heat exchanger configuration, size, construction materials, safety precautions, etc. Despite the environmental concerns of its disposal, chemical cleaning remains the best method to clean heat exchangers. By using either a chelant or an acid, a high level of fouling could be eliminated in short time achieving the restoration of operation performance.

Chemical cleaning discussion in this chapter will be limited to the acid cleaning due to its widespread application compared to the use of chelants that are considered as more costly. The handling of this subject is usually a multidisciplinary task including chemist, process and material specialist and HSE[1] engineer. However, a brief discussion of the chemistry aspect will be considered and the main aim of this chapter is to analyze the material integrity aspect.

---

[1] Health Safety and Environment

A. A. Meroufel (✉)
Desalination Technologies Research Institute, Saline Water Conversion Corporation, Jubail, Saudi Arabia
e-mail: ameroufel@swcc.gov.sa

© Springer Nature Switzerland AG 2020
V. S. Saji et al. (eds.), *Corrosion and Fouling Control in Desalination Industry*,
https://doi.org/10.1007/978-3-030-34284-5_10

This chapter begins with a general description of the main heat exchangers present in thermal desalination plants. Then, the fouling chemistry is described according to the operating conditions of the main heat exchangers. Following this, acid cleaning procedures are discussed based on desalination plant manufacturer's recommendations. Corrosion prevention practices are detailed with associated limited studies. Finally, the practical conclusions and future needed research or engineering needed efforts are outlined.

Due to the interchangeably use of scaling and fouling within scientific community and to be inline with the book authors, fouling in this chapter correspond to the inorganic part usually called scale.

10.2 Thermal Desalination Heat Exchanger Configurations

The heat exchangers in thermal desalination plants can be classified based on different criteria such as the purpose of use, fluid side, operating conditions, etc. Hou et al. [1] mentioned additional classification parameters such as construction, transfer processes, degree of compactness, flow arrangements, phases of process fluids and heat transfer mechanisms.

Among the different types of heat exchangers, shell and tube heat exchangers with one pass tube-side configuration are the most considered in thermal desalination and designed as per specific industrial standards.

Under thermal MSF desalination, three main heat exchangers are considered: heat rejection, heat recovery and heat input (brine heater). However, for the thermal MED desalination only two heat exchangers will be considered i.e. MED effect and final condenser. There are other heat exchangers but with reduced operation importance which justify our limitation to the above-mentioned list. Table 10.1 summarizes the operating conditions affecting fouling, typical construction materials used for the fouled side of the heat exchangers and the fluids employed.

From Table 10.1, it appears that the variation of materials chemistry and operating conditions induces variations in the fouling rate and chemistry. The most fixed

Table 10.1 The main heat exchangers in thermal desalination plants

Heat exchanger name	Operating conditions (max. temperature, °C/ velocity, m/s)	Typical construction materials in contact with corrosive fluid	Tube side/shell side fluid
MSF desalination			
Heat rejection condenser	40/2	Ti/modified7030/AlBr/NAB	Seawater/ steam
Heat recovery condenser	100/2	Ti/CuNi(9010 or 7030)/AlBr	Brine/steam
Brine heater	110/2	Ti/CuNi(9010 or 7030)/AlBr	Brine/steam
MED desalination			
MED effect	65/2	Ti/CuNi 9010	Steam/ seawater
Final condenser	35/2	CuNi (9010 or 7030)/AlBr	Seawater/ steam

operating parameter is the seawater/brine velocity which is optimized at 2 m/s for optimum heat transfer and minimal erosion/impingement risk on tubes. In addition, the circulation of the cleaning solution is usually ensured by existing pumps such as the brine recycle pump for MSF and brine blow-down for MED evaporator. Consequently, pump materials should be considered such as austenitic cast iron (Ni-resist) and duplex or austenitic stainless steel grades.

Different efforts are made to minimize the fouling on heat transfer either during the design phase and/or during operation. For instance, a huge quantity of chemicals is adopted to minimize fouling growth during operation. According to Hou et al. [1], 40% of the chemicals used by industry are for scaling control in cooling towers, boilers and other heat exchangers. Even though, the offline chemical cleaning remains the most powerful and well-established method to restore the heat exchanger performance.

The ease of heat exchangers cleaning is one of their critical design criteria. Within thermal desalination, the integration of a chemical cleaning system to the MSF or MED evaporator was considered by most of the manufacturers.

10.3   Fouling Chemistry and the Descalants

Many authors from different disciplines including chemist, microbiologists, electrochemists and materials scientists extensively studied heat exchanger fouling. The variation of the fouling chemical composition and related deposition mechanisms are attracting a lot of research efforts taking benefit from sophisticated tools [2–5]. The details of mechanisms involved in both inorganic and biofouling are described in Chaps. 14 and 15, respectively. Compared to the heat exchanger fouling, RO membranes fouling in fact attracted more research interest from the desalination community.

In terms of inorganic fouling, the soft alkaline scale is the most dominant form that has been characterized in both MSF and MED thermal desalination [6, 7]. This corresponds to $CaCO_3$ and $Mg(OH)_2$ components which coexist with proportions that are dependent on pH and temperature. For instance, it has been observed that $Mg(OH)_2$ dominates the scale above 87 °C and below this temperature, it is the $CaCO_3$ which dominates [8]. In fact, it is important to mention that the value of this transition temperature, which induces a modification of the scale chemistry, depends on seawater composition and velocity. In high-temperature MSF with TBT of 112 °C, the hard scale in the form of anhydrite $CaSO_4$ was also characterized at stage 4 [9]. However, $CaSO_4$ presence in MSF was limited to supersaturated seawater with high concentration factor in the range of 1.8–2 [10]. The scale chemistry in MED desalination plant where conditions are mild was found identical to that of MSF [11].

Malik et al. [12] studied the solubility of different scale samples from the brine heater and MSF heat recovery tubes. Three different acids were considered i.e. hydrochloric, sulfuric and sulfamic where tests were conducted at ambient temperature and under laboratory stirring. Depending on the scale chemistry, the acid descaling properties vary where hydrochloric acid was the most effective for $CaCO_3$ and sulfuric acid for $Mg(OH)_2$ scale. To fit the compromise between descaling performance and corrosion risk of the base metal, the authors suggested hydrochloric acid with a corrosion inhibitor. Phosphoric acid was also suggested for seawater cooling heat exchangers used in the shipping industry [13, 14]. However, the concern was more focused on the restoration of the metal substrate protection layer.

In addition to the above-mentioned acids, Tahir et al. [15] studied the descaling performance of nitric and citric acids at a concentration between 1 and 5%. According to their results, use of hydrochloric acid allows a saving of 50% in terms of circulation time compared to nitric acid. In rare cases, $CaSO_4$ anhydrite was found in MSF heat exchangers where a different descaling agent was employed. Kwong et al. [16] studied the descaling performance of three methods: sodium glycollate, ethylenediaminetetraacetic acid (EDTA) and sodium carbonate added hydrochloric acid. The most successful and adopted method was the alternation between sodium carbonate at 80 °C and inhibited hydrochloric acid. Such an approach seems impractical when considering the necessity to heat which induces an additional cost.

10.4 Acid Cleaning Procedures

Depending on the conditions of the desalination plant performance, different acid cleaning options are considered. For instance, the whole MSF evaporator can be cleaned excluding some parts such as brine heater or heat rejection. Due to the low fouling level, the cleaning of the heat rejection in MSF or the final condenser in MED is non-frequent.

Criteria for acid cleaning depend on various operation parameters related to the production and efficiency. For instance, the necessity to clean an MSF brine heater is considered if the shell internal temperature exceeds a particular setting value in spite that the plant performance ratio is higher than a designed value.

Typical acid cleaning procedure for both MSF and MED is illustrated in Fig. 10.1 where most of the practices adopt the use of acidified seawater instead of distilled water. The consideration of temperature to accelerate descaling is proposed especially when using sulfuric or sulfamic acids. In practice, a temperature of 50–60 °C is adopted for enhanced descaling with these acids.

In terms of acid cleaning duration and continuity decision, manufacturers provide different parameters including the return pH and copper ion concentration. Copper ions are considered as an indicator of heat exchanger tube corrosion. For instance, when the pH of the circulating cleaning solution reaches 1.5–2, the cleaning should be stopped within 1–2 h at this pH. Such over-conservative criteria aim

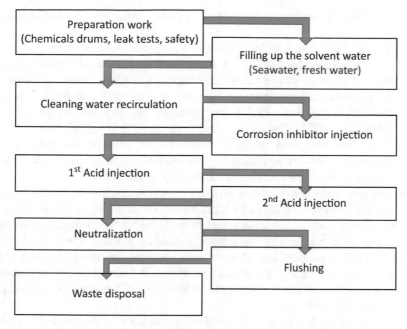

Fig. 10.1 MSF and MED acid cleaning typical procedure

to protect the base metal from any unexpected attack. In other cases, a qualitative criterion is provided such as an abnormal increase of copper ions in the cleaning solution. However, the practical experience showed that the simultaneous monitoring of a set of parameters can be more efficient and rational including parameters related to the scale (calcium and magnesium), acidity (pH) and the base metal (iron and copper elements).

10.5   Corrosion Prevention during Acid Cleaning

10.5.1   Acid Choice

Corrosion prevention of the base metal during descaling starts by the acid choice. Although the descaling properties of the used acid are important to operators, the compromise with metal resistance to the acid remains a primordial condition. For these reasons, most thermal desalination manufacturers recommend the use of either diluted sulfuric or sulfamic acid. However, with experience, many operators moved to the use of hydrochloric acid due to the excellent descaling properties and short time for desalination unit restoration. From a corrosion point of view, acid cleaning using diluted or weak acids with a concentration that does not exceed 5% should not be aggressive. However, in the present section, available data on the behavior of the different concerned alloys in both sulfuric and hydrochloric acids will be presented.

Corrosion of different desalination metallic alloys in the concerned diluted mineral acids is not fully discussed in the open literature [17–21]. Most of the references focused on concentrated acids with temperature effect. However, all studies agree that corrosion in mineral acids where the pH is below 5 is dependent on three parameters: hydrogen ion activity, the oxidizing capacity of the solution and counter-ions present [22]. For instance, iron corrosion rate in sulfuric acid with a pH range from 0 to 4 was found limited by diffusion of protons and saturation concentration of iron sulfate ($FeSO_4$) [23]. By increasing the concentration of sulfuric acid, the phenomenon becomes sensitive to the flowing conditions.

However, in hydrochloric acid a different mechanism was suggested where the corrosion rate of iron is rapid at pH values below 3 with strong participation of the chloride counter-ions [24]. It is important to recall that usually sulfuric acid is considered as an oxidizing acid while hydrochloric acid is a reducing acid that induces different corrosion mechanisms. However, some authors claim that sulfuric acid can alternate between oxidizing and reducing characters depending on its concentration and the temperature of the medium [25]. In low concentration range, sulfuric acid would behave as a reducing acid.

In the case of hydrochloric acid, steel corrosion leads to the formation of iron chloride compound which is very soluble in the acid, offering very low protection as mentioned by Richardson [26]. For both ferritic and austenitic steels, both alloy composition and the acid concentration must be considered simultaneously to predict the behavior in mineral acids. The behavior can be under activation control for ferritic and sensitive to the flow for austenitic grades depending on acid concentration [27, 28]. With hydrochloric acid, the impact is detrimental especially to austenitic steels where the risk of localized attack is higher [29]. This situation is similar for austenitic stainless steels where hydrochloric acid affects the passivity through a significant decrease of the critical breakdown potential [30]. Krawczyk et al. [31] studied the impact of temperature and microstructure on the corrosion type (uniform, localized or mixed). Corrosion type diagram was developed depending on the acid concentration and temperature. From another perspective, cold rolling up to 20% did not show significant impact on corrosion rate and type. Isocorrosion chart shown in Fig. 10.2 indicates the corrosion risk on typical low grades of stainless steel in diluted hydrochloric acid at ambient temperature. According to this diagram, it is necessary to select higher grades with higher nickel and molybdenum content in contact with hydrochloric acid solution. However, in sulfuric acid duplex stainless steel can be used up to 40% of the acid concentration depending on the temperature as shown in Fig. 10.3.

Hydrochloric acid is also detrimental to titanium alloy where the metal corroded in a sudden manner [32]. The corrosion rate of commercial titanium named grade 2 and other titanium alloys in hydrochloric acid was found to be affected by the presence of ferric ions. Figure 10.4 shows the Isocorrosion chart where we can see that grade 2 can be used in hydrochloric acid up to an acid concentration of 5% at ambient temperature. In sulfuric acid, commercial titanium can be safely used up to

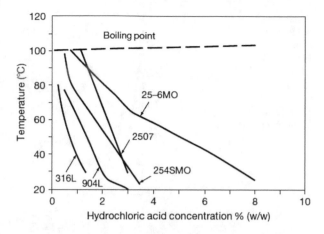

Fig. 10.2 Isocorrosion chart (0.1 mm/yr) for different stainless steels in hydrochloric acid. Reproduced with permission from Ref. [26]; Copyright © 2010 Elsevier

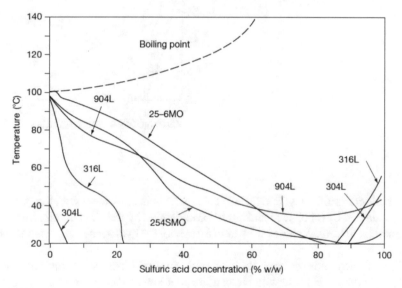

Fig. 10.3 Isocorrosion chart (0.1 mm/yr) for carbon steel and stainless steels in sulfuric acid. Reproduced with permission from Ref. [25]; Copyright © 2010 Elsevier

a concentration of 5% and a temperature of 50 °C based on the Isocorrosion chart in naturally aerated sulfuric acid [33].

From another perspective, Andijani et al. [34] studied the risk of hydrogen absorption on titanium tubes. Indeed, the galvanic coupling between titanium tubes, iron sacrificial anodes and cupronickel alloy cladding on water box in MSF evaporator would induce a role of the cathode for titanium due to its high potential. The authors confirmed the absence of significant hydrogen absorption in commer-

Fig. 10.4 Isocorrosion
chart (0.13 mm/yr) for
different titanium and its
alloys in hydrochloric acid
with and without the
presence of ferric ions.
Reproduced with
permission from Ref. [26];
Copyright © 2010 Elsevier

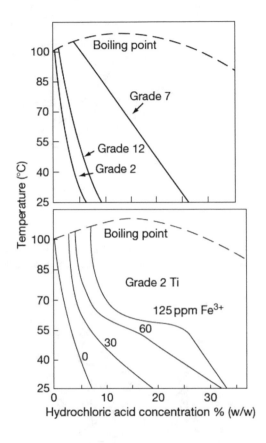

cial titanium in sulfuric acid at 50 °C. This result is in agreement with the investigation conducted by Satoh [35] and could be explained by the alpha microstructure of commercial titanium known for its low diffusivity for hydrogen as reported by Been et al. [36].

Copper alloys are successfully used with non-oxidizing acids, such as sulfuric acid and hydrochloric acid with the condition of low acid concentration, deaeration and low ferric ion concentration. However, in air-saturated acidic solution, corrosion can reach up to 1.2 mm/yr. and would increase in hydrochloric acid [37].

It is important to notice that all the above-mentioned data are for a stagnant diluted acidic solution where the solvent is distilled or freshwater. However, in flowing acidified seawater there is a scarcity of published studies to assess the real corrosion risks. This emphasize the need to use of corrosion inhibitors as discussed in the following section.

10.5.2   Corrosion Inhibitors

Among the different types of corrosion inhibitors, acid corrosion inhibitors will be the focus in this section. The selection criteria for acid cleaning of MSF unit is not limited to the corrosion inhibition performance but also include the safety of use, economic limitations, chemical stability (interaction with other chemicals during cleaning), and environmental impact.

Corrosion inhibitor selection starts with the understanding of the corrosion mechanisms as well as how the inhibitor will behave on the different materials in the acid cleaning operation. Finding one corrosion inhibitor for various metallurgy represents one of the remaining technical challenges for the corrosion community. In the literature, acid corrosion inhibitors usually cover two metallic alloys as maximum.

On the other hand, the optimum concentration of the corrosion inhibitor depends on the environment parameters such as the pH and presence of corrosive constituents (halides) [38]. Many authors reported that below a particular critical concentration value of the inhibitor, inhibition turns into corrosion stimulation [39]. In addition, it is accepted that for certain critical pH value where we shift into slightly acidic condition, oxygen reduction has to be considered with hydrogen reduction that accelerates the kinetics of corrosion [39]. However, below that critical pH value, hydrogen reduction is the dominant cathodic process as in the situation of acid cleaning.

The presence of halides adsorbent (iodide, bromide and chlorides) was found to enhance the physical adsorption of the organic inhibitor and surface coverage on the steel metal surface offering better corrosion inhibition [40]. This synergy was variable between halides with a debated order between iodide, bromide and chloride. In the same manner, Asefi et al. [41] obtained a slight increase and stabilization of corrosion inhibitor efficiency with chloride concentration up to 0.1 mol/L. On the other side, chlorides are known for their simultaneous dissolution and localized corrosion stimulation in acidic medium for cupronickel and stainless steel alloys respectively [42, 43]. Alvarez et al. [44] reported typical polarization curves of iron, nickel and their alloys in 1 M sulfuric acid for various chlorides concentrations where the passive region found disappeared with an increase of chloride concentration. The impact of chlorides present in the seawater in acid cleaning merit more investigation and can vary according to the metal substrate.

In the following section, the main corrosion inhibitors will be discussed for each alloy then a summary of the inhibitors that can be used for more than one metal/ alloy will be mentioned. For detailed information on corrosion inhibitors in acidic solution, the reader are advised to refer other references [45, 46].

Acid corrosion inhibitors on steel and its alloys are the most studied with an important focus on organic compounds with multiple heteroatoms containing polar functional groups to enhance the efficiency as reported by Revie et al. [47]. Within the context of acidic corrosion, organic corrosion inhibitors showed better passivation of the steel substrate compared to the inorganic inhibitors. Amine, aniline and amino acids and their derivatives based inhibitors are used with success in acidic corrosion due to their non-bonding π–electrons that allow the formation of coordinating bonds with metal surface [48–50]. The presence of oxide layer on the steel was found to enhance the bonding by the creation of hydrogen bond [51]. On the other hand, several Schiff base inhibitors were also studied demonstrating excellent inhibition efficiencies [52, 53]. Their excellent inhibition raise from their known combination of lone pair of electrons from both heteroatoms and π-electrons. For cast iron, Rajeswari et al. [54] studied the performance of three Schiff bases in the corrosion inhibition in 1 N hydrochloric acid solution. The efficiency was dependent on the inhibitor concentration, exposure time and inhibitor substituent molecules.

Similarly, organic corrosion inhibitors dominate the protection of copper alloys in descaling conditions. Azoles, amines and amino acid-types are major copper corrosion inhibitors [55–57]. The mechanism of corrosion inhibition is the same as for carbon steels and cast iron.

For passive alloys such as stainless steels and titanium, the story is different where the main risk to be mitigated is the localized corrosion in the form of pitting. Corrosion inhibitors suggested for stainless steel included both synthetic and natural compounds. For instance, Caliskan et al. [58] suggested the use of pyrimidine derivatives for austenitic stainless steel in 0.5 M sulfuric acid. Using different techniques, the authors could demonstrate the mixed protection mode of the inhibitor as well as the spontaneous adsorption mechanism. Sanni et al. [59] used an eco-friendly waste (eggshell powder) as a corrosion inhibitor for 316 L stainless steel in 0.5 M sulfuric acid. Very high corrosion inhibition was obtained with a mixed inhibition mechanism due to the amino acids and proteins present in this inhibitor.

Most of the published works during the last decade on acid corrosion inhibitors focused on steel and hydrochloric acid. This is explained by the application target such as the pickling process during the manufacturing processes. The presence of chlorides and stirring were rarely considered where most authors studied in distilled water acidified by hydrochloric acid at concentrations ranging between 0.5–2 mol/L. Therefore, the obtained results are irrelevant for thermal desalination acid cleaning conditions. The growing interest to the green corrosion inhibitor use was also noticed but it is facing a serious challenge of scale-up and adoption by chemicals manufacturers.

Malik et al. [60] were among the rare authors studying the corrosion behavior of thermal MSF desalination alloys in flowing acidified (by sulfuric and hydrochloric acids) gulf seawater. The authors focused on the corrosion resistance of tubing materials such as cupronickel and titanium alloys. They observed an initial high corrosion rate followed by a decline to negligible corrosion rate in the presence of a commercial amine-based corrosion inhibitor. This was attributed to the gradual formation of the passive layer. Following this study, Malik et al. [60] investigated the impact of the presence of oxygen where they could demonstrate the efficiency of commercial inhibitors in both aerated and deaerated dynamic acid solutions. Figure 10.5 shows the results obtained for immersion conditions for 6 h and at room temperature.

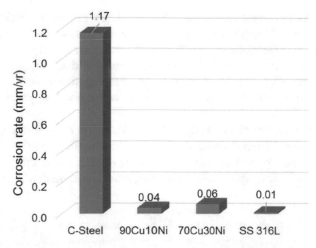

Fig. 10.5 Corrosion rate of MSF alloys in dynamic acidified (2% hydrochloric acid) seawater at room temperature in the presence of a commercial corrosion inhibitor

Deaerating the acid cleaning solution increases the cost of cleaning and that can be avoided by other alternatives. On the other hand, we can notice the low inhibitor protectiveness for active alloys such as carbon steel and Ni-resist (austenitic cast iron) in dynamic aerated conditions. These alloys are usually adopted for MSF flash chamber bottom, sacrificial anode and brine recycle pump components. Since acid cleaning duration exceeds 6 h, it is advised to explore the corrosion inhibitor efficiency for a longer duration.

Another practical experience comes from the field data presented by Abu Dayyeh et al. [61] on MSF acid cleaning using hydrochloric acid at Ghubra plant (Oman). The authors focused on the copper alloy corrosion and could estimate a very low thickness loss on different MSF evaporator tubes. Usually, MSF acid cleaning in the presence of corrosion inhibitors induces a copper concentration in the range of 60–150 ppm. When considering the vast cupronickel alloy area, the corresponding calculated corrosion rate will be negligible as estimated by Abu Dayyeh et al. [61].

Meroufel et al. [62] studied the performance of a commercial amine-based corrosion inhibitor to protect different MSF alloys in stagnant acidified Gulf seawater at 40 °C for different durations. Figure 10.6 shows the decline of the general corrosion rate with immersion time in the presence of the commercial corrosion inhibitor. When comparing with the results of Fig. 10.5, it appears that the flow reduced the corrosion rate significantly.

It is important to mention that the determination of acid corrosion inhibitor efficiency was discussed in NACE TM 0193 standard [63]. It seems that due to the variety of chemical cleaning applications, there is no universal test method and the corrosion severity obtained on clean metal coupons is not the same experienced during real acid cleaning operation. On the other hand, the duplication of field acid cleaning conditions is quite difficult in laboratory tests. Especially when considering the ratio of sample surface area to the solution volume, exposure duration, seawater salinity, velocity and interaction between scale and corrosion inhibitors.

Thus, the assessment of corrosion inhibitor performance during real acid cleaning process is the most efficient way to fix the corrosion risks. Figure 10.7 shows a typical evolution of copper ions during an acid cleaning of real MSF unit using 2% hydrochloric acid acidified seawater with a commercial approved corrosion inhibitor.

We can observe, the slope change of the copper ion concentration due to the action of corrosion inhibitor which slows down the corrosion rate. This later can be obtained from the slope of copper evolution curve after considering the exposed surface area. Then, the slope is converted into corrosion current density as performed

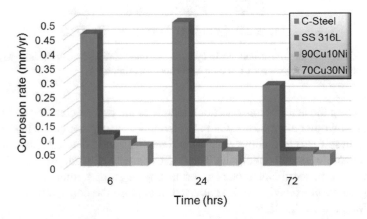

Fig. 10.6 Commercial corrosion inhibitor efficiency to protect MSF alloys in stagnant acidified (2% hydrochloric acid) seawater at 40 °C

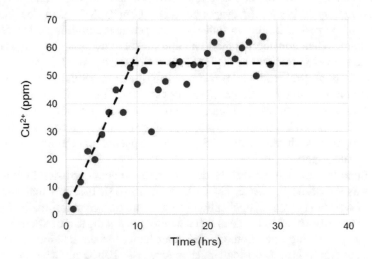

Fig. 10.7 Copper evolution during acid cleaning of MSF unit

by Amin et al. [64]. This method was found to be well correlated with weight loss method if the copper ions concentration is accurately determined.

10.6   Conclusions and Outlook

In this chapter, acid cleaning of heat exchangers in thermal desalination plant has been discussed from a corrosion point of view with a brief indication of descaling performance. The domination of hydrochloric acid due to its super descaling properties raise the concern of corrosion risk on different alloys. This risk is mitigated with a suitable corrosion inhibitor. Limited published studies were observed relative to the acid cleaning conditions and materials of thermal desalination alloys. The absence of industrial standards and the over conservatism of both desalination and materials manufacturers raise the technical and scientific need for more corrosion research on the mechanisms and inhibition possibilities.

Whereas the corrosion inhibitor concentration adopted in practice is below 2%, the selection of suitable corrosion inhibitors for simultaneous protection for multiple alloys remains the main research question. Indeed, the commercially available corrosion inhibitors are usually performing for one or two alloys. Another pending research question is the ability of commercial corrosion inhibitor to mitigate pitting corrosion of low stainless steel grades such as 316 L.

On the other hand, many efforts are conducted to avoid acid cleaning of heat exchangers including the adoption of physical (pneumatic) methods using a hydromechanical system. However, this kind of techniques is limited to one heat exchanger and cannot be applied to large MSF condensers or MED effects because of the time-consuming aspect.

References

1. T.K. Hou, S. Kazi, A. Mahat, C.B. Teng, A. Al-Shamma'a, A. Shaw, Industrial heat exchangers: Operation and maintenance to minimize fouling and corrosion. in *Heat Exchangers – Advanced Features and Applications*, ed. by S.M.S. Murshed, M.M. Lopes, (Intechopen, 2017), ISBN: 978-953-51-3092-5
2. J. Cowan, D. Weintritt, *Water-Formed Scale Deposits*, (Gulf Publishing, Houston, 1976), ISBN: 978-0872018969
3. T.R. Bott, *Fouling of Heat Exchangers*, (Elsevier, Amsterdam, 1995), ISBN: 9780444821867
4. M.M. Awad, Fouling of heat transfer surfaces. in *Heat Transfer: Theoretical Analysis, Experimental Investigations and Industrial Systems*, ed. by A. Belmiloudi, (Intechopen, 2011), ISBN: 978-953-307-226-5
5. J. Zhao, M. Wang, H.M.S. Lababidi, H. Al-Adwani, K.K. Gleason, A review of heterogeneous nucleation of calcium carbonate and control strategies for scale formation in multi stage flash (MSF) desalination plants. Desalination **442**, 75–88 (2018)
6. M.A.K. Al-Sofi, Fouling phenomena in multi stage flash (MSF) distillers. Desalination **126**, 61–76 (1999)

7. K. Krömer, S. Will, K. Loisel, S. Nied, J. Detering, A. Kempter, H. Glade, Scale formation and mitigation of mixed salts in horizontal tube falling film evaporators for seawater desalination. Heat Transf. Eng. **36**, 750–762 (2015)

8. A.M.S. El-Din, R.A. Mohammed, Brine and scale chemistry in MSF distillers. Desalination **99**, 73–111 (1994)

9. E1Din AMS, E1-Dahshan ME, R.A. Mohammed, Scale formation in flash chambers of high-temperature MSF distillers. Desalination **177**, 241–258 (2005)

10. E. Ghiazza, A.M. Ferro, *The Scaling of Tubes in MSF Evaporators: A Critical Review across 20 Years of Operational Experience* (IDA world congress, Manama, 2002). https://www.fisiait.com/static/upload/ghi/ghiazza_05.pdf

11. P. Budhiraja, A.A. Fares, Studies on scale formation and optimization of antiscalant dosing in multi-effect thermal desalination units. Desalination **220**, 313–325 (2008)

12. A.U. Malik, I. Andijani, A. Al-Mubayaed, Comparative study on the use of sulfamic, sulfuric and hydrochloric acid as descalants for brine heater and heat recovery tubes. SWCC Technical Report No. 36, 1995

13. T.M. Wolejsza, E.J. Lemieux, K.E. Lucas, *Evaluation of a Phosphoric Acid Based Descaling Solvent for Shipboard Cleaning of Seawater Heat Exchangers*. NACE-03261, Corrosion 2003, NACE International, Houston, 2003

14. E.J. Lemieux, R. Bayles, T.M. Newbauer, K.E. Lucas, Evaluation of descaling chemicals for use with titanium, alloy 625 and nickel copper alloy heat exchanger and piping system components. Corrosion 2006, NACE-06289, NACE International, Houston, 2006

15. M.S. Tahir, M. Saleem, Experimental study of chemical de-scaling – effect of acid concentration. J. Faculty Eng. Technol, 1–9 (2007-2008). https://pdfs.semanticscholar.org/51ba/4d1797de617ff85a85212e14f2a178d68dac.pdf

16. C.Y. Kwong, L.K. Kuen, Chemical removal of calcium sulphate scale-plant experience of Lok on Pai desalter. Desalination **30**, 359–371 (1979)

17. A.M. Shams El-Din, H.A. El Shayeb, F.M. Abdel Wahab, Stability of titanium tubes toward acid wash. J. Electroanal. Chem. **214**, 567–587 (1986)

18. W.T. Hanbury, T. Hodgkiess, K. Al-Omari, Aspects of acid cleaning operations in MSF plants. Desalination **158**, 1 (2003)

19. T. Hodgkiess, K. Al-Omari, K. Bontems, B. Lesiak, Acid cleaning of thermal desalination plant: Do we need to use corrosion inhibitors? Desalination **183**, 209–216 (2005)

20. T.M.H. Saber, M.K. Tag El Din, A.M. Shams El Din, Dibutyl thiourea as corrosion inhibitor for acid washing of multistage flash distillation plant. Br. Corros. J. **27**(2), 139–143 (1992)

21. J. Kish, N. Stead, D. Singbeil, Corrosion of stainless steel in sulfamic acid cleaning solutions. Corrosion 2006, NACE-07205, NACE International, Houston, 2007

22. D.C. Silverman, Aqueous Corrosion, in *Corrosion: Fundamentals, Testing, and Protection*, ed. by S. D. Cramer, B. S. Covino, vol. 13A., ASM Handbook, (ASM International, 2003), pp. 190–195

23. B.T. Ellison, W.R. Schmea, Corrosion of steel in concentrated sulfuric acid. J. Electrochem. Soc. **125**, 524–531 (1978)

24. R.J. Chin, K. Nobe, Electrodissolution kinetics of iron in chloride solutions. J. Electrochem. Soc. **119**, 1457–1461 (1972)

25. J.A. Richardson, Corrosion in sulfuric acid, in *Shreir's Corrosion*, (Elsevier, 2010), pp. 1226–1249

26. J.A. Richardson, Corrosion in hydrogen halides and hydrohalic acids, in *Shreir's Corrosion*, (Elsevier, 2010), pp. 1207–1225

27. B. Alexandre, A. Caprani, J.C. Charbonnier, M. Keddam, P.H. Morel, The influence of chromium on the mass transfer limitation of the anodic dissolution of ferritic steels Fe-Cr in molar sulfuric acid. Corros. Sci. **21**, 765–780 (1981)

28. The corrosion resistance of nickel containing alloys in sulfuric acid and related compounds, CEB-1, International Nickel Company, Inc., (1983), https://www.nickelinstitute.org/

media/1829/thecorrosionresistanceofnickel_containingalloysinsulphuricacidandrelatedcompounds_1318_.pdf

29. Resistance of nickel and high nickel alloys to corrosion by hydrochloric acid, hydrogen chloride, and chlorine, CEB-3, International nickel company, Inc., (1974), https://www.nickelinstitute.org/media/1776/resistanceofnickelandhigh_nickelalloystocorrosionbyhydrochloricacid_hydrogenchlorideandchlorine_279_.pdf

30. G. Bianchi, F. Mazza, S. Torchio, Stress-corrosion cracking of austenitic stainless steel in hydrochloric acid media at room temperature. Corros. Sci. **13**, 165–173 (1973)

31. B. Krawczyk, P. Cook, J. Hobbs, D.L. Engelberg, Corrosion behavior of cold rolled type 316L stainless steel in HCl-containing environments. Corrosion **73**, 1346–1358 (2017)

32. D.C. Silverman, Derivation and application of EMF-pH diagrams. in *Electrochemical Techniques for Corrosion Engineers*, ed. by R. Baboian, (NACE International, 1986)

33. B. Craig, D. Anderson, *Handbook of Corrosion Data*, (ASM international, 1995)

34. I. Andijani, S. Ahmad, A.U. Malik, Corrosion behavior of titanium metal in the presence of inhibited sulfuric acid at 50 °C. Desalination **129**, 45–51 (2000)

35. H. Satoh, Hydrogen absorption and its prevention of titanium in a simulated descaling environment in a desalination plant. Desalination **97**, 45–51 (1994)

36. J. Been, J.S. Grauman, Titanium and titanium alloys, in *Uhlig Corrosion Handbook*, (Wiley, 2011), p. 861

37. Corrosion, *Metals Handbook*, Vol 13, (ASM Handbook, 1987)

38. P.A. Schweitzer, *Fundamentals of Corrosion: Mechanisms, Causes, and Preventative Methods*. (CRC Press, Taylor and Francis, 2010), ISBN 9781420067705, p. 318

39. R. Hausler, Corrosion inhibition and inhibitors. in *Corrosion Chemistry,* ed. by G.R. Brubaker, P.B.P. Phipps, *ACS Symposium Series*, 89, 262, (American Chemical Society, 1979)

40. A. Khamis, M.M. Saleh, M.I. Awad, B.E. El-Anadouli, Enhancing the inhibition action of cationic surfactant with sodium halides for mild steel in 0.5 M H_2SO_4. Corros. Sci. **74**, 83–91 (2013)

41. D. Asefi, M. Arami, N.M. Mahmoodi, Electrochemical effect of cationic gemini surfactant and halide salts on corrosion inhibition of low carbon steel in acid medium. Corros. Sci. **52**, 794–800 (2010)

42. F.K. Crundwell, The anodic dissolution of 90% copper-10% nickel alloy in hydrochloric acid solutions. Electrochim. Acta **36**, 2135–2141 (1991)

43. H. Kaiser, G.A. Eckstein, Corrosion of alloys, in *Corrosion and Oxides Films*, ed. by M. Stratmann, G. S. Frankel, (Wiley-VCH, 2003)

44. M.G. Alvarez, J.R. Galvele, Pitting corrosion, in *Shreir's Corrosion*, (Elsevier, 2010), pp. 772–800

45. Hart E (2016) Corrosion Inhibitors: Principles, Mechanisms and Applications. Nova Science publisher, ISBN: 978-1-63485-791-8

46. V.S. Sastri, *Green Corrosion Inhibitors – Theory and Practice* (Wiley, 2011). 9780470452103

47. R.W. Revie, H.H. Uhlig, *Corrosion and corrosion control*, 4th edn. (Wiley, 2007)

48. M.A. Quraishi, R. Sardar, Di-thiazolidines- A new class of heterocyclic inhibitors for prevention of mild steel corrosion in hydrochloric acid solution. Corrosion **58**, 103–107 (2002)

49. A.L.D. Baddini, S.P. Cardoso, E. Hollauer, J.A.D.P. Gome, Statistical analysis of a corrosion inhibitor family on three steel surfaces (duplex, super-13 and carbon) in hydrochloric acid solutions. Electrochim. Acta **53**, 434–446 (2007)

50. H. Ashassi-Sorkhabi, M.R. Majidi, K. Seyyedi, Investigation of inhibition effect of some amino acids against steel corrosion in HCl solution. Appl. Surf. Sci. **225**, 176–185 (2004)

51. M.A. Amin, K.F. Khaled, Q. Mohsen, H.A. Arida, A study of the inhibition of iron corrosion in HCl solutions by some amino acids. Corros. Sci. **52**, 1684–1695 (2010)

52. M. Lashgari, M.R. Arshadi, S. Miandari, The enhancing power of iodide on corrosion prevention of mild steel in the presence of a synthetic soluble Schiff-base: Electrochemical and surface analyses. Electrochim. Acta **55**, 6058–6063 (2010)

53. M.A. Hegazy, A.M. Hasan, M.M. Emara, M.F. Bakr, A.H. Youssef, Evaluating four synthesized Schiff base as corrosion inhibitors on the carbon steel in 1M hydrochloric acid. Corros. Sci. **65**, 67–76 (2012)
54. V. Rajeswari, D. Kesavan, M. Gopiraman, P. Viswanathamurthi, Inhibition of cast iron corrosion in acid, base and neutral media using Schiff Base derivatives. J. Surfact. Deterg. **16**, 571–580 (2013)
55. E.M. Sherif, S.M. Park, Effect of 2-Amino-5-ethylthio-1,3,4-thiadiazole on copper corrosion as a corrosion inhibitor in aerated acidic pickling solutions. Electrochim. Acta **51**, 6556–6562 (2006)
56. E.M. Sherif, S.M. Park, Inhibition of copper corrosion in acidic pickling solutions by N-phenyl-1,4-phenylenediamine. Electrochim. Acta **51**, 4665–4673 (2006)
57. J.B. Matos, L.P. Pereira, S.M.L. Agostinho, O.E. Barcia, G.O. Cordeiro, E. D'Elia, Effect of cysteine on the anodic dissolution of copper in sulfuric acid medium. J. Electroanal. Chem. **570**, 91–94 (2004)
58. N. Caliskan, E. Akbas, The inhibition effect of some pyrimidine derivatives on austenitic stainless steel in acidic media. Mater. Chem. Phys. **126**, 983–988 (2011)
59. O. Sanni, A.P.I. Popoola, O.S.I. Fayomi, Enhanced corrosion resistance of stainless steel type 316 in sulphuric acid solution using eco-friendly waste product. Results Phys. **9**, 225–230 (2018)
60. A.U. Malik, I. Andijani, A. Al-Mubayed, Comparative study of sulfamic acid, sulfuric acid and hydrochloric acid as descalants for brine heater and heat recovery tubes. SWCC-RDC Technical report No. 36, 1995
61. A.R. Abu Dayyeh, R. Hamdan, W.F. Zaki, A.S. Abutalib, Cleaning of the cooling tubes in the MSF evaporators and its effect in maintaining the plant performance. IDA World Congress, Paper No. 135, 8–13 March 2002, Manama
62. A. Meroufel, A. Al-Enazi, Green corrosion development for MSF acid cleaning. Unpublished results
63. NACE, TM0193, Laboratory corrosion testing of metals in static chemical cleaning solutions at temperatures below 93 °C. NACE, Houston
64. M.A. Amin, K.F. Khaled, S.A. Fadl-Allah, Testing validity of Tafel extrapolation and monitoring corrosion of cold rolled steel in HCl solutions- experimental and theoretical studies. Corros. Sci. **52**, 140–151 (2010)

Chapter 11
Advanced Corrosion Prevention Approaches: Smart Coating and Photoelectrochemical Cathodic Protection

Viswanathan S. Saji

11.1 Introduction

Corrosion is a natural process causing materials damage and components failures. The worldwide corrosion cost is projected to be approximately US$ 2.5 trillion, which is equivalent to 3.4% of the global gross domestic product (GDP) [1–3]. Appropriate modifications in the material (e.g. corrosion-resistant alloys), environment (e.g. corrosion inhibitors) or material's surface (e.g. coatings) is requisite for effective corrosion protection. Corrosion can also be prevented by cathodic (by sacrificial anode or impressed current methods) and anodic methods [4, 5]. In several industrial applications, a combination of surface coatings and cathodic protection (CP) can deliver the most effective corrosion mitigation. Coatings alone may not be fully adequate to provide corrosion protection for the underlying metal in aggressive environments such as seawater as they are accompanied by microscopic pores, and the coating defects increase with the increase in service time. Conversely, CP alone is generally not cost-effective.

Seawater desalination has emerged as a viable alternative water supply in the modern world, predominantly in arid countries. Research and development over the last five decade in this area bring about many advancements leading to more efficient and cost-effective desalination systems [6, 7]. However, corrosion remains a major threat in seawater desalination systems. The average salinity of seawater is ~ 3.5%, and the oxygen concentration reaches a maximum at this salt level. Combined with the high oxygen and the salt content and the presence of various other aggressive species, microbes, fouling agents and pollutants make seawater extremely corrosive to structural materials.

V. S. Saji (✉)
King Fahd University of Petroleum and Minerals, Center of Research Excellence in Corrosion, Dhahran 31261, Saudi Arabia
e-mail: saji.viswanathan@kfupm.edu.sa

© Springer Nature Switzerland AG 2020
V. S. Saji et al. (eds.), *Corrosion and Fouling Control in Desalination Industry*,
https://doi.org/10.1007/978-3-030-34284-5_11

225

The primary objective of this chapter is to provide a concise description of two selected cutting-edge corrosion control approaches, one each from the surface coating and the CP, which can be employed in desalination industries for more effective corrosion protection. The first section of the chapter deals with smart coatings, an attractive coating approach that can deliver controlled active protection with self-healing abilities. The smart coating approaches are explained with reference to supramolecular interactions. The second section explains impending solar energy assisted CP method; namely photoelectrochemical cathodic protection (PECP).

11.2   Smart Coatings

A corrosion protective surface coatings in general works by passive protection or active protection. The passive protection depends entirely on the coating's barrier properties, whereas the active protection depends on the incorporated corrosion inhibitors in the coating matrix [8–10]. The smart coating concept makes use of both the active and the passive protection intelligently via the controlled on-demand release of inhibitors.

In general, a smart coating approach relies on a specialized resin, pigment, or other material that can respond to an environmental stimulus and react to it. Instead of direct doping of corrosion inhibitors in the coating matrix, a smart coating utilizes approaches such as encapsulation of inhibitors in nano/microcontainers that are dispersed homogeneously in the host coating [11, 12]. The immobilization/encapsulation method can overcome the problems with the direct doping approach, such as early inhibitor leakage and undesirable inhibitor-matrix interactions [13]. When the coating is damaged, the incorporated containers releases the pre-loaded inhibitors to repair the defects. A perfect nano/microcontainer is projected to be featured with the following attributes [14]:

 (i)   Chemical and mechanical stability
 (ii) Compatibility with the coating matrix
 (iii) Adequate loading capability
 (iv) An impervious covering wall to avoid leakage of the active matter
 (v)   Skill to sense corrosion initiation
 (vi) Release of the active matter on response.

The smart coating relies on supramolecular chemical approaches [15]. Supramolecular chemistry, otherwise called 'chemistry of molecular assemblies' or 'the chemistry of the intermolecular (non-covalent) bond', are branded by the three-dimensionally arranged supramolecular species with intermolecular interactions (e.g., ion-ion, ion-dipole, dipole-dipole, hydrogen bonding, π–π, etc.). Supramolecular compounds can be conveniently categorized into (i) molecular self-assemblies, (ii) host-guest complexes, and (iii) molecules built to specific shapes (rotaxanes, dendrimers etc.). In the host-guest chemistry-based systems, a host (a large molecule, aggregate or a cyclic compound having a cavity) binds a guest molecule making a host-guest supramolecule. More details on supramolecular compounds can be found elsewhere [15–20].

Structural alterations of supramolecular assemblies permit stimuli responsiveness. The nanocontainers disseminated in coatings can discharge the captured inhibitors in a regulated manner, which can be accustomed to occur along with the corrosion instigation or local environment changes (temperature, pH or pressure changes, radiation, mechanical action etc.) by using reversible covalent bonds or feebler non-covalent interactions. Smart coatings can be divided into several types based on the functionalization, and that can have several functions such as self-cleaning, self-healing, anti-biofouling and anti-corrosion [12, 21–29]. Several research efforts are presently ongoing in fabricating porous, hollow, or layer structured nanocontainers. Here, we describe recent research advancements in nano/microcontainers where supramolecular approaches are utilized in crafting more proficient smart coatings.

11.2.1 Polymer-Based Nano/Microcapsules

Polymeric capsules have been explored extensively as a carrier of core components in smart coating applications. Here, the active constituent (core) is embedded within another material (shell) to protect the core from the direct surroundings. Either monomers/pre-polymers (by methods such as emulsion polymerization, interfacial polycondensation etc.) or polymers (by suspension cross-linking, solvent extraction etc.) can be employed for the production of nano/microcontainers. Polyurea (PUA)-formaldehyde and polyurethane (PU)-based containers were among the first strategies developed in this area [30].

Huang et al. prepared PU microcapsules encircling hexamethylene diisocyanate via interfacial polymerization of methylene diphenyl diisocyanate prepolymer and 1,4-butanediol in oil-in-water emulsion [31]. Latnikova et al. showed that the type of polymer used had a noteworthy influence on the structure of the capsules formed. The PU and PUA loaded with the inhibitors were found to be evenly dispersed in the polymer matrix when compared with polyamide containers [32]. Gite et al. on their studies on quinoline incorporated PUA microcapsules showed that well defined, thermally stable spherical microcapsules with mean diameters of 72 and 86 μm were displayed excellent thermal and storage stabilities up to 200 °C [33]. All these studies presented considerably enhanced corrosion resistance for the self-healing coating after incorporating an optimized amount of microcapsules. For example, a low corrosion rate of 0.65 mm year^{-1} was obtained in 5 wt.% HCl for mild steel coated with PU with the incorporation of 4 wt.% of microcapsules loaded quinoline corrosion inhibitor [33]. Li et al. described pH-responsive polystyrene (PS) nano-container where benzotriazole (BTA) inhibitors were encapsulated in the PS matrix during polymerization, and that was followed by adsorption of a highly branched polyethylenimine as a controlled release system of BTA [34]. Maia et al. showed that incorporation of PUA microcontainers loaded with 2-mercaptobenzothiazole

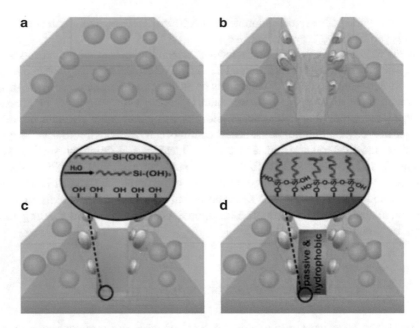

Fig. 11.1 Schematic of a self-healing smart coating. (**a**) Coating encompassing microcapsules loaded with alkoxysilanes on a metal substrate. (**b**) The release of encapsulated alkoxysilanes on coating damage. (**c**) The reaction of alkoxysilanes with H_2O from the ambient environment. (**d**) The damaged surface becomes passive and hydrophobic. Reproduced with permission from Ref. 36; Copyright 2011 © Royal Society of Chemistry

(MBT) in sol-gel coating (on 2024 Al alloy) was beneficial to improve the coating's barrier properties as well as coating adhesion [35]. Latnikova et al. described an intersting concept by the synergistic combination of the passivation and hydrophobicity (Fig. 11.1) where alkoxysilanes with long hydrophobic tails were captured in PU microcapsules. On coating damage, the released alkoxysilanes form covalent interaction with the hydroxyl groups on the metal surface developing an inactive electrochemical film. The highly hydrophobic properties ascribed to the long hydrocarbon tails enhanced the barrier properties by precluding water/aggressive ions infiltration [36]. Further details on polymeric nanocontainers can be found elsewhere [30]. Several studies investigated polymer nano/microgels-based novel carriers [37, 38].

11.2.2   Host-Guest Chemistry-Based

Stimulus feed-back anti-corrosion coatings, made-up by the dissemination of 'guest' materials (inhibitor loaded intelligent nanocontainers) into 'host' matrix have attracted considerable research attention [12, 26, 27, 39, 40]. The expected properties of the guest materials are:

   (i) High loading capability
   (ii) Good compatibility
  (iii) Stimulus-responsiveness
  (iv) Controlled release property
   (v) Zero premature release
  (vi) Prompt high sensitivity

Here, we provides a few examples of host-guest chemistry-based systems of mesoporous silica and cyclodextrins (CDs).

Mesoporous silica nanocontainers (MSNs) have attracted considerable research attention due to their large specific surface area, easy surface functionalization, biocompatibility, high stability, and controllable pore diameter [41–43]. Borisov et al. showed that too low concentrations (0.04 wt.%) of embedded nanocontainers led to good coating barrier properties, however, reduced active corrosion inhibition due to insufficient inhibitor availability. In contrast, higher concentrations of the nanocontainers (0.8–1.7 wt.%) degraded the coating integrity, leading to diffusion paths for aggressive species and reduced barrier effect. A coating with 0.7 wt.% inhibitor loaded MSNs exhibited the best corrosion protection [44, 45]. Zheludkevich et al. described a layer by layer (LBL) assembly where pH-sensitive silica nanoparticles were covered with polyelectrolyte layers, and layers of BTA and that was successively assimilated into a hybrid sol-gel coating ($\sim$ 95 mg of inhibitor /1 g of SiO_2). The enhancement of corrosion resistance was credited to the better adhesion, improved barrier effect, and placid inhibitor release. When corrosion commenced, the local pH changes and that opens the polyelectrolyte shell of the nanocontainers releasing BTA. After corrosion suppressed, the local pH gets recovered, closing the polyelectrolyte shell and terminating further inhibitor release [21]. Chen and Fu fabricated hollow mesoporous silica nanocontainers (diameter 0.5–08 μm, shell thickness $\sim$ 100 nm) using poly(vinylpyrrolidone) and cetyltrimethylammonium bromide templates, and subsequently, pH-sensitive supramolecular nanovalves comprising of ucurbit [6] uril (CB [6]) rings and bisammonium stalks were anchored to the nanocontainers's outer surface. At neutral pH, the CB [6] incorporated the bisammonium stalks firmly, closing the nanopore orifices effectively, and once the pH augmented, the stalks became deprotonated and CB [6] de-threaded from the stalks, opening the nanovalves with inhibitor release [46, 47]. Ding et al. explored an innovative organic silane-based corrosion potential-stimulus feedback active coating (CP-SFAC) for Mg alloys, based on redox-responsive self-healing concept [48, 49] where the surface potential reduction after localized corrosion initiation was exploited as a trigger. The design utilized supramolecular self-assembly technique where corrosion potential-stimulus responsive nanocontainers (CP-SNCs) that were composed of $Fe_3O_4@mSiO_2$ as magnetic nanovehicles and bipyridinium $\subset$ water-soluble pillar [5] arenes (BIPY $\subset$ WP [5]) assemblies as gatekeepers/disulfide linkers. The $Fe_3O_4@mSiO_2$ produced by the reverse microemulsion method displayed a core-shell structure with $\sim$ 50 nm thick amorphous SiO_2 shell. Figure 11.2 schematically shows the synthesis procedure of 8-hydroxyquinoline (8-HQ) encapsulated CP-SNCs from $Fe_3O_4@mSiO_2$. The fabricated CP-SNCs were successively

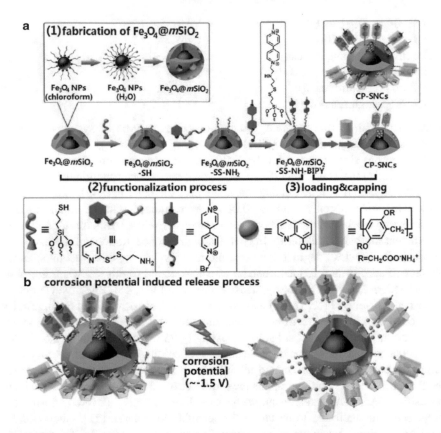

Fig. 11.2 Schematic diagrams of (**A**) Synthetic method of CP-SNCs [(1) Fabrication of Fe$_3$O$_4$@ mSiO$_2$, (2) Functionalization process, and (3) Loading & Capping]; and (**B**) Corrosion potential-induced release. Reproduced with permission from Ref. 49; Copyright 2017 © American Chemical Society

assimilated into the hybrid sol-gel coating to construct CP-SFAC and coated on the AZ31B alloy [49]. The supramolecular assemblies with high binding affinity efficiently blocked the captured corrosion inhibitor (8-HQ), within the mesopores of Fe$_3$O$_4$@mSiO$_2$. At corrosion potentials (-1.5 V *vs* SHE, Mg alloy), the inhibitor gets released instantly because of the cleavage of disulfide linkers and the removal of the supramolecular assemblies (Fig. 11.2) [49].

An efficient inhibition strategy for early leakage through organosilyl modification at the outlet of mesopores were reported by Zheng et al. The nanocontainers assimilated coating showed good barrier effect, long-term self-healing and stimuli responsiveness to local pH change [50]. Zhao et al. fabricated hollow silica nanocapsules with Mg(OH)$_2$ precipitation in shells (HSNs) through an inverse emulsion polymerization where BTA has been immobilized in HSNs. Ultraviolet-visible (UV–vis) spectroscopic release studies at different pH values displayed that the amount of released BTA from the salt-precipitated HSNs in acidic solution was considerably greater than that of HSNs without salt-precipitation [51]. Liang et al. reported an acid and alkali dual-stimuli responsive SiO_2-imidazoline nanocomposites (SiO_2–IMI) with high inhibitor loading capacity. The SiO_2-IMI were homogeneously disseminated into the hydrophobic SiO_2 sol, and a superhydrophobic surface was realized on Al alloy by dip-coating [52]. Wang et al. reported redox-triggered-smart-nanocontainers, assembled by mounting β CD functionalized with ferrocene onto the exterior surface of MSNs [53].

CDs are one of the foremost supramolecular hosts that can deliver diverse self-assembled structures and functionalities [17, 54, 55]. They are attractive in terms of availability, biocompatibility and cost factor. CDs are cyclic oligomers of glucose consisting of 6–8 D-glucopyranoside units connected edge to edge, with faces pointing inwards, headed for a central hydrophobic cavity wherein the interior cavity can accommodate guest molecules (Fig. 11.3). Amiri and Rahimi used α, β and γ forms of CD-based inclusion complex containing organic inhibitors to create on-demand release coatings. Salt spray test revealed that larger nanocontainer's size, higher nanocontainer content, and thicker coating afforded better corrosion protection. Corrosion initiation of the coated Al substrate happened after 6 days of salt spray on the bare sample, but the samples with nanocontainer-based coatings did not rust even after 1000 h [56, 57].

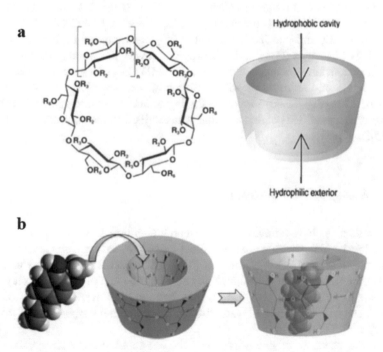

Fig. 11.3 (a) Molecular structure of CDs. (b) Schematic showing the formation of inclusion complex with guest molecule. Reproduced with permission from Ref. 56 & 57; Copyright 2015 & 2014 © Springer

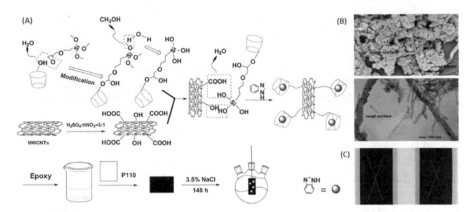

Fig. 11.4 (**A**) Synthesis of nanocontainers and fabrication sequence of the coating. (**B**) SEM and TEM images of MWCNTs/β CD. (**C**) Scarification test results (10 wt.% NaCl solution, 12 h) for steel specimens coated with (left) epoxy resin mixed with 3 wt.% β CD/MWCNTs loaded with inhibitor and (right) without corrosion inhibitor. Reproduced with permission from Ref. 58; Copyright 2016 © Elsevier

He et al. fabricated a nanoreservoir based on multi-walled carbon nanotubes (MWCNTs) and β CDs through a modest chemical way (Fig. 11.4). As evident from the SEM image, there was a noteworthy enlargement of MWCNTs after being amended by β CD, and that indicated that β CD was mounted on the surface of MWCNTs. TEM images (Fig. 11.4b) showed that it was an assembly of individual MWCNTs with heterogeneous layers. The β CD/MWCNTs were then loaded with a benzimidazole inhibitor and blended into epoxy. The scarification test in NaCl solution for 12 h (Fig. 11.4c) supported the better anti-corrosion and self-healing capability of the modified coating and that was ascribed to the controlled inhibitor release [58].

11.2.3 Inorganic Clay-Based

The inorganic clays have attracted much research attention as a low-cost nanocontainer. Halloysite clay nanotubes are the most widely investigated one. Halloysites appear in different morphologies, such as platy, spheroidal, and tubular, where the most prevalent morphology is tubular. The chemical formula of halloysite clay can be represented as $Al_2(OH)_4Si_2O_5 \cdot nH_2O$. The hydrated halloysite ($n = 2$), called 'halloysite-10 (Å)', has a layer of water molecules between the adjacent clay layers (Fig. 11.5) [64].

The hollow structure of halloysite nanotubules can adequately encapsulate the active agent. The strong surface charges can permit multilayer-nanoscale-assemblies through LBL adsorptions, and that can be enclosed with polyelectrolytes to form

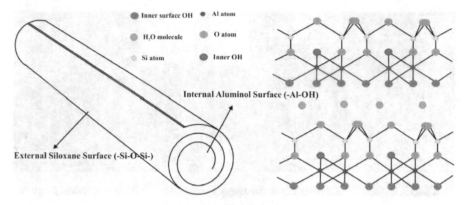

Fig. 11.5 Structure of halloysite clay particle. Reproduced with permission from Ref. 64; Copyright 2017 © Elsevier

nano-shells, overcrowding the tube ends and realizing controlled release of encapsulated inhibitors against an external stimulus [59]. Figure 11.6 shows scanning electron microscopy (SEM) images of halloysite nanotubes before and after loading a corrosion inhibitor (MBT). The initial halloysites comprise of well-defined nanotubes with sizes at the range of 1–15 μm. The maximum MBT quantity loaded was 5 wt.%. To achieve controlled release properties the nanotubes surfaces were modified by two polyelectrolyte bilayers [poly(styrene sulfonate) (PSS)/poly(allylamine hydrochloride) (PAH)] (Fig. 11.6) resulting in an inhibitor/halloysite/PAH/PSS/PAH/PSS layered structure. The opening of the shell can be realized by changing the system pH to acidic or alkali levels due to the onset of corrosion [22].

More details on inorganic clay nanocontainers can be found elsewhere [22, 60–64]. Several reports on anionic clays are available [65–67].

11.2.4   Polyelectrolyte-Based

Polyelectrolyte multilayers are widely investigated for smart coating applications. Polyelectrolytes (polymer + electrolyte) comprising of the fraction of monomeric units with ionized functional groups (cationic/anionic polyelectrolytes) can be assembled on the surface of nanoparticles via LBL approach. The incorporated inhibitors will be released in a controlled way when the confirmation of the polyelectrolyte molecules alters due to changes in pH or mechanical impact of the surrounding media [30, 68–71]. Polyelectrolytes can provide smart anti-corrosion protection due to different ways [70]:

- Buffering activity of the polyelectrolyte multilayers
- Self-curing due to mobility of the polyelectrolyte multilayers
- Polyelectrolyte multilayers as a carrier for corrosion inhibitors

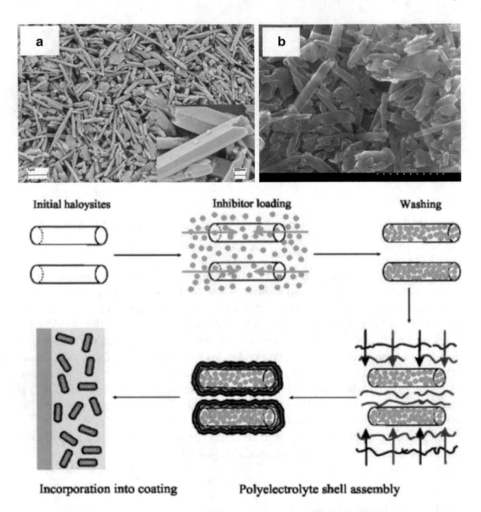

Initial haloysites   Inhibitor loading   Washing

Incorporation into coating   Polyelectrolyte shell assembly

Fig. 11.6 SEM images of (**A**) as-formed halloysite nanotubes and (**B**) halloysites nanotubes doped with inhibitor (MBT) and coated with PAH/PSS/PAH/PSS polyelectrolyte layers (scale bar in B is 1.5 μm). Schematic illustration of the fabrication of MBT-loaded halloysite/polyelectrolyte nanocontainers is provided below. Reproduced with permission from Ref. 22; Copyright 2008 © American Chemical Society

A schematic showing the mechanism of action of smart self-healing polyelectrolyte coating with incorporated inhibitors is presented in Fig. 11.7. More details can be found elsewhere [68–71]

The design and preparation of novel supramolecular nanocontainers with multi-functionalities is anticipated to bring excellent prospects for developing new generation corrosion prevention coatings. In spite of the substantial extent of research efforts in this direction, there are only a few commercially offered products [27]. Many reported works available in the recent literature have the potential for industrial-scale applicability.

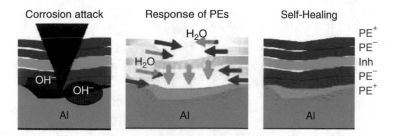

Fig. 11.7 Schematic mechanism of self-healing action of a smart polyelectrolyte anti-corrosion coating. Corrosion attack causes pH change of the system stimulating response of the polyelectrolyte coating: pH buffering, rearrangement of polymer chains and release of corrosion inhibitors (PE$^+$ – positively charged polyelectrolyte; PE$^-$ – negatively charged polyelectrolyte; Inh – corrosion inhibitor). Reproduced with permission from Ref. 70; Copyright 2010 © American Chemical Society

11.3   Photoelectrochemical Cathodic Protection

Cathodic protection (sacrificial anode and impressed current methods) is an extensively employed corrosion prevention method. CP impedes corrosion by altering the active (anodic) sites on the metallic structure to passive (cathodic) sites by providing electrical current. The technique has a wide range of industrial applications; to mention a few, water/fuel pipelines, storing tanks, boat/ship hulls, offshore platforms, onshore well casings, and reinforcing steels. However, the cost factor occasionally limits the application.

While the sacrificial anode cathodic protection (SACP) works by the sacrificial current from an anode (e.g. Al, Mg); the impressed current (ICCP) method makes use of direct current from a rectifier to cathodically protect the steel. Both methods have advantages and disadvantages of their own. ICCP is not economic in terms of continuous usage of current whereas SACP is not attractive in terms of expensive anode materials. SACP is not considered environmentally friendly and the electrodes get consumed during the operation [72].

A recent environmentally friendly approach in this arena is photoelectrochemical cathodic protection (PECP) that utilizes renewable energy (solar energy) to generate electricity using a semiconductor photoelectrode (photoanode). On illumination, electrons in a semiconductor can be excited to conduction band yielding holes in valence band. The electrons produced can be used for protecting a metallic structure by altering the metal potential cathodically (to more negative values) than the potential at which corrosion occurs (Fig. 11.8). The photoanode does not get used up during the operation and is environmentally friendly. The method is economical as it does not consume electrical energy or waste anode material. The technique can be worthy for areas where power supplies are limited (Fig. 11.9).

TiO$_2$ is the extensively researched semiconducting transition metal oxide for photoanode fabrication. The material has attracted widespread research attention

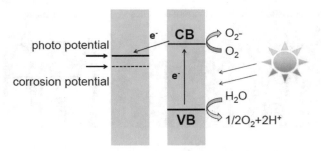

Metal   TiO₂ semiconductor

Fig. 11.8 The model of PECP of metal by semiconductor photoanode. Reproduced with permission from Ref. 73; Copyright 2017 © Creative Commons Licence

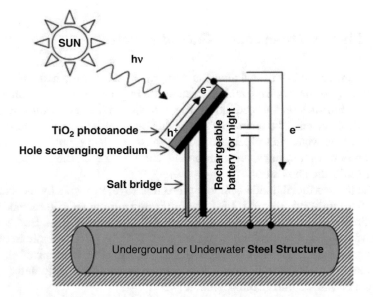

Fig. 11.9 Schematic diagram of the proposed photocathodic metal protection system using a TiO₂ photoanode and solar light. Reproduced with permission from Ref. 74; Copyright 2002 © American Chemical Society

because of its interesting photocatalytic properties, environmental friendliness, chemical stability and economic feasibility. The idea of PECP for metals by TiO_2 coatings under UV illumination was firstly reported in the year 1995 [75]. Since then, several studies have been reported on corrosion prevention of metals using TiO_2-based surface coatings [76, 77]. Park et al. further advanced the PECP technique motivated by the concept of employing a photoanode as an alternative to the sacrificial anode for prevention of steel corrosion. On UV illumination, a TiO_2 photoanode in contact with a hole scavenging medium can supply photocurrent to an electrically joined steel cathode, shifting the cathode potential to protective passive values [74, 78]. Since then, PECP by TiO_2 photoanodes has received significant research attention [79–103]. In addition to the use of commercial TiO_2, different cost-effective approaches were employed for the preparation of TiO_2 such as sol-gel [79], hydrothermal [80], and liquid phase deposition [81]. Anodization of metallic Ti was also widely employed [82]. Among these methods, the liquid phase deposition (LPD) method is perhaps the most economical and can produce crystalline metal oxide semiconductor films at lower temperatures [81, 83, 84].

It is the wide bandgap (3.0 and 3.2 eV respectively for rutile and anatase phases) that limits photocatalytic properties of TiO_2 to UV region (wavelength < 390 nm). Different approaches were investigated to augment the photoactivity of TiO_2 from UV to the visible part, for example, metal/non-metal doping, dye sensitization, and semiconductor coupling. Among them, non-metal doping (with N, C, S etc.) [85, 86] has attracted significant recent research attention. It has been shown that N and F-doping eased the formation of surface oxygen vacancies, with a higher number of Ti^{3+} ions [85]. Sol-gel solvothermal synthesized N-F-doped TiO_2 presented excellent photocatalytic properties than a commercial-grade P25 TiO_2 [87]. Similarly, hydrothermally synthesized N-F-doped anatase TiO_2 from $(NH_4)_2TiF_6$ and NH_3 aqueous solutions showed strong absorption at 400–550 nm wavelength range [88]. Lei et al. have employed N-F-doped TiO_2 surface coating for the PECP of steel, that showed excellent visible light response at 600–750 nm wavelength range [81]. Although doping can reduce the band-gap, it has disadvantages such as indirect doping level with low absorption efficiency, and introduction of new recombination centers [73]. As an alternative to doping, semiconductor coupling was extensively researched.

Coupling TiO_2 with other semiconductors with appropriate energy levels is an accepted way to ease the separation of photoelectron-hole pairs. Small bandgap semiconductors, such as CdS [89], CdSe [90], PbS [91], PbSe [92] etc. have been introduced into mesoporous TiO_2 films to improve the visible light response. Zhang et al. reported a highly efficient CdSe/CdS co-sensitized TiO_2 nanotube films for PECP of stainless steel (SS) [93]. Li et al. reported an electrodeposited CdS-modified ordered anodized TiO_2 nanotubes photoelectrodes. The composite electrode efficiently harvested solar light in the UV and visible region and that was credited to the improved charge separation and electron transport efficiencies. The electrode potentials of 304 SS coupled with the composite photoanode shifted negatively for about 246 mV and 215 mV under UV and white light irradiation, respectively [94]. Several reports are available on carbon composite TiO_2 and polymer-modified TiO_2 photoanodes. More details can be found elsewhere [73].

Among the several alternatives reported for TiO_2, the graphitic carbon nitride (g-C_3N_4) is a recently reported promising material. However, the fast secondary recombination of photogenerated charge carriers and the amphoteric property of g-C_3N_4 are negatives. A nanocomposite with structured carbons, for example, graphene can improve the electrical conductivity and the charge separation efficiency of g-C_3N_4. Jing et al. reported a secondary reduced graphene oxide modified g-C_3N_4 (R-rGO-C_3N_4) that was prepared by a thermal polycondensation method and a subsequent reduction process (Fig. 11.10 A). An electrophoretic deposition was utilized

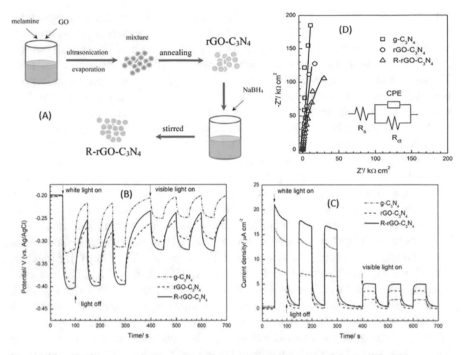

Fig. 11.10 (**A**) The schematic illustration of the preparation process of the R-rGO-C₃N₄ composite. The variations of the potentials (**B**) and current densities (**C**) of the 304 SS electrode coupled with the g-C₃N₄, rGO-C₃N₄ and R-rGO-C₃N₄ thin film photoanodes under intermittent light illumination. (**D**) Nyquist plots of the g-C₃N₄, rGO-C₃N₄ and R-rGO-C₃N₄ thin film photoanodes measured in the dark. The inset is the equivalent circuit used to fit the EIS results. Reproduced with permission from Ref. 95. Copyright 2017 © The Electrochemical Society

to fabricate the photoanode. The variations of the potentials and the current densities of the galvanic coupling under intermittent white light and visible light illumination are shown in Fig. 11.10 B & C. In dark, the potentials of all the three materials were close to the corrosion potential of 304 SS in 3.5 wt.% NaCl. The obvious shift of potential on light illumination corresponds to the cathodic polarization of steel. The photoinduced potential drops caused by rGO-C₃N₄ (190 mV) and R-rGO-C₃N₄ (200 mV) were larger than that of g-C₃N₄ (120 mV), demonstrating the better PECP performance of R-rGO-C₃N₄. The stability in potentials corresponds to stability of the prepared photoanodes. Positive current densities (Fig. 11.10 B) are detected for all the three galvanic couplings under light illumination, signifying successful electron transfer from the photoanode to the cathode. The photoinduced current densities of the g-C₃N₄, rGO-C₃N₄ and R-rGO-C₃N₄ photoanodes under white light illumination were approximately 7.4, 13.5 and 17.8 μA.cm⁻², respectively. Under visible light illumination, these values correspondingly decreased to 1.8, 3.5 and 5.1 μA.cm⁻². The charge transfer resistance (R_{ct}) was evaluated by performing electrochemical impedance spectroscopy (EIS) studies in the dark at the open circuit potential (OCP) in 0.1 mol L^{-1} Na$_2$SO$_4$ (Fig. 11.10 D). The R_{ct} (obtained by curve fitting) of R-rGO-C$_3$N$_4$ (1.32 × 10^6 Ω.cm^2) was smaller than rGO-C$_3$N$_4$ (5.86 × 10^6 Ω.cm^2) and g-C$_3$N$_4$ (1.30 × 10^7 Ω.cm^2). The results displayed that rGO modification enhanced the electrical conductivity of g-C$_3$N$_4$, and promoted the transfer and separation of photogenerated charge carriers [95].

Another main issue in terms of commercialization of the technique is the absence of photoelectrochemical response of the photoanode in dark. In this direction, many research attempts are going on in fabricating novel photoanodes with extended energy (photoelectron) storage capacities in addition to PECP. The idea is to store surplus energy in the photoelectrode during day time so that it can be used at night. Several semiconductor transition metal oxides such as MoO$_3$ [96] and WO$_3$ [76, 97] were investigated along with TiO$_2$ in this direction. Bu and Ao classified the works reported in this domain as below:

- WO$_3$-TiO$_2$ composite photoanodes
- SnO$_2$-TiO$_2$ composite photoanodes
- CeO$_2$-TiO$_2$ composite photoanodes
- Porous structured TiO$_2$ photoanodes

Oxides of Mo and W are famous for their higher pseudocapacitance. Saji and Lee has investigated in detail the potential and pH-dependent pseudocapacitance of Mo-surface oxides [98–100]. A nanosized TiO$_2$/WO$_3$ bilayer coating on the surface of 304 SS was reported by Zho et al. and found that this coating could maintain the CP effect for 6 h in the dark after switching off the light [101]. Park et al. measured the galvanic current between the TiO$_2$/WO$_3$ photoelectrode and the coupled steel electrode and verified that the photoinduced electrons generated by TiO$_2$ transferred to WO$_3$ and then overflowed to the coupled steel during the light illumination, and the stored electrons in WO$_3$ flowed to the coupled steel directly or through TiO$_2$ after the light was switched off [102]. Liang et al. synthesized WO$_3$/TiO$_2$ nanotube composite film coated photoanodes where TiO$_2$ nanotubes were synthesized on Ti by electrochemical anodization, followed by annealing at 450 °C and WO$_3$ nanoparticles were deposited on the film via potentiostatic electrodeposition [97]. Other than coupling with semiconductor, TiO$_2$ films with special porous structures were also found to have certain electron storage properties [73].

Li et al. fabricated CdSe/RGO/TiO$_2$ photoanodes by electrochemical deposition (Fig. 11.11A). Under visible-light illumination, the OCP of the photoanode coupled with 304 SS shifted cathodically indicative of efficient PECP. The photocurrent of the composite electrode (Curve d, Fig. 11.11 b) is ~ 8 times greater than that of pure TiO$_2$ (Curve b). The enhanced performance was credited to its fast electron transfer, large surface area and ordered mesoporous structure. The composite photoanode provided satisfactory CP even in the dark (Fig. 11.11B). The potentiodynamic polarization curves (Fig. 11.11 C) clearly showed the better performance of the composite electrode. On illumination, the corrosion potential of both CdSe/TiO$_2$ and CdSe/RGO/TiO$_2$ electrodes exhibited a significant cathodic shift compared

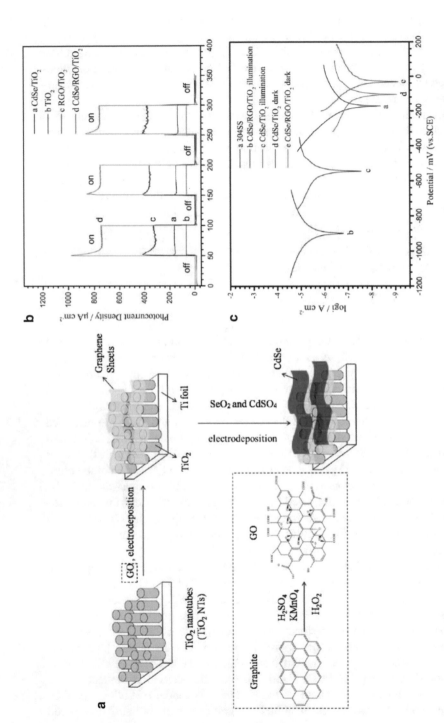

Fig. 11.11 (**A**) Schematic diagram of procedures for preparing the CdSe/RGO/TiO$_2$ composites. (**B**) Photocurrent-time curves of different photoanodes under intermittent illumination with visible light (λ > 400 nm). The electrolyte was 0.1 M Na$_2$SO$_4$ solution. The immersion time was 0.5 h. (**C**) Comparison of the polarization curves of bare 304 SS, and 304 SS coupled with CdSe/RGO/TiO$_2$ and CdSe/TiO$_2$ electrodes with or without visible-light illumination. The electrolyte in the corrosion cell was 3.5 wt.% NaCl solution, whereas in the photoanode cell was 0.1 M NaOH and 0.1 M Na$_2$S solution. Reproduced with permission from Ref. 103; Copyright 2015 @ Elsevier

with that of pure 304 SS in the dark. The corrosion potential shift was much more negative for CdSe/RGO/TiO$_2$ than CdSe/TiO$_2$ electrode. The corrosion current density exhibited a noteworthy increase when steel was coupled with CdSe/RGO/TiO$_2$ under illumination and that was attributed to the photogenerated electrons [103].

Table 11.1 shows a summary of PECP performance of TiO$_2$-based photoanodes [73]. Several works not presented in Table 11.1 are available in the current literature that includes Bi$_2$O$_3$/TiO$_2$ [104], g-C$_3$N$_4$/TiO$_2$ [105], SrTiO$_3$/TiO$_2$ [106], In$_2$O$_3$/TiO$_2$ [107], ZnO/In$_2$S$_3$ [108], NiO/TiO$_2$ [109], WO$_3$/TiO$_2$ [110], and RGO/WO$_3$/TiO$_2$ [111]. In spite of the fact the PECP technology has a high potential for industrial-scale application, more research efforts need to be put forward to enhance the protection efficiency and photoelectron storage capability.

Table 11.1 A summary of PECP performance of TiO$_2$-based photoanodes. Reproduced with permission from Ref. 73; Copyright 2017 © Creative Commons Licence

Photoanode	Method	Photo source	OCP (mV)	Metal
Pure TiO$_2$	Sol–gel	UV light	600	Cu
Pure TiO$_2$	Spray pyrolysis	UV light	250	304 SS
Pure TiO$_2$	LPD	White light	655	304 SS
Pure TiO$_2$	Anodization	White light	354	304 SS
Pure TiO$_2$	Hydro thermal	White light	262	316L SS
Pure TiO$_2$	Sol–gel	White light	525	304 SS
Pure TiO$_2$	Sol–gel	UV light	439	316L SS
Pure TiO$_2$	Sol–gel, hydrothermal	White light	560	403 SS
Ni–TiO$_2$	Sol–gel	Visible light	300	304 SS
Cr–TiO$_2$	Sol–gel	Simulated sunlight	230	316L SS
Fe–TiO$_2$	LPD	White light	405	304 SS
N–TiO$_2$	Anodization	Visible light	400	316L SS
N–TiO$_2$	Hydro thermal	UV light	470	316L SS
N–F–TiO$_2$	LPD	Visible light	515	304 SS
CdS/TiO$_2$	Anodization	UV light	246	304 SS
	Electrochemical deposition	White light	215	304 SS
ZnS/CdS@TiO$_2$	Anodization	White light	900	403 SS
	Electrochemical deposition			
Ag/SnO$_2$/TiO$_2$	Photo-reduction deposition, sol–gel	Visible light	550	304 SS
MWCNT/TiO$_2$	Sol–gel	UV light	400	304 SS
GR/TiO$_2$	Sol–gel	UV light	400	304 SS
Sodiumpolyacrylate/TiO$_2$	LPD	White light	710	304 SS

11.4 Conclusions and Outlook

The chapter provides a brief account of two promising corrosion prevention strategies that can be applied to desalination systems. In the first section, a short description of controlled on-demand release smart coatings with encapsulated inhibitors are provided. The section explains various nano/microcontainers that can be incorporated into the coating matrix for active protection. Among the four different types described, most of the recent works focused on mesoporous silica and cyclodextrin-based host-guest systems. Systematic practices need to be done in this area to optimize a particular nanocontainer in a coating, where the containers do not decline but augment the mechanical strength, adhesion and barrier properties of the coating. Further research is requisite in realizing sustained-release nanocontainers.

In the second section, the environmentally friendly solar energy-assisted cathodic protection, the photoelectrochemical cathodic protection is briefed. This concept is different from the photovoltaic powered cathodic protection (solar cells used as the external power source). Here, a semiconductor photoanode is utilized to generate electricity from solar light instead of using solar cell as a power source. These methods are worthy for areas where power supplies are limited. Non-metal doping of TiO_2 and semiconductor coupling of TiO_2 with charge storing transition metal oxides are highlighted in this section. In spite of having a few promising research, more works need to be performed in this area for its potential large scale application. Future research can be focused on combined systems of photoelectrochemical cathodic protection with traditional cathodic protection methods.

Acknowledgement The authors would like to acknowledge the support and fund provided by King Fahd University of Petroleum & Minerals through Project No. SR171021 under the Deanship of Research.

References

1. V.S. Saji, R. Cook, *Corrosion Control and Prevention Using Nanomaterials*, (Elsevier, 2012), ISBN 978-1-84569-949-9
2. G. Koch, J. Varney, N. Thompson, O. Moghissi, M. Gould, J. Payer, International measures of prevention, application, and economics of corrosion technologies study, *In* NACE International Impact, ed. by G. Jacobson, March 2016, Houston
3. L.T. Popoola, A.S. Grema, G.K. Latinwo, B. Gutti, A.S. Balogun, Corrosion problems during oil and gas production and its mitigation. Inter. J. Ind. Chem. **4**, 1–15 (2013)
4. M.G. Fontana, N.D. Greene, *Corrosion Engineering*, 3rd edn., (Mcgraw Hill, 1985), ISBN 978-0070214637
5. H.H. Uhlig, R.W. Review, (2009) *Corrosion and Corrosion Control: An Introduction to Corrosion Science and Engineering*, 4th edn., (Wiley, 2009), ISBN 978-0-471-73279-2
6. J. Kucera, Desalination – Water from Water, (Scrivener Publishing LLC, Massachusetts, Wiley, 2014), ISBN: 9781118208526
7. M. Elimelech, W.A. Phillip, The future of seawater desalination: Energy, technology, and the environment. Science **333**, 712–717 (2011)

8. A.E. Hughes, I.S. Cole, T.H. Muster, R.J. Varley, Designing green, self-healing coatings for metal protection. NPG Asia Mater. **2**, 143–151 (2010)
9. V.S. Saji, J. Thomas, Nano-materials for corrosion control. Corr. Sci. **92**, 51–55 (2007)
10. V.S. Raja, A. Venugopal, V.S. Saji, K. Sreekumar, S. Nair, M.C. Mittal, Electrochemical impedance behavior graphite dispersed electrically conducting acrylic coating on AZ 31 Mg alloy in 3.5 wt.% NaCl solution. Prog. Org. Coat. **67**, 12–19 (2010)
11. M. Jakab, J. Scully, On-demand release of corrosion-inhibiting ions from amorphous Al–co–Ce alloys. Nat. Mater. **4**, 667–670 (2005)
12. A.S.H. Makhlouf, *Handbook of Smart Coatings for Materials Protection*, (Elsevier, 2014), ISBN 9780857096883
13. S.V. Lamaka, M.L. Zheludkevich, K.A. Yasakau, R. Serra, S. Poznyak, M. Ferreira, Nanoporous titania interlayer as reservoir of corrosion inhibitors for coatings with self-healing ability. Prog. Org. Coat. **58**, 127–135 (2007)
14. Y. Feng, *Intelligent Nanocoatings for Corrosion Protection of Steels*, PhD Thesis, University of Calgary, (2017)
15. V.S. Saji, Supramolecular concepts and approaches in corrosion and biofouling prevention. Corr. Rev. **37**, 187–230 (2019)
16. J.M. Lehn, *Supramolecular Chemistry: Concepts and Perspectives*, (Wiley, 1995), ISBN 978-3527293117
17. J. Szejtli, T. Osa, *Comprehensive Supramolecular Chemistry*, Vol. 3, (Pergamon, 1999), ISBN 9780080912844
18. J.W. Steed, J.L. Atwood, *Supramolecular Chemistry*, (Wiley, 2000), ISBN 978-0-470-51234-0
19. J.M. Lehn, Towards complex matter: Supramolecular chemistry and self-organization. Proc. Natl. Acad. Sci. USA. **99**, 4763–4768 (2002)
20. J. Atwood, G.W. Gokel, L. Barbour, *Comprehensive Supramolecular Chemistry II*, 2nd edn., (Elsevier, 2017), ISBN 9780128031995
21. M.L. Zheludkevich, D.G. Shchukin, K.A. Yasakau, H. Möhwald, M.G.S. Ferreira, Anticorrosion coatings with self-healing effect based on nanocontainers impregnated with corrosion inhibitor. Chem. Mater. **19**, 402–411 (2007)
22. D.G. Shchukin, S.V. Lamaka, K.A. Yasakau, M.L. Zheludkevich, M.G.S. Ferreira, H. Möhwald, Active anticorrosion coatings with halloysite nanocontainers. J. Phys. Chem. C **112**, 958–964 (2008)
23. J. Baghdachi, Smart coatings, in *ACS Symposium Series*; (ACS, Washington, 2009), https://doi.org/10.1021/bk-2009-1002.ch001
24. M.L. Zheludkevich, J. Tedim, M.G.S. Ferreira, Smart coatings for active corrosion protection based on multi-functional micro and nanocontainers. Electrochim. Acta **82**, 314–323 (2012)
25. D.G. Shchukin, H. Möhwald, A coat of many functions. Science **341**, 1458–1459 (2013)
26. M.F. Montemor, Functional and smart coatings for corrosion protection: A review of recent advances. Surf. Coat. Technol. **258**, 17–37 (2014)
27. A. Stankiewicz, M.B. Barker, Development of self-healing coatings for corrosion protection on metallic structures. Smart Mater. Struct. **25**, 084013 (2016)
28. A.A. Nazeer, M. Madkour, Potential use of smart coatings for corrosion protection of metals and alloys: A review. J. Mol. Liq. **253**, 11–22 (2018)
29. F. Zhang, P. Ju, M. Pan, D. Zhang, Y. Huang, G. Li, X. Li, Self-healing mechanisms in smart protective coatings: A review. Corros. Sci. **144**, 74–88 (2018)
30. H. Wei, Y. Wang, J. Guo, N.Z. Shen, D. Jiang, X. Zhang, X. Yan, J. Zhu, Q. Wang, L. Shao, H. Lin, S. Wei, Z. Guo, Advanced micro/nanocapsules for self-healing smart anticorrosion coatings. J. Mater. Chem. A **3**, 469–480 (2015)
31. M. Huang, J. Yang, Facile microencapsulation of HDI for self-healing anticorrosion coatings. J. Mater. Chem. **21**, 11123–11130 (2011)
32. A. Latnikova, D.O. Grigoriev, H. Möhwald, D.G. Shchukin, Capsules made of cross-linked polymers and liquid core: Possible morphologies and their estimation on the basis of Hansen solubility parameters. J. Phys. Chem. **116**, 8181–8187 (2012)

33. V.V. Gite, P.D. Tatiya, R.J. Marathe, P.P. Mahulikar, D.G. Hundiwale, Microencapsulation of quinoline as a corrosion inhibitor in polyurea microcapsules for application in anticorrosive PU coatings. Prog. Org. Coat. **83**, 11–18 (2015)
34. G.L. Li, M. Schenderlein, Y. Men, H. Möhwald, D.G. Shchukin, Monodisperse polymeric core-shell nanocontainers for organic self-healing anticorrosion coatings. Adv. Mater. Interfaces **1**, 1300019 (2014)
35. F. Maia, K.A. Yasakau, J. Carneiro, S. Kallip, J. Tedim, T. Henriques, A. Cabral, J. Venâncio, M.L. Zheludkevich, M.G.S. Ferreira, Corrosion protection of AA2024 by sol-gel coatings modified with MBT-loaded polyurea microcapsules. Chem. Eng. J. **283**, 1108–1117 (2016)
36. A. Latnikova, D.O. Grigoriev, J. Hartmann, H. Möhwald, D.G. Shchukin, Polyfunctional active coatings with damage-triggered water-repelling effect. Soft Matter **7**, 369–372 (2011)
37. M. Cai, Y. Liang, F. Zhou, W. Liu, Functional ionic gels formed by supramolecular assembly of a novel low molecular weight anticorrosive/antioxidative gelator. J. Mater. Chem. **21**, 13399–13405 (2011)
38. V.C. Cécile, B. Fouzia, S. Pierre, J. Loïc, Surface-assisted self-assembly strategies leading to supramolecular hydrogels. Angew. Chem. Int. Ed. **57**, 1448–1456 (2018)
39. D.G. Shchukin, H. Möhwald, Self-repairing coatings containing active nanoreservoirs. Small **3**, 926–943 (2007)
40. A. Popoola, O.E. Olorunniwo, O.O. Ige, Corrosion resistance through the application of anti-corrosion coatings, in *Developments in Corrosion Protection*, (Intech, 2014). https://doi.org/10.5772/57420
41. B. Chang, J. Guo, C. Liu, J. Qian, W. Yang, Surface functionalization of magnetic mesoporous silica nanoparticles for controlled drug release. J. Mater. Chem. **20**, 9941–9947 (2010)
42. F. Maia, J. Tedim, A.D. Lisenkov, A.N. Salak, M.L. Zheludkevich, M.G.S. Ferreira, Silica nanocontainers for active corrosion protection. Nanoscale **4**, 1287–1298 (2012)
43. T.D. Nguyen, K.C.F. Leung, M. Liong, C.D. Pentecost, J.F. Stoddart, J.I. Zink, Construction of a pH-driven supramolecular nanovalve. Org. Lett. **8**, 3363–3366 (2006)
44. D. Borisova, H. Möhwald, D.G. Shchukin, Influence of embedded nanocontainers on the efficiency of active anticorrosive coatings for aluminum alloys part I: Influence of nanocontainer concentration. ACS Appl. Mater. Interfaces **4**, 2931–2939 (2012)
45. D. Borisova, H. Möhwald, D.G. Shchukin, Influence of embedded nanocontainers on the efficiency of active anticorrosive coatings for aluminum alloys part II: Influence of nanocontainer position. ACS Appl. Mater. Interfaces **5**, 80–87 (2013)
46. T. Chen, J.J. Fu, pH-responsive nanovalves based on hollow mesoporous silica spheres for controlled release of corrosion inhibitor. Nanotechnology **23**, 235605 (2012)
47. T. Chen, J.J. Fu, An intelligent anticorrosion coating based on pH-responsive supramolecular nanocontainers. Nanotechnology **23**, 505705 (2012)
48. C.D. Ding, Y. Liu, M.W. Wang, T. Wang, J.J. Fu, Self-healing, superhydrophobic coating based on mechanized silica nanoparticles for reliable protection of magnesium alloys. J. Mater. Chem. A **4**, 8041–8052 (2016)
49. C.D. Ding, J.H. Xu, L. Tong, G.C. Gong, W. Jiang, J. Fu, Design and fabrication of a novel stimulus-feedback anticorrosion coating featured by rapid self-healing functionality for the protection of magnesium alloy. ACS Appl. Mater. Interfaces **9**, 21034–21047 (2017)
50. Z. Zheng, M. Schenderlein, X. Huang, N.J. Brownbill, F. Blanc, D. Shchukin, Influence of functionalization of nanocontainers on self-healing anticorrosive coatings. ACS Appl. Mater. Interfaces **7**, 22756–22766 (2015)
51. D. Zhao, D. Liu, Z. Hu, A smart anticorrosion coating based on hollow silica nanocapsules with inorganic salt in shells. J. Coat. Technol. Res. **14**, 85–94 (2017)
52. Y. Liang, M.D. Wang, C. Wang, J. Feng, J.S. Li, L.J. Wang, J.J. Fu, Facile synthesis of smart nanocontainers as key components for construction of self-healing coating with superhydrophobic surfaces. Nanoscale Res. Lett. **11**, 231 (2016)
53. T. Wang, L.H. Tan, C.D. Ding, M.D. Wang, J.H. Xu, J.J. Fu, Redox-triggered controlled release systems-based bi-layered nanocomposite coating with synergistic self-healing property. J. Mater. Chem. A **5**, 1756–1768 (2017)

54. J. Szejtli, Introduction and general overview of cyclodextrin chemistry. Chem. Rev. **98**, 1743–1754 (1998)
55. W. Saenger, Cyclodextrin inclusion compounds in research and industry. Angew. Chem. Int. Ed. Engl. **19**, 344–362 (1980)
56. S. Amiri, A. Rahimi, Synthesis and characterization of supramolecular corrosion inhibitor nanocontainers for anticorrosion hybrid nanocomposite coatings. J. Polym. Res. **22**, 66 (2015)
57. S. Amiri, A. Rahimi, Preparation of supramolecular corrosion-inhibiting nanocontainers for self-protective hybrid nanocomposite coatings. J. Polym. Res. **21**, 566 (2014)
58. Y. He, C. Zhang, F. Wu, Z. Xu, Fabrication study of a new anticorrosion coating based on supramolecular nanocontainers. Synth. Met. **212**, 186–194 (2016)
59. E. Abdullayev, R. Price, D. Shchukin, Y. Lvov, Halloysite tubes as nanocontainers for anticorrosion coating with benzotriazole. ACS Appl. Mater. Interfaces **1**, 1437–1443 (2009)
60. D.G. Shchukin, H. Möhwald, Surface-engineered nanocontainers for entrapment of corrosion inhibitors. Adv. Funct. Mater. **17**, 1451–1458 (2007)
61. D. Fix, D.V. Andreeva, Y.M. Lvov, D.G. Shchukin, H. Möhwald, Application of inhibitor-loaded halloysite nanotubes in active anti-corrosive coatings. Adv. Funct. Mater. **19**, 1720–1727 (2009)
62. M.L. Zheludkevich, S.K. Poznyak, L.M. Rodrigues, D. Raps, T. Hack, L.F. Dick, T. Nunes, M.G.S. Ferreira, Active protection coatings with layered double hydroxide nanocontainers of corrosion inhibitor. Corros. Sci. **52**, 602–611 (2010)
63. J.M. Falcón, T. Sawczen, I.V. Aoki, Dodecylamine-loaded halloysite nanocontainers for active anticorrosion coatings. Front. Mater. **2**, 69 (2015)
64. K.A. Zahidah, S. Kakooei, M.C. Ismail, P.B. Raja, Halloysite nanotubes as nanocontainer for smart coating application: A review. Prog. Org. Coat. **111**, 175–185 (2017)
65. D. Álvarez, A. Collazo, M. Hernández, X.R. Nóvoa, C. Pérez, Characterization of hybrid sol-gel coatings doped with hydrotalcite-like compounds to improve corrosion resistance of AA2024-T3 alloys. Prog. Org. Coat. **67**, 152–160 (2010)
66. N.C. Rosero-Navarro, L. Paussa, F. Andreatta, Y. Castro, A. Durán, M. Aparicio, L. Fedrizzi, Optimization of hybrid sol-gel coatings by combination of layers with complementary properties for corrosion protection of AA2024. Prog. Org. Coat. **69**, 167–174 (2010)
67. J. Tedim, S.K. Poznyak, A. Kuznetsova, D. Raps, T. Hack, M.L. Zheludkevich, M.G.S. Ferreira, Enhancement of active corrosion protection via combination of inhibitor-loaded nanocontainers. ACS Appl. Mater. Interfaces **2**, 1528–1535 (2010)
68. I. Dewald, A. Fery, Polymeric micelles and vesicles in polyelectrolyte multilayers: Introducing hierarchy and compartmentalization. Adv. Mater. Interfaces **4**, 1600317 (2017)
69. D.O. Grigoriev, K. Köhler, E. Skorb, D.G. Shchukin, H. Möhwald, Polyelectrolyte complexes as a smart depot for self-healing anticorrosion coatings. Soft Matter **5**, 1426–1432 (2009)
70. D.V. Andreeva, E.V. Skorb, D.G. Shchukin, Layer-by-layer polyelectrolyte/inhibitor nanostructures for metal corrosion protection. ACS Appl. Mater. Interfaces **2**, 1954–1962 (2010)
71. M. Biesalski, J. Ruehe, R. Kuegler, W. Knoll, Polyelectrolytes at solid surfaces: Multilayers and brushes, *In Handbook of Polyelectrolytes and their Applications*, ed. by S.K. Tripathy, J. Kumar, H.S. Nalwa, Vol. 1, (2002), 39–63
72. W. von Baeckmann, W. Schwenk, W. Prinz, *Handbook of Cathodic Corrosion Protection*, 3rd edn., ISBN: 9780884150565, (Elsevier, 1997)
73. Y. Bu, J.P. Ao, A review on photoelectrochemical cathodic protection semiconductor thin films for metals. Green Energy Environ. **2**, 331–362 (2014)
74. H. Park, K.Y. Kim, W. Choi, Photoelectrochemical approach for metal corrosion prevention using a semiconductor photoanode. J. Phys. Chem. B **106**, 4775–4781 (2002)
75. J. Yuan, S. Tsujikawa, Characterization of sol-gel-derived TiO_2 coatings and their photoeffects on copper substrates. J. Electrochem. Soc. **142**, 3444–3450 (1995)
76. T. Tatsuma, S. Saitoh, Y. Ohko, A. Fujishima, TiO_2-WO_3 photoelectrochemical anticorrosion system with an energy storage ability. Chem. Mater. **13**, 2838–2842 (2001)

77. Y. Ohko, S. Saitoh, T. Tatsuma, Fujishima, A Photoelectrochemical anticorrosion and self-cleaning effects of a TiO$_2$ coating for type 304 stainless steel. J. Electrochem. Soc **148**, B24–B28 (2001)

78. H. Park, K.Y. Kim, W. Choi, A novel photoelectrochemical method of metal corrosion prevention using a TiO$_2$ solar panel. Chem. Commun., 281–282 (2001)

79. R. Subasri, T. Shinohara, Investigations on SnO$_2$-TiO$_2$ composite photoelectrodes for corrosion protection. Electrochem. Commun. **5**, 897–902 (2003)

80. Y.F. Zhu, R.G. Du, W. Chen, H.Q. Qi, C.J. Lin, Photocathodic protection properties of three-dimensional titanate nanowire network films prepared by a combined sol–gel and hydrothermal method. Electrochem. Commun. **12**, 1626–1629 (2010)

81. C.X. Lei, Z.D. Feng, H. Zho, Visible-light-driven photogenerated cathodic protection of stainless steel by liquid-phase-deposited TiO$_2$ films. Electrochim. Acta **68**, 134–140 (2012)

82. J. Li, C.J. Lin, C.G. Lin, A photoelectrochemical study of highly ordered TiO$_2$ nanotube arrays as the photoanodes for cathodic protection of 304 stainless steel. J. Electrochem. Soc. **158**, C55–C62 (2011)

83. C.X. Lei, H. Zhou, Z.D. Feng, Effect of liquid-phase-deposited parameters on the photogenerated cathodic protection properties of TiO$_2$ films. J. Alloys Compd **542**, 164–169 (2012)

84. T.P. Niesen, M.R. De Guire, Review: Deposition of ceramic thin films at low temperatures from aqueous solutions. J. Electroceram. **6**, 169–207 (2001)

85. D. Li, N. Ohashi, S. Hishita, T. Kolodiazhnyi, H. Haneda, Origin of visible-light-driven photocatalysis: A comparative study on N/F-doped and N–F-codoped TiO$_2$ powders by means of experimental characterizations and theoretical calculations. J. Solid State Chem. **178**, 3293–3302 (2005)

86. H. Yun, J. Li, H.B. Chen, C.J. Lin, A study on the N-, S- and Cl-modified nano-TiO$_2$ coatings for corrosion protection of stainless steel. Electrochim. Acta **52**, 6679–6685 (2007)

87. D.G. Huang, S.J. Liao, J.M. Liu, Z. Dang, L. Petrik, Preparation of visible-light responsive N-F-codoped TiO$_2$ photocatalyst by a sol–gel-solvothermal method. J. Photochem. Photobiol. A Chem. **184**, 282–288 (2006)

88. T. Yamada, Y. Gao, M. Nagai, Hydrothermal synthesis and evaluation of visible light active photocatalysts of (N, F)-codoped anatase TiO$_2$ from an F-containing titanium chemical. J. Ceram. Soc. Jpn. **116**, 614–618 (2008)

89. B. Jiang, X.L. Yang, X. Li, D.Q. Zhang, J. Zhu, G.S. Li, Core–shell structure CdS/TiO$_2$ for enhanced visible-light-driven photocatalytic organic pollutants degradation. J. Sol-Gel Sci. Technol. **66**, 504–511 (2013)

90. L.L. Su, J. Lv, H.E. Wang, L.J. Liu, G.Q. Xu, D.M. Wang, Z.X. Zheng, Y.C. Wu, Improved visible light photocatalytic activity of CdSe modified TiO$_2$ nanotube arrays with different intertube spaces. Catal. Lett. **144**, 553–560 (2014)

91. R. Plass, S. Pelet, J. Krueger, M. Gratzel, U. Bach, Quantum dot sensitization of organic–inorganic hybrid solar cells. J. Phys. Chem. B **106**(2002), 7578–7580 (2002)

92. R.D. Schaller, V.I. Klimov, High efficiency carrier multiplication in PbSe nanocrystals: Implications for solar energy conversion. Phys. Rev. Lett. **92**, 186601 (2004)

93. J. Zhang, R.G. Du, Z.Q. Lin, Y.F. Zhu, Y. Guo, H.Q. Qi, L. Xu, C.J. Lin, Highly efficient CdSe/CdS co-sensitized TiO$_2$ nanotube films for photocathodic protection of stainless steel. Electrochim. Acta **83**, 59–64 (2012)

94. J. Li, C.J. Lin, J.T. Li, Z.Q. Lin, A photoelectrochemical study of CdS modified TiO$_2$ nanotube arrays as photoanodes for cathodic protection of stainless steel. Thin Solid Films **519**, 5494–5502 (2011)

95. J. Jing, M. Sun, Z. Chen, J. Li, F. Xu, L. Xu, Enhanced photoelectrochemical cathodic protection performance of the secondary reduced graphene oxide modified graphitic carbon nitride. J. Electrochem. Soc. **164**, C822–C830 (2017)

96. Q. Liu, J. Hu, Y. Liang, Z.C. Guan, H. Zhang, H.P. Wang, R.G. Du, Preparation of MoO$_3$/TiO$_2$ composite films and their application in photoelectrochemical anticorrosion. J. Electrochem. Soc. **163**, C539–C544 (2016)

97. Y. Liang, Z.C. Guan, H.P. Wang, R.G. Du, Enhanced photoelectrochemical anticorrosion performance of WO_3/TiO_2 nanotube composite films formed by anodization and electrodeposition. Electrochem. Commun. **77**, 120–123 (2017)
98. V.S. Saji, C.W. Lee, Molybdenum, molybdenum oxides and their electrochemistry. ChemSusChem **5**, 1146–1161 (2012)
99. V.S. Saji, C.W. Lee, Reversible redox transition and pseudocapacitance of molybdenum/surface molybdenum oxides. J. Electrochem. Soc. **160**, H54–H61 (2013)
100. V.S. Saji, C.W. Lee, Potential and pH-dependent pseudocapacitance of Mo/Mo oxides – An impedance study. Electrochim. Acta **137**, 647–653 (2014)
101. M. Zhou, Z. Zeng, L. Zhong, Photogenerated cathode protection properties of nano-sized TiO_2/WO_3 coating. Corros. Sci. **51**, 1386–1391 (2009)
102. H. Park, A. Bak, T. Jeon, S. Kim, W. Choi, Photo-chargeable and dischargeable TiO_2 and WO3 heterojunction electrodes. Appl. Catal. B Environ. **115-116**, 74–80 (2012)
103. H. Li, X. Wang, L. Zhang, B. Hou, Preparation and photocathodic protection performance of CdSe/reduced graphene oxide/TiO_2 composite. Corros. Sci. **94**, 342–349 (2015)
104. Z.-C. Guan, H.-P. Wang, X. Wang, J. Hu, R.-G. Du, Fabrication of heterostructured β-Bi2O3-TiO2 nanotube array composite film for photoelectrochemical cathodic protection applications. Corros. Sci. **136**, 60–69 (2018)
105. D. Ding, Q. Hou, Y. Su, Q. Li, L. Liu, J. Jing, B. Lin, Y. Chen, g-C3N4/TiO2 hybrid film on the metal surface, a cheap and efficient sunlight active photoelectrochemical anticorrosion coating. J. Mater. Sci. Mater. Electron. **30**, 12710–12717 (2019)
106. Y.-F. Zhu, L. Xu, J. Hu, J. Zhang, R.-G. Du, C.-J. Lin, Fabrication of heterostructured SrTiO3/TiO2 nanotube array films and their use in photocathodic protection of stainless steel. Electrochim. Acta **121**, 361–368 (2014)
107. M. Sun, Z. Chen, Enhanced photoelectrochemical cathodic protection performance of the in O/TiO composite. J. Electrochem. Soc. **162**, C96–C104 (2014)
108. J. Jing, Z. Chen, Y. Bu, Visible light induced photoelectrochemical cathodic protection for 304 SS by In2S3-sensitized ZnO nanorod array. Int. J. Electrochem. Sci. **10**, 8783–8796 (2015)
109. C. Han, Q. Shao, J. Lei, Y. Zhu, S. Ge, Preparation of NiO/TiO2 p-n heterojunction composites and its photocathodic protection properties for 304 stainless steel under simulated solar light. J. Alloys Compd. **703**, 530–537 (2017)
110. S.Q. Yu, Y.H. Ling, R.G. Wang, J. Zhang, F. Qin, Z.J. Zhang, Constructing superhydrophobic WO3@TiO2 nanoflake surface beyond amorphous alloy against electrochemical corrosion on iron steel. Appl. Surf. Sci. **436**, 527–535 (2018)
111. W. Liu, T. Du, Q. Ru, S. Zuo, Y. Cai, C. Yao, Preparation of graphene/WO3/TiO2 composite and its photocathodic protection performance for 304 stainless steel. Mater. Res. Bull. **102**, 399–405 (2018)

Part III
Fouling in Desalination

Chapter 12
Inorganic Scaling in Desalination Systems

Khaled Touati, Haamid Sani Usman, Tiantian Chen, Nawrin Anwar,
Mahbuboor Rahman Choudhury, and Md. Saifur Rahaman

12.1 Introduction

Access to freshwater is becoming a major challenge in many places around the world. This trend is expected to continue with the increasing global population and industrial developments. To address the water scarcity problem, seawater desalination has been considered as one of the promising solutions that meet the rising demand for freshwater. According to the International Desalination Association (IDA) report, there are over 16,000 desalination plants in over 150 countries, with a total production capacity of 90 million m^3/day [1, 2]. The pursuit of freshwater has motivated the research in several desalination technologies such as Reverse Osmosis (RO), Forward Osmosis (FO), and several thermal-based processes (Membrane Distillation (MD), Multi-stage flashing (MSF), Multi-effects Distillation (MED)). RO is considered as the most efficient and reliable seawater desalination process, and despite the promising potential in producing freshwater, is still affected by fouling issues.

The prominent types of fouling in desalination can be classified into organic fouling, inorganic scaling, and biofouling (Fig. 12.1) [1]. The inorganic scaling, also known as scaling, refers to the accumulation of calcium, magnesium, carbonate, sulfate, and phosphorus on the surface of the membrane. It occurs by the precipitation and crystallization processes where ions in the supersaturated solution

K. Touati · H. S. Usman · T. Chen · N. Anwar · M. S. Rahaman (✉)
Department of Building, Civil and Environmental Engineering, Concordia University,
Montreal, QC, Canada
e-mail: saifur.rahaman@concordia.ca

M. R. Choudhury
Department of Building, Civil and Environmental Engineering, Concordia University,
Montreal, QC, Canada

Civil and Environmental Engineering Department, Manhattan College, Bronx, NY, USA

© Springer Nature Switzerland AG 2020
V. S. Saji et al. (eds.), *Corrosion and Fouling Control in Desalination Industry*,
https://doi.org/10.1007/978-3-030-34284-5_12

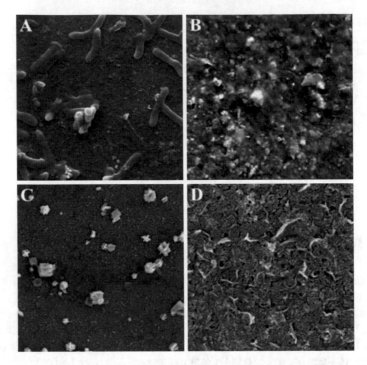

Fig. 12.1 Scanning electron microscopy (SEM) images of four fouling types on membrane surfaces. (**A**) Biofouling. (**B**) Organic fouling. (**C**) Inorganic scaling. (**D**) Colloidal fouling. Reproduced with permission from Ref. [8]; Copyright 2017@ Elsevier

crystallize on the membrane [3, 4, 5]. This chapter describes inorganic scaling in desalination plants with special emphasis on membrane fouling. Scaling mechanisms, scaling prediction, and membrane scaling control strategies are also described.

12.2   Scaling in Seawater Desalination Processes

Scale deposition is a problem encountered when water containing ions of sparingly soluble salts are used. Scale precipitation occurs whenever process conditions lead to the creation of supersaturation as it relates to one or more of the sparingly soluble salts. Another problem associated with membrane scaling is the adherent deposits of scale on the internal walls of industrial or domestic equipment that causes severe technical problems with a tremendous economic impact. The nature of the scale formed depends on the desalination process (pressure-driven or thermal desalination) due to the difference in the operating conditions (see below) [6–9].

The scaling phenomenon affects the performance of the membrane by decreasing the permeate flux rate as scalant will attach to the membrane surface, thus, blocking the membrane pores. The membrane pores will be continuously and increasingly clogged until a stop of service is needed to clean/replace the membrane. Scaling further result in the decline of permeate flux, which necessitates additional cost through increasing the operational pressures to maintain a constant permeate flux through a decreasing amount of available membrane pores. Other effects of scaling include deterioration of membrane (membrane decay), additional costs of labor, and additional cost in chemical cleaning (which also may cause membrane decay) [6, 7]. The combined consequences of scaling will lead to an increased price of water. Principally, research topics are focusing on developing mitigation and preventive measures that focus on reducing the effect of scaling on membranes by developing enhanced materials that improve membrane performance [7].

12.2.1   Pressure Driven Processes

Pressure driven processes for seawater desalination are well-defined as systems that use hydraulic pressure (i.e., RO) or osmotic pressure (i.e., FO) on either side of a semi-permeable membrane to desalinate seawater. RO is the most used desalination process worldwide due to several facts, most importantly, the energy efficiency of the process, the high productivity, and the constant investments into improving the process. FO is still not commercialized and published results studying the effectiveness are only coming from laboratory-scale tests. As mentioned previously, scale formation drastically reduces the performance of membrane-based desalination processes by [10–12]:

(1) Blocking the active membrane layer where the water-salts separation occurs, reducing the separation performance of the membrane, decreases the productivity, and damages the active membrane layer.
(2) Depositing on the pipes which may cause damage to the system.
(3) Occurring during the pre-treatment steps, limiting its efficiency.
(4) Depositing on the rejected brine side due to the high salt concentration.

The types of scaling in RO membranes include alkaline (e.g., calcium carbonate), non-alkaline (e.g., calcium sulfate) or silica-based. Calcium carbonate ($CaCO_3$) is the most common scale-forming mineral, which originates in the form of calcium and bicarbonate ions in industrial water, seawater, or groundwater sources. When the water temperature increases, the solubility of $CaCO_3$ decreases resulting in precipitation onto heated surfaces. Heated surfaces are not always required for calcium salts to form scale as it can, occur whenever the solubility concentration is exceeded [13, 14]. $CaCO_3$ is usually the leading primary precipitate in seawater RO and crystallizes in three different crystal forms: calcite, aragonite, and vaterite. Calcite is classified as a hard scale, whereas aragonite and vaterite are classified as softer types of scale that are more easily removed. $CaCO_3$ deposition occurs depending on

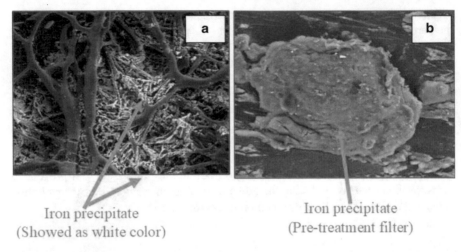

Iron precipitate
(Showed as white color)

Iron precipitate
(Pre-treatment filter)

Fig. 12.2 SEM images of (**A**): Iron precipitates on RO active layer (150X), (**B**): Iron precipitates in pre-filters (600 X). Reproduced with permission from Ref. [19]; Copyright 2019 @ Springer

operating parameters such as salt concertation and temperature, becoming a problem for the RO and FO process alike. Lab-scale research has verified the deposition of $CaCO_3$ and gypsum in FO desalination systems [15, 16].

Scaling by calcium sulfate is one of the common problems in seawater desalination. While the precipitation of calcium carbonate scale can often be minimized by adjusting the feed water to acidic conditions, calcium sulfate is insensitive to pH [17]. Iron precipitation may also occur in desalination processes [18]. This type of scale has a severe impact on the membrane (Fig. 12.2A) and the efficiency of the pre-treatment, as it can precipitate in pre-treatment as well (Fig. 12.2B). Ongoing research is being completed to find new techniques for iron removal such as limestone filter treatment, ash [19], adsorption techniques [20], and oxidation using potassium and permanganate. Aeration, oxidation, and separation are the most popular anti-foulant processes. Another major foulant for RO membranes is silica. Several investigative studies of fouled membranes have found silica, together with biofilms, to be the most egregious foulant [21]. Moreover, scale inhibitors or antiscalant that are successful in controlling other mineral types are usually ineffective against silica because of the latter's tendency to form an amorphous solid instead of a crystalline one [22]. Compared to other scaling types, silica scaling is considered a complex process and not well understood. In RO feed waters, silica usually exists in the following different forms [22]:

(1) The monomeric form, also known as silicic acid [$Si(OH)_4$];
(2) The polymeric form, also known as colloidal silica; or,
(3) Granular/ particulate silica.

Moreover, when operating with high recovery (especially with brackish water desalination), condensation reactions result in the silicic acid monomers forming siloxane bonds to produce various silica species such as dimers, trimers, polymers, and particles. Due to more favorable thermodynamics, these species usually deposit and polymerize on the membrane surface, a phenomenon known as heterogeneous crystallization.

12.2.2 Thermal Desalination Processes

Contrary to pressure-driven processes, thermal desalination is based on water evaporation and condensation. Due to its high recovery, these processes may suffer from severe mineral precipitation. Like RO, scaling can occur in different parts of thermal processes, namely during pre-treatment, distillation, and post-treatment. In distillation processes, the scales formed can be divided into two categories: (a) alkaline scales, which are mainly calcium carbonate and magnesium hydroxide and are considered as soft scales; and (b) sulfate scales, which are mainly calcium sulfate. Calcium sulfate scales are considered hard scaling and generally form at temperatures above 121 °C when higher concentrations of calcium and sulfate ions are prevalent. As an example, in MSF processes, scaling occurs at the orifices of the flash chambers and inside the tubes of the heat exchangers and the brine heaters [24]. The scale species are mainly $CaCO_3$ and $Mg(OH)_2$ that are formed due to the decomposition of bicarbonate and the increased pH of the brine. High-temperature MSF processes (108–112 °C) often lead to magnesium scaling (i.e., $Mg(OH)_2$ dominant) in the first few stages, while $CaCO_3$ is mainly found in early stages at low-temperature units (95 °C).

The operational temperatures of MD, which are higher than those of other membrane processes, alter the solubility and scaling behaviors of the sparingly soluble salts in the feed solutions. The solubility of NaCl and silica increases with the increase of temperature [25], whereas $CaCO_3$, $CaSO_4$, and Na_2SO_4 exhibit inverse solubility with temperature [25]. Consequently, a higher temperature enhances the scaling propensity of $CaCO_3$ and $CaSO_4$, as the solubility rises, but enables the stability of silica at higher concentrations. Temperature polarization, which reduces the feedwater temperature at the membrane surface, facilitates membrane scaling by silica, but this phenomenon mitigates the precipitation of $CaCO_3$ and $CaSO_4$. Additionally, temperature polarization in partially wetted pores facilitates the precipitation of silica at the solution-vapor interface [26, 27].

12.3 Scaling Mechanism

A proper understanding of scaling mechanisms is required to improve the performance of desalination processes. Understanding the different interactions between membrane surfaces and scaling constituents can help guide the scaling mitigation program, provide more precise scaling projections, and effectively increase the efficiencies of desalination processes.

12.3.1 Nucleation

Nucleation is a process in which the free or dissolved ions or molecules gather and organize to form new structures. The appearance of the new structures, or crystals, can only be achieved when a degree of supersaturation is reached. Three stages can be distinguished during the precipitation of $CaCO_3$: nucleation, dehydration, and crystal growth [28–31]. The process begins with the ions that agglomerate forming a "cluster." The grouping of these aggregates forms a colloidal nucleus which grows and produces a stable crystal. Nucleation can be separated into primary nucleation and secondary nucleation, where primary nucleation is related to the appearance of the first (or the creation at any time) crystal and the secondary nucleation relies upon crystals pre-existing within the system. Primary nucleation can further be divided into homogeneous and heterogeneous nucleation, as shown in Fig. 12.3, are distinguished. Homogeneous nucleation is obtained when the nucleus develops only in the liquid phase and is not influenced by the presence of impurities in the reaction medium. The number of molecules in a stable crystal can vary from about 100 to several thousand. The actual formation of such a nucleus results from the simultaneous collision of the number of necessary molecules. This is a highly unlikely phenomenon. The process of forming such a crystal involves specific energy called the free energy of activation of nucleation. Nucleation is said to be heterogeneous when the nucleus develops by the support or in the presence of impurities. These impurities can act as an accelerator as well as an inhibitor [32]. It is generally accepted that heterogeneous nucleation can be initiated at a lower degree of super-saturation than homogeneous nucleation. The variation in free enthalpy associated with heterogeneous nucleation is lower than that associated with homogeneous nucleation [33, 34]. The secondary nucleation, in a weakly supersaturated solution, is induced and supported by seed crystals of the same nature as the precipitate. In this type of nucleation, the affinity between the crystals of the solute formed and the seed crystals (solid surface) is complete, which corresponds to the critical energy of zero nucleation [35].

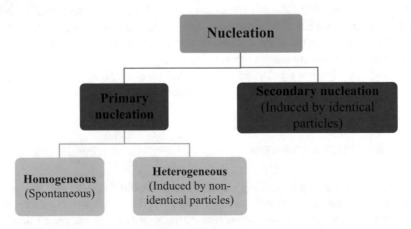

Fig. 12.3 Schematic of nucleation mechanisms

12.3.2 Scale Formation

A high concentration of inorganic salts in the water when in contact with the membrane surfaces results in the formation of scale. The high concentration of salts may exceed 4–10 times the average solubility, which prompts the nucleation process. As discussed in previous sections, the main constituents causing scaling are inorganic salts, including $CaSO_4$, $CaCO_3$, SiO_2, and $BaSO_4$ [36].

The two primary mechanisms of inorganic scaling are bulk crystallization (cake layer formation) and surface crystallization (surface blockage) [37]. For bulk crystallization, which is more predominant in high pressure and moderate cross-flow velocity systems, occurs due to the homogenous growth of particles in bulk phase [38]. For surface crystallization, surface blocking of the membrane occurs mainly due to the convective transportation of colloidal particulate in a solution [39] and is more common in high pressure and low cross velocity operating conditions [40]. The principal stages of scale deposition in RO membrane (Fig. 12.4) begins with dissolving of mineral in the feed water, and continues through the saturation in rejected brine and early nucleation before the final stages of micro-crystal growth and scale formation.

Studies have been conducted to determine the impact of this scaling on membrane surface/performance, from the perspective of both heterogeneous and homogeneous crystallization, and the direct impact it plays on permeate flux decline [8, 41–43]. When the size of the colloid is close to membrane pore size, pore blockage occurs, whereas when the size of the colloids is larger than membrane pore size, cake formation happens [31]. The dynamic formation of the cake on the surface of the

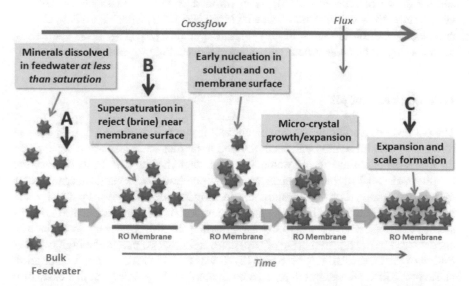

Fig. 12.4 Schematic illustration of the critical steps in scale formation onto the RO membrane surface over time. Reproduced with permission from Ref. [1]; Copyright 2018 @ Elsevier

membrane occurs at different stages. The first step is the pore blockage, followed by cake formation, while the final step is the cake compression. As a result, the effect of pore blockage at every stage is the reduction of flux.

12.3.3   Factors Affecting Scale Formation

The scaling in membranes is related to the feed water composition, water chemistry (pH, cation concertation, ionic strength), membrane properties (membrane morphology), mode of operations, feed water temperature, as well as the hydrodynamic characteristics of a membrane [31]. Each factor will contribute in the scaling formation in membranes, and acquiring information on all these factors will lead mitigation processes, scaling monitoring and to the general improvement in the performance of RO systems in seawater desalination [44].

12.3.3.1   Effect of Temperature

The temperature of the feed is one of the underlying reasons for scale formation as it impacts the separation performance of the membrane [45, 46]. Higher temperatures (> 23 °C) along the membrane create a growth environment for bacteria, which leads to biofilm and other formations of scaling [47]. The high temperature in the feed water further leads to salt diffusion which increases the viscosity, causing swelling of pores in the membrane network and leads to cake formations on the surface of the membrane [1, 14]. As mentioned previously, with the increase of temperature, the solubility increases with NaCl and silica, whereas $CaCO_3$, $CaSO_4$, and Na_2SO_4 exhibit inverse solubility with temperature [26] (Fig. 12.5). Increasing the solubility of any constituent will enhance the scaling propensity.

12.3.3.2   Effect of pH

The pH level of a solution is a significant factor that affects scaling in desalination plants. For that, all scaling indices developed to predict scale formation depends on the measured pH of the treated water. Several studies showed that the increase of the solution pH (>8.3 at 25 °C) enhances the precipitation of several inorganic salts such as calcium carbonate, gypsum, and iron hydroxides [48, 49]. Practically, the increase in the solution pH leads to the acceleration of calcium carbonate and gypsum nucleation, which leads to crystal growth [50, 51]. This case is frequently encountered in the rejected brine of two-stage RO process, due to the high concentration of ions and relatively high pH [10]. It was also shown that iron precipitation could occur in the desalination process by increasing the pH [19]. In this case, iron hydroxides precipitate even with a low concentration of iron ions. As for thermal desalination, calcium carbonate, magnesium hydroxide, and calcium sulfate are the

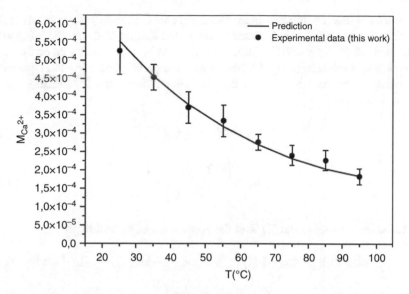

Fig. 12.5 Influence of the temperature on the $CaCO_3$ solubility (P = 1 bar). Reproduced with permission from Ref. [23]; Copyright 2012 @ Elsevier

main scale-forming salts [52], and only the first two scale-former salts depend on pH. Since thermal processes are based on increasing the water temperature, carbon dioxide (CO_2) exhausts the solution at high temperatures, which leads to an increase in pH. Therefore, magnesium hydroxide and calcium carbonate may precipitate. For that, the choice of the initial operating pH and pH adjustment during the operation are crucial steps to avoid scale formation in desalination plants [53].

12.3.3.3 Effect of Ionic Strength

Another factor for the formation of scaling is the effect of ionic strength. High ionic strength results in aggressive cake formation on membranes. The repulsive energy, which acts as a barrier between molecules, is nullified with the increase of ionic strength leading to the compression of the electrical double layer of counter ions. Through this, the vander Waals forces tend to dominate, contributing to colloidal instability (resulting in aggressive cake formation) [39, 54], to higher diffusion, and higher pore adsorption.

12.4 Scaling Prediction

Three scaling indices are usually used to assess the scaling tendency in seawater desalination processes. While the Langelier Saturation Index (LSI) (Langelier 1936) is often applied for brine solutions of a TDS not exceeding 10 g/L [55, 56], the Stiff and Davis Saturation Index (S&DSI) (Stiff & Davis 1952) [57] and MLSI (Elfil & Roques 2004) [58] are used to assess the scaling tendency of highly saline solutions. Most of the stability indices used in the literature are based on the determination of saturation pH (pH_s). pH_s is formulated by combining the equation of the solubility product, K_s, with the equilibrium relation of the 2nd ionization of the carbonic acid, K_2:

$$Ks = \left(Ca^{2+}\right)\left(CO_3^-\right) \tag{12.1}$$

$$K_2 = \frac{\left(CO_3^{2-}\right)\left(H_3O^+\right)}{\left(HCO_3^-\right)} \tag{12.2}$$

Combining Eqs. (1) and (2) with the application of logarithm gives:

$$pH_s = pK_2 - pK_s - \log\gamma_{Ca^{2+}} - \log\gamma_{HCO_3^-} - \log\left[Ca^{2+}\right] - \log\left[Alc\right] \tag{12.3}$$

where [Alc] is the concentration of the alkalinity which is related to the concentration of HCO_3^- at solution pH < 9 ([Alc] $\approx \left[HCO_3^-\right]$). The LSI index is then defined as follows:

$$LSI = pH_{mes} - pH_s \tag{12.4}$$

where pH_{mes} is the measured pH of the treated solution, and pH_s is the pH of saturation for calcite. pH_s is defined as:

$$pH_s = pK_2' - pK_s' - \log\left[Ca^{2+}\right] - \log\left[Alc\right] - \log\left[1 + 2.10^{pK_s' - pK_2'}\right] \tag{12.5}$$

pK_2' and pK_s' are defined as the apparent constants of the second dissociation of carbonic acid and the solubility product of calcite at a given ionic strength. LSI uses the concentrations instead of activities. When pH < 9.5, the term $1 + 2.10^{pK_s' - pK_2'}$ is neglected. Then, the LSI expression becomes:

$$pH_s = pK_2' - pK_s' - \log\left[Ca^{2+}\right] - \log\left[Alc\right] \tag{12.6}$$

Similar to the LSI index formula, the S & DSI is presented as follows:

$$S \& DSI = pH_{mes} - pH_s \tag{12.7}$$

where:

$$pH_s = -\log\left[Ca^{2+}\right] - \log\left[Alc\right] + K \tag{12.8}$$

K is an empirical constant which is associated with the ionic strength. According to LSI calculations, three situations can be expected:

- LSI < 0: Water is under-saturated for CaCO3. The under-saturated water tends to remove existing calcium carbonate protective coatings in pipelines and equipment.
- LSI = 0: Water is considered to be neutral; neither scale-formation nor scale removal.
- LSI > 0: Water is supersaturated with respect to CaCO3, and scale formation may occur.

For S & DSI scaling index, a positive value of S & DSI indicates the tendency to form $CaCO_3$ scale. $CaCO_3$ hydrated forms (Amorphous Calcium Carbonate: ACC, and Monohydrate Calcium Carbonate: MCC) are crucial precursors for the $CaCO_3$ spontaneous nucleation [59]. The solubility products of the MCC and the ACC have shown to be the lower limits for spontaneous and instantaneous nucleation, respectively, in the temperature interval ranging between 20 °C and 60 °C [60]. According to these findings, in temperature ranges between 25 °C and 60 °C, the MCC solubility product constitutes a lower limit that must be exceeded to obtain spontaneous nucleation [61]. The saturation pH is calculated using the calcium carbonate monohydrated ($CaCO_3.H_2O$) solubility product instead of LSI and S & DSI cases. For a pH ranging between 6 and 9, the saturation pH is described as:

$$pH_{s/MMC} = pK_2 - pK_{s/MMC} - \log\gamma_{Ca^{2+}} - \log\gamma_{HCO_3^-} - \log\left[Ca^{2+}\right] - \log\left[Alc\right] \quad (12.9)$$

where pH_S/MCC and pK_S/MCC are respectively the MCC saturation pH and solubility product. Similar to the previous indices, the MLSI is then defined as follows [58]:

$$MLSI = pH - pH_{S/MCC} \quad (12.10)$$

The activity coefficients are calculated by simple models, such as "Modified Debye & Hückel" for solutions with an ionic strength of less than 0.2 M (or TDS <10 g / L). For more saline waters, such as seawater desalination by RO, more complex models are valid as Simplified Pitzer (Ionic force <2 M). According to MLSI values, four situations are expected:

- MLSIinst < MLSI; Water is highly scaling, and the phenomenon is instantaneous.
- 0 < MLSI < MLSIinst; Water is scaling; the phenomenon depends on the wall nature and will be slow for MLSI values close to 0.
- MLSIagr < MLSI <0; There is no spontaneous precipitation; water is at a calco-carbonic equilibrium.
- MLSI < MLSIagr; Water is undersaturated with respect to calcite and can be considered as aggressive.

The MLSI$_{inst}$ is given by the same Eq. 12.9 when substituting pH$_S$/MCC by the ACC saturation pH (pH$_S$/ACC). Surpassing this value is an indication of a highly scaling behavior as the ACC was shown to be a precursor for CaCO$_3$ nucleation at high supersaturation [58]. MLSI$_{agr}$ is also given by Eq. (9) when replacing the pHs/MCC by the calcite saturation pH, pH$_S$/calcite, which is exactly the saturation pH defined by Langelier.

12.5 Membrane Scaling Control Strategies

Membrane scaling prevention and control is a complicated process which adopts several different methods. The methodology adopted is dependent on the nature of the scaling, type of foulants, and the mode of operation. Some of the most widely adopted techniques of scale prevention and control are categorized into three groups, namely [8]:

(a) Development of novel RO membranes or membrane surface modifications,
(b) Membrane cleaning strategies and
(c) Pre-treatment strategies and processes.

12.5.1 Surface Modifications and Novel Membrane Materials

Membrane smoothness and hydrophilicity have a considerable impact on scaling propensities [62]. Studies on surface modification demonstrated the relationship between the smoothness, hydrophilicity, and hydrophobicity of a membrane to scaling showing that the smoother and more hydrophilic a membrane surface is, the lower tendency it has to scale. Similarly, the rougher and more hydrophobic a membrane surface, the higher the chances of scaling [63, 64]. The surface modification increases the anti-scaling performance of a membrane by changing its properties such as surface charge, morphology, chemical groups, as well as hydrophilicity [10, 65, 66].

Polyamide thin-film composites (TFC) [67, 68] has been used to develop novel RO membranes and to modify membrane surfaces. Other novel materials used are carbon nanotubes (CNTs), nanophorous graphene, and metal oxide nanoparticles [69–74]. Surface modifications are further introduced to improve anti-scaling performance.

12.5.2 Physical and Chemical Cleaning

It is one of the more conventional methods for the mitigation and control of scaling. It involves the use of chemical facilitated by physical activities such as backwash to remove scaling and other foulants. In order to further improve the cleaning efficiency, some of the physical cleaning techniques require the removal and soaking of the membrane. In a recent development, a method referred to as Chemical Enhanced Backwashing (CEB) [75–77] combines both physical and chemical cleaning process for the enhancement of scaling removal in the RO system. In this method, a soaking period of 5–20 min is adopted to allow for chemical reaction to take place. As an alternative, the membrane can be soaked for several hours, during which air sparging can use [54].

Similarly, at an industrial scale, the cleaning process is commonly used to recover the membrane performance. The practice referred to as Cleaning-in-Place (CIP) comprises of different stages which comprise emptying the filtration rack of both feed and permeate side, the addition of chemical agents and finally, disinfection/ rinsing [54, 78, 79].

12.5.3   Pre-treatment Method

This method, considered as a preventive, scaling control technique, is popular and widely adopted in most RO systems to improve feed water quality [80, 81]. Enhancing the pre-treatment impact on membrane scaling has gained popularity in research fields with a vast interest in desalination systems. Using pre-treatment for the feed water improves the reliability of RO systems while prolonging the life span of the membrane [81]. Pre-treatment requirements are based upon the feed water composition. For example, water with a high level of hardness will require a pre-treatment process to soften the water [8, 80, 82]. Some of the widely RO pre-treatment processes that were adopted in the last 10 years for the prevention of scaling includes coagulation/flocculation, ultrafiltration, microfiltration, scale inhibitors, amongst others [83–85]. Furthermore, a disinfection stage in the pre-treatment process is becoming common.

The common pretreatment technologies for RO (Fig. 12.6) includes disinfection, coagulation/flocculation, filtration, microfiltration, and ultrafiltration amongst others [86]. The selection and the application of the pre-treatment technologies mostly depend on the type, quality as well as the characteristics of feedwater. The conventional pretreatment processes are widespread and are typical in seawater desalination [87]. However, to achieve higher feed quality, the non-conventional pretreatment processes (membrane) are being adopted as the pretreatment process in most RO desalination plants [86].

12.6   Conclusions and Outlook

Scaling is a major obstacle that limits the effectiveness of thermally based desalination as it results in a reduction of heat transfer, and increases temperature polarization, which leads to a reduction in water flux. For a membrane thermally based

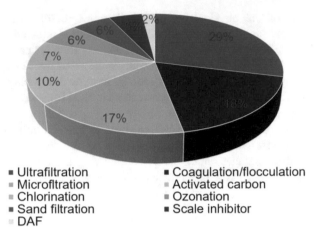

- Ultrafiltration   - Coagulation/flocculation
- Microfltration   - Activated carbon
- Chlorination   - Ozonation
- Sand filtration   - Scale inhibitor
- DAF

Fig. 12.6 Common studied RO pre-treatment technologies in the past 10 years. Reproduced with permission from Ref. [8]; Copyright 2017 @ Elsevier

process such as MD, the effect of scaling on the membrane performance is significant as it results in membrane wetting. Scaling is a major gridlock in pressure driven processes such as RO by pore blockage and increase of the energy consumption. This chapter highlighted the importance of controlling scaling as it affects the cost of operation and maintenance of the system. Membrane scaling is related to the feed water composition, water chemistry (pH, cation concertation, ionic strength), membrane properties (membrane morphology), mode of operations, feed water temperature, as well as the hydrodynamic characteristics of the process. Membrane scaling prevention and control is also a complicated process which adopts several procedures. As earlier discussed, the methodology adopted is dependent on the nature of the scalants, physical-chemical characteristics of the water, and the mode of operation. For this reason, finding a lasting solution to the challenge of scaling requires more research, especially in the optimization of membranes through the advancement of anti-scaling materials as well as improved pretreatment techniques. These, when achieved, will result in a favorable outcome that will lead to a significant increase in the efficiency and the performance of desalination processes.

References

1. P.S. Goh, W.J. Lau, M.H.D. Othman, A.F. Ismail, Membrane fouling in desalination and its mitigation strategies. Desalination **425**, 130–155 (2017)
2. A. Antony, J.H. Low, S. Gray, A.E. Childress, P. Le-Clech, G. Leslie, Scale formation and control in high-pressure membrane water treatment systems: A review. J. Memb. Sci. **383**, 1–16 (2011)
3. H. Lin, M. Zhang, F. Wang, F. Meng, B. Liao, H. Hong, J. Chen, W. Gao, A critical review of extracellular polymeric substances (EPSs) in membrane bioreactors: Characteristics, roles in membrane fouling and control strategies. J. Memb. Sci. **460**, 110–125 (2014)

4. M. David, J. Swaminathan, E. Guillen, H.A. Arafat, J.H. Scaling, C. Link, Scaling and fouling in membrane distillation for desalination applications: A review. Desalination **356**, 294–313 (2015)
5. A.S. Al-amoudi, Factors affecting natural organic matter (NOM) and scaling fouling in NF membranes: A review. Desalination **259**, 1–10 (2010)
6. S. Shirazi, C. Lin, D. Chen, Inorganic fouling of pressure-driven membrane processes – A critical review. Desalination **250**, 236–248 (2010)
7. B. Mi, M. Elimelech, Silica scaling and scaling reversibility in forward osmosis. Desalination **312**, 75–81 (2013)
8. S. Jiang, Y. Li, B.P. Ladewig, Science of the Total environment a review of reverse osmosis membrane fouling and control strategies. Sci. Total Environ. **595**, 567–583 (2017)
9. A.A. Alanezi, A. Altaee, I. Ibrar, O. Naji, A. Sharif, A. Malekizadeh, A. Alhawari, A review of fouling mechanisms, control strategies and real-time fouling monitoring techniques in forward osmosis. Water **11**, 695 (2019)
10. K. Touati, M. Hila, K. Makhlouf, H. Elfil, Study of fouling in two-stage reverse osmosis desalination unit operating without an inlet pH adjustment: Diagnosis and implications. Water Supply **17**, 1682–1693 (2017)
11. K. Touati, H. Cherif, N. Kammoun, M. Jendoubi, H. Elfil, Inhibition of calcium carbonate scaling by precipitation using secondary nucleation coupled to degassing with atmospheric air. J. Water Process Eng. **22**, 258–264 (2018)
12. K. Touati, E. Alia, H. Zendah, H. Elfil, A. Hannachi, Sand filters scaling by calcium carbonate precipitation during groundwater reverse osmosis desalination. Desalination **430**, 24–32 (2018)
13. Y. Zarga, H.B. Boubaker, N. Ghaffour, H. Elfil, Study of calcium carbonate and sulfate co-precipitation. Chem. Eng. Sci. **96**, 33–41 (2013)
14. B. Tomaszewska, M. Bodzek, Desalination of geothermal waters using a hybrid UF-RO process. Part II: Membrane scaling after pilot-scale tests. Desalination **319**, 107–114 (2013)
15. J. Zhanga, W. Lay, C. Loong, S. Chou, C. Tang, R. Wang, A.G. Fane, Membrane biofouling and scaling in forward osmosis membrane bioreactor. J. Membr. Sci. **403–404**, 8–14 (2012)
16. M. Zhang, J. Shan, C.Y. Tang, Gypsum scaling during forward osmosis process – A direct microscopic observation study. Desalin. Water Treat. **57**, 3317–3327 (2016)
17. D.L. Shaffer, M.E. Tousley, M. Elimelech, Influence of polyamide membrane surface chemistry on gypsum scaling behavior. J. Membr. Sci. **525**, 249–256 (2017)
18. E. Melliti, K. Touati, H. Abidi, H. Elfil, Iron fouling prevention and membrane cleaning during reverse osmosis process. Int. J. Environ. Sci. Tech. **16**, 3809–3818 (2019)
19. B. Das, P. Hazarika, G. Saikia, H. Kalita, D.C. Goswami, H.B. Das, S.N. Dube, R.K. Dutta, Removal of iron by groundwater by ash: A systematic study of a traditional method. J. Hazard. Mater. **141**, 834–841 (2007)
20. S.S. Tahir, N. Rauf, Removal of Fe^{2+} from the waste water of a galvanized pipe manufacturing industry by adsorption onto bentonite clay. J. Environ. Manag. **73**, 285–292 (2004)
21. A. Antony, J.H. Low, S. Gray, A.E. Childress, P. Le-Clech, G. Leslie, Scale formation and control in high pressure membrane water treatment systems: A review. J. Membr. Sci. **383**, 1–16 (2011)
22. A. Matin, F. Rahman, H.Z. Shafi, S.M. Zubair, Scaling of reverse osmosis membranes used in water desalination: Phenomena, impact, and control; future directions. Desalination **455**, 135–157 (2019)
23. B. Coto, C. Martosa, J.L. Pe˜na, R. Rodríguez, G. Pastor, Effects in the solubility of $CaCO_3$: Experimental study and model description. Fluid Phase Equilibr **324**, 1–7 (2012)
24. R.C. Newton, C.E. Manning, Evidence for SiO_2-NaCl complexing in H_2O-NaCl solutions at high pressure and temperature. Geofluids **16**, 342–348 (2016)
25. D.M. Warsinger, E.W. Tow, J. Swaminathan, J.H. Lienhard, Theoretical framework for predicting inorganic fouling in membrane distillation and experimental validation with calcium sulfate. J. Membr. Sci. **528**, 381–390 (2017)

26. D.M. Warsinger, J. Swaminathan, E. Guillen-Burrieza, H.A. Arafat, J.H. Lienhard, Scaling and fouling in membrane distillation for desalination applications: A review. Desalination **356**, 294–313 (2015)

27. A.A. Alanezi, A.I. Ibrar, O. Naji, A. Sharif, A. Malekizadeh, A. Alhawari, A review of fouling mechanisms, control strategies and real-time fouling monitoring techniques in forward osmosis. Water **11**, 695 (2019)

28. Y. Zhan, H. Dong, F. Yi, C. Yue, The combined process of paper filtration and ultrafiltration for the pre-treatment of the biogas slurry from swine manure. Intl J. Environ. Res. Public Health **15**, 894 (2018)

29. W. Guo, H. Ngo, J. Li, Bioresource technology- a mini-review on membrane fouling. Bioresour. Technol. **122**, 27–34 (2012)

30. G. Montes-Hernandez, Calcite precipitation from CO_2-H_2O-$Ca(OH)_2$ slurry under high pressure of CO_2. J. Cryst. Growth **308**, 228–236 (2007)

31. M. Kellermeier, A. Picker, A. Kempter, H. Cölfen, D. Gebauer, A straightforward a treatment of activity in aqueous $CaCO_3$ solutions and the consequences for nucleation theory. Adv. Mater. **26**, 752–757 (2014)

32. M. Kellermeier, P. Raiteri, J.K. Berg, A. Kempter, J.D. Gale, D. Gebauer, Entropy drives calcium carbonate ion association. ChemPhysChem **17**, 3535–3541 (2016)

33. T.A. Hoang, Mechanisms of scale formation and inhibition, in *Mineral Scales and Deposits*, ed. by Z. Amjad, K. D. Demadis, (Elsevier, Amsterdam, 2015), pp. 47–83. ISBN: 978-0-444-63228-9

34. M. Davoody, J. Lachlan, W. Graham, J. Wu, P.J. Witt, S. Madapusi, R. Parthasarathy, Mitigation of scale formation in unbaffled stirred tanks-experimental assessment and quantification. Chem. Eng. Res. Design **146**, 11–21 (2019)

35. T. Lee, Y.J. Choi, Y. Cohen, Gypsum scaling propensity in semi-batch RO (SBRO) and steady-state RO with partial recycle (SSRO-PR). J. Membr. Sci. **588**, 117106 (2019)

36. Y. Wang, Composite fouling of calcium sulfate and calcium carbonate in a dynamic seawater reverse osmosis unit. Msc Thesis, The University of New South Wales Sydney, Australia 2005, (2019), http://unsworks.unsw.edu.au/fapi/datastream/unsworks:1099/SOURCE1

37. A.J. Karabelas, Critical assessment of medic series of R & D reports project. Chemical Process Engineering Research Institute Greece. 98-BS-034. August, 2003. (2019), https://medrc.org/jdownloads/Alumni%20Research%20Projects/98-BS-034.pdf

38. N.T.K. Thanh, N. Maclean, S. Mahiddine, Mechanisms of nucleation and growth of nanoparticles in solution. Chem. Rev. **114**, 7610–7630 (2014)

39. T.M. Pääkkönen, M. Riihimäki, E. Puhakka, E. Muurinen, C.J. Simonson, R.L. Keiski, Crystallization fouling of $CaCO_3$ – Effect of bulk precipitation on mass deposition on the Heat transfer surface. Proceedings of the International Conference Heat Exchanger Fouling and Cleaning, pp. 209–216. (2009), http://heatexchanger-fouling.com/papers/papers2009/30_Paeaekkoenen_F.pdf

40. I.M.A. Elsherbiny, A.S.G. Khalil, M. Ulbricht, Influence of surface micro-patterning and hydrogel coating on colloidal silica fouling of polyamide thin-film composite membranes. Membranes **9**, 67 (2019)

41. M. Nergaard, C. Grimholt, An introduction to scaling causes, problems and solutions. NTNU – Institutt for petroleumsteknologi og anvendt geofysikk. TPG 4140 – Natural Gas. November (2010), https://pdfs.semanticscholar.org/5f20/dbd659b1ef869d2a54795145e46075611826.pdf

42. V.A. Grover, Adsorption of divalent metals to metal oxide nanoparicles: Competitive and temperature effects (Order No. 1492085). *Available from ProQuest Dissertations & Theses Global.* (868328325). (2011), Retrieved from https://lib-ezproxy.concordia.ca/login?

43. Y. Liu, B. Mi, Combined fouling of forward osmosis membranes: Synergistic foulant interaction and direct observation of fouling layer formation. J. Membr. Sci. **407-408**, 136–144 (2012)

44. Y.M. Kim, S.J. Kim, Y.S. Kim, S. Lee, I.S. Kim, J. Ha, Overview of systems engineering approaches for a large-scale seawater desalination plant with a reverse osmosis network. Desalination **238**, 312–332 (2009)

45. H. Li, Y. Lin, P. Yu, Y. Luo, L. Hou, FTIR study of fatty acid fouling of reverse osmosis membranes: Effects of pH, ionic strength, calcium, magnesium and temperature. Sep. Purif. Technol. **77**, 171–178 (2011)
46. J. Wu, A.E. Contreras, Q. Li, Studying the impact of RO membrane surface functional groups on alginate fouling in seawater desalination. J. Membr. Sci. **458**, 120–127 (2014)
47. C. Jarusutthirak, G. Amy, Role of soluble microbial products (SMP) in membrane fouling and flux decline. Environ. Sci. Technol. **40**, 969–974 (2006)
48. P. Sahachaiyunta, T. Koo, R. Sheikholeslami, Effect of several inorganic species on silica fouling in RO membranes. Desalination **144**, 373–378 (2002)
49. S.G. Yiantsios, D. Sioutopoulos, A.J. Karabelas, Colloidal fouling of RO membranes: An overview of key issues and efforts to develop improved prediction techniques. Desalination **183**, 257–272 (2005)
50. Y. Jin, Y. Ju, H. Lee, S. Hong, Fouling potential evaluation by cake fouling index: Theoretical development, measurements, and its implications for fouling mechanisms. J. Membr. Sci. **490**, 57–64 (2015)
51. Y. Liao, A. Bokhary, E. Maleki, B. Liao, A review of membrane fouling and its control in algal-related membrane processes. Bioresour. Technol. **264**, 343–358 (2018)
52. L. Zheng, D. Yu, G. Wang, Z. Yue, C. Zhang, Y. Wang, Characteristics and formation mechanism of membrane fouling in a full-scale RO wastewater reclamation process: Membrane autopsy and fouling characterization. J. Membr. Sci. **563**, 843–856 (2018)
53. Y. Yu, S. Lee, S. Hong, Effect of solution chemistry on organic fouling of reverse osmosis membranes in seawater desalination. J. Membr. Sci. **351**, 205–213 (2010)
54. X. Shi, G. Tal, N.P. Hankins, V. Gitis, Fouling and cleaning of ultrafiltration membranes: A review. J. Water Process. Eng **1**, 121–138 (2014)
55. W.F. Langelier, The analytic control of anti-corrosion water treatment. J. Am. Water Works Asso. **28**, 1500–1521 (1926)
56. R.J. Ferguson, Scaling indices: Types and applications, in *Mineral Scales and Deposits*, ed. by Z. Amjad, K. D. Demadis, (Elsevier, Amsterdam, 2015), pp. 721–735. ISBN: 978-0-444-63228-9
57. H.A. Stiff, L.E. Davis, A method for predicting the tendency of oil field water to deposit calcium carbonate. Trans. Metall. Soc. Am. Inst. Min., Metall. Pet. **195**, 213–216 (1952)
58. H. Elfil, H. Roques, Prediction of the limit of the metastable zone in the "CaCO$_3$–CO$_2$–H$_2$O" system. AICHE J. **50**, 1908–1916 (2004)
59. J.Y. Gal, Y.H. Fovet, N. Gache, Mechanisms of scale formation and carbon dioxide partial pressure influence, Part I: Elaboration of an experimental method and a scaling model. Water Res. **36**, 755–763 (2002)
60. H. Elfil, A. Hannachi, Reconsidering water scaling tendency assessment. AICHE J. **52**, 3583–3591 (2006)
61. H. Elfil, H. Roques, Role of hydrate phases of calcium carbonate on the scaling phenomenon. Desalination **137**, 177–186 (2001)
62. L.F. Dumée, Towards integrated anti-microbial capabilities: Novel bio-fouling resistant membranes by high velocity embedment of silver particles. J. Membr. Sci. **475**, 552–561 (2015)
63. K.P. Lee, T.C. Arnot, D. Mattia, A review of reverse osmosis membrane materials for desalination – development to date and future potential. J. Membr. Sci. **370**, 1–22 (2011)
64. L. Ni, J. Meng, X. Li, Y. Zhang, Surface coating on the polyamide TFC RO membrane for chlorine resistance and antifouling performance improvement. J. Membr. Sci. **451**, 205–215 (2014)
65. J. Xue, Z. Jiao, R. Bi, R. Zhang, X. You, F. Wang, L. Zhou, Y. Su, Z. Jiang, Chlorine-resistant polyester thin film composite nanofiltration membranes prepared with β-cyclodextrin. J. Membr. Sci. **584**, 282–289 (2019)
66. J. Lee, R. Wang, T. Bae, A comprehensive understanding of co-solvent e ffects on interfacial polymerization: Interaction with trimesoyl chloride. J. Membr. Sci. **583**, 70–80 (2019)
67. M. Gyu, S. Park, S. Jin, H. Kwon, J. Bae, J. Lee, Facile performance enhancement of reverse osmosis membranes via solvent activation with benzyl alcohol. J. Membr. Sci. **578**, 220–229 (2018)

68. S. Ho, S. Kwak, B. Sohn, T. Hyun, Design of TiO$_2$ nanoparticle self-assembled aromatic polyamide thin-film-composite (TFC) membrane as an approach to solve biofouling problem. J. Membr. Sci. **211**, 157–165 (2003)

69. G. Kang, Y. Cao, Development of antifouling reverse osmosis membranes for water treatment: A review. Water Resour. **46**, 584–600 (2011)

70. J. Wang, R. Xu, F. Yang, J. Kang, Y. Cao, M. Xiang, Probing in fluences of support layer on the morphology of polyamide selective layer of thin film composite membrane. J. Membr. Sci. **556**, 374–383 (2018)

71. M. Safarpour, A. Khataee, V. Vatanpour, Thin film nanocomposite reverse osmosis membrane modified by reduced graphene oxide/TiO$_2$ with improved desalination performance. J. Membr. Sci. **489**, 43–54 (2015)

72. H. Chae, J. Lee, C. Lee, I. Kim, P. Park, Graphene oxide-embedded thin-film composite reverse osmosis membrane with high flux, anti-biofouling, and chlorine resistance. J. Membr. Sci. **483**, 128–135 (2015)

73. A. Inurria, P. Cay-Durgun, D. Rice, H. Zhang, D. Seo, M.L. Lind, F. Perreault, Polyamide thin-film nanocomposite membranes with graphene oxide nanosheets: Balancing membrane performance and fouling propensity. Desalination **451**, 139–147 (2019)

74. V. Vatanpour, S. Siavash, R. Moradian, S. Zinadini, B. Astinchap, Fabrication and characterization of novel antifouling nanofiltration membrane prepared from oxidized multiwalled carbon nanotube/polyethersulfone nanocomposite. J. Membr. Sci. **375**, 284–294 (2011)

75. H. Zarrabi, M. Ehsan, V. Vatanpour, A. Shockravi, M. Safarpour, Improvement in desalination performance of thin film nanocomposite nano filtration membrane using amine-functionalized multiwalled carbon nanotube. Desalination **394**, 83–90 (2016)

76. E. Lee, J. Kwon, H. Park, W. Hyun, H. Kim, A. Jang, Influence of sodium hypochlorite used for chemical enhanced backwashing on biophysical treatment in MBR. Desalination **316**, 104–109 (2013)

77. T. Zsirai, P. Buzatu, P. Aerts, S. Judd, Efficacy of relaxation, back flushing, chemical cleaning and clogging removal for an immersed hollow fibre membrane bioreactor. Water Resour. **46**, 4499–4507 (2012)

78. S.K. Lateef, B.Z. Soh, K. Kimura, Bioresource technology direct membrane filtration of municipal wastewater with chemically enhanced backwash for recovery of organic matter. Bioresour. Technol. **150**, 149–155 (2013)

79. C. Brepols, K. Drensla, A. Janot, M. Trimborn, N. Engelhardt, Strategies for chemical cleaning in large scale membrane bioreactors. Water Sci. Technol. **57**, 457–463 (2008)

80. S. Jamaly, N.N. Darwish, I. Ahmed, S.W. Hasan, A short review on reverse osmosis pretreatment technologies. Desalination **354**, 30–38 (2014)

81. A. Bennett, Desalination: Developments in pre-treatment technology. Filtr. Separat. **49**, 16–20 (2012)

82. S. Fatima, R. Hashaikeh, N. Hilal, Reverse osmosis pre-treatment technologies and future trends: A comprehensive review. Desalination **452**, 159–195 (2018)

83. M. Badruzzaman, N. Voutchkov, L. Weinrich, J.G. Jacangelo, Selection of pre-treatment technologies for seawater reverse osmosis plants: A review. Desalination **449**, 78–91 (2019)

84. S. Hee, B. Tansel, Novel technologies for reverse osmosis concentrate treatment: A review. J. Environ. Manag. **150**, 322–335 (2015)

85. A. Subramani, J.G. Jacangelo, Treatment technologies for reverse osmosis concentrate volume minimization: A review. Sep. Purif. Technol. **122**, 472–489 (2014)

86. P. Noka, L. Qi-Feng, K. Seung-Hyun, Pre-treatment strategies for seawater desalination by reverse osmosis system. Desalination **249**, 308–316 (2009)

87. B. Mohammad, V. Nikolay, W. Lauren, G.J. Joseph, Selection of pretreatment technologies for seawater reverse osmosis plants: A review. Desalination **449**, 78–91 (2019)

Chapter 13
Biofouling in RO Desalination Membranes

Nawrin Anwar, Liuqing Yang, Wen Ma, Haamid Sani Usman,
and Md. Saifur Rahaman

13.1 Introduction

The global freshwater crisis is considered as one of the most critical challenges currently faced by the international community. Wastewater reuse and seawater desalination have been considered as highly feasible ways to alleviate global water scarcity. In desalination processes, the cost of water production by seawater reverse osmosis (SWRO) is reported to be one-half to one-third of the cost of thermal distillation [1]. Reverse osmosis (RO) has proven to be a competitive technology for various types of wastewater reclamation and brackish/seawater desalination. RO is preferred for its superior efficiency in the removal of small-sized contaminants (salt, metal ions, pharmaceuticals, organic colloid, etc.), smaller footprints [2] as well as for lower capital and operating costs when compared to traditional treatment methods e.g., thermal distillation [3].

RO is a high-pressure membrane-based process which utilizes a dense membrane to separate water from molecular-sized contaminants such as dissolved organic compounds, colloids, and monovalent ions (e.g., Na^+, Cl^-) [4]. In microfiltration (MF) and ultrafiltration (UF) processes, membrane pore structures are

The original version of this chapter was revised. The correction to this chapter is available at
https://doi.org/10.1007/978-3-030-34284-5_18

N. Anwar · L. Yang · H. S. Usman · M. S. Rahaman (✉)
Building, Civil and Environmental Engineering Department, Concordia University,
Montreal, QC, Canada
e-mail: saifur.rahaman@concordia.ca

W. Ma
Building, Civil and Environmental Engineering Department, Concordia University,
Montreal, QC, Canada

Chemical and Environmental Engineering Department, Yale University,
New Haven, CT, USA

© Springer Nature Switzerland AG 2020, Corrected Publication 2021
V. S. Saji et al. (eds.), *Corrosion and Fouling Control in Desalination Industry*,
https://doi.org/10.1007/978-3-030-34284-5_13

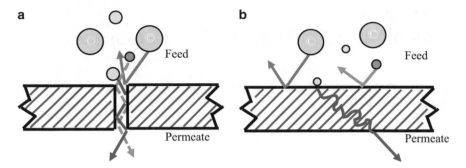

Fig. 13.1 Contaminants transport via (**a**) size exclusion (microfiltration, ultrafiltration, etc.) and (**b**) solution diffusion (Nanofiltration, Reverse Osmosis) mechanisms in a membrane process

designed to remove contaminants based upon their size (size exclusion mechanism) [5], and only the contaminants which are larger than membrane pore sizes are retained as shown in Fig. 13.1 (a). The pore size of RO membrane is about 0.1 nm where "solution-diffusion" is the main mechanism [6, 7] for water transport as shown in Fig. 13.1 (b).

The semi-permeable membrane is the core component of the RO process. The first generation of commercially available RO membranes were developed in the 1960s using cellulose acetate (CA) [8]. In the early 1980s, a polyamide (PA) casted membrane with a thin film composite structure (TFC) was introduced by the Film Tec corporation [9]. TFC membranes displayed higher water permeability, operated at higher temperatures, and operated at higher pressures than CA membranes, while also using wider range of pH values. TFC is still regarded as a "state-of-art" material in RO process [8]. PA is a widely used material for the fabrication of commercial TFC membranes. PA active layers can be formed through cross-linking between trimesoyl chloride (TMC) and m-phenylene diamine (MPD) [10].

Despite having a high-quality permeate product, one of the major limitations that hinders the widespread application of RO is membrane fouling. Feed water in RO systems generally contains four main types of contaminants: inorganic compounds (salts, metal hydroxide, metal carbonate, etc.), natural organic matter (NOM), gel-colloids, and microorganisms that can cause four different categories of membrane fouling: inorganic fouling/scaling, organic fouling, particulate fouling, and biofouling respectively. During long operational periods, these contaminants may reside on the membrane surface and form an additional fouling layer, which jeopardizes membrane performance (Fig. 13.2). Periodic physical/chemical cleaning is necessary [11] to maintain the desired flux of the RO membrane. The cleaning actions, especially by chemicals, shortens the membranes life. The membrane fouling has significant economic impact on RO plant operation as it accounts for about 50% of the total costs [12] via cleaning, loss of operation due to cleaning, and increasing pressures to maintain constant flux through the clogged membrane. RO desalination is extensively used in Middle East, and around 70% of these desalination plants experience biofouling [12, 13] due to the water in the Gulf region having high organic content, considerable amount of microorganisms, and total dissolved solids (TDS) content. As biofouling account up to 35–45% of all fouling in the RO process [14], the comprehensive under-

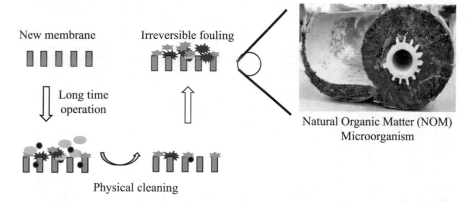

New membrane   Irreversible fouling

Long time
operation

Natural Organic Matter (NOM)
Microorganism

Physical cleaning

Fig. 13.2 Formation of irreversible fouling on polyamide membrane. (Reproduced the inset image with permission from Dr. Florian Beyer; Copyright @ Dr. Florian Beyer)

standing of biofouling of RO membranes is crucial for effective biofouling management. This chapter will provide an insight into the mechanism of biofouling, formation of biofilms, role of EPS, and critical factors affecting the biofilms. Also, biofouling impact on permeate water flux and salt rejection is further discussed along with performance degradation mechanism and energy consumption.

13.2   Biofouling

The unwanted deposition/growth of microorganisms on or within the membrane surface results in biofouling of the membrane. Generally, any fouling resulting from microbial colonization and biofilm formation is considered biofouling. Very few organisms present in the feed water can lead to significant biofouling of the membrane. Even membranes using pre-treated influent with substantial removal of microbe is susceptible to significant biofouling [14, 15]. Biofouling causes membrane flux decline, membrane biodegradation, enhanced salt passage, increased differential and feed pressure, permeate quality degradation, the necessity of frequent cleaning which eventually would result in high treatment cost, and often process failure. Bacteria, fungi, and yeasts are the primary microorganisms causing biofouling. The bacteria can tolerate a wide range of pH (0.5–13) and temperature (-12–110 °C) while being able to colonize on all membrane surfaces in RO plants in varying conditions [16]. Table 13.1 represents frequently observed microorganism on membranes in RO plants.

13.2.1   Mechanism of Biofilm Formation

Accumulation of microbial cells on/within the membrane surface along with a matrix of extracellular polymeric substances (EPS) is known as biofilm. In general, microbes are abundant in all water systems and can colonize rapidly on favourable

Table 13.1 List of frequently observed microorganism on RO membranes [16, 17]; Adapted with permission from [14]; Copyright 2016 © Springer, Creative Commons CC BY

Bacteria	Fungi	Yeasts
Pseudomonas	Penicillium	Occasionally present in substantial quantity
Bacillus	Trichoderma	
Mycobacterium	Mucor	
Corynebacterium	Fusarium	
Flavobacterium	Aspergillus	
Arthrobacte		
Acinetobacter		
Cytophaga		
Moraxella		
Micrococcus		
Serratia		
Lactobacillus		
Aeromonas		

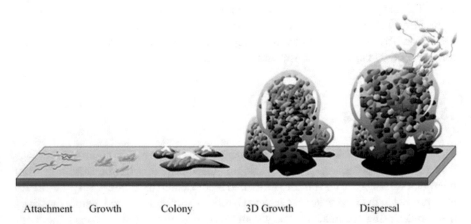

Attachment   Growth   Colony   3D Growth   Dispersal

Fig. 13.3 Different stages of biofilm development. Adapted with permission from [19]; Copyright 2007 © Don Monroe, Creative Commons CC BY

surfaces. The microbes get attached on the membrane surface and grow due to the presence of nutrients in the feed/influent. Eventually, the microbes excrete EPS, in which they are embedded, and form further biofilm.

Biofilm formations are usually a complex multistage process that can be reversible or irreversible. The development of biofilms on membrane surfaces usually follows the following steps [18], as illustrated in Fig. 13.3:

(1) Adsorption and attachment of cells on the membrane surface altering the membrane properties and forming the conditioning film.
(2) Aggregation and growth of new cells which are controlled by different chemical and physical factors with hydrophobic and non-polar surfaces enhancing the irreversible attachment along with excretion of EPS.

(3) Formation of microbial colonies and biofilm development and maturation with the continuous production of EPS.
(4) Three-dimensional growth and further maturation of the biofilm, and
(5) Detachment or release of cells from the matrix of biofilm to form new colonies on new locations.

During the initial induction phase, the attachment can occur on membranes in as early as 2 h. A logarithmic microbial growth phase occurs after the adhesion and primary colonization of microbes from the initial induction phase. The growth phase is subsequently followed by the nutrient controlled plateau phase where the membrane is covered by biofilm [12]. The plateau phase attains a balance between biofilm growth and cell detachment. Biofilm growth and cell detachment are governed by nutrient concentration, the resultant growth rate, the mechanical stability of the biofilm, and the effective shear force on the biofilm.

Biological substances are unavoidable in any water treatment environment. Even if 99.9% of the bacteria are destroyed in the pre-treatment process [20], those entering the RO system still deposit on the membrane surface and start the formation of a biofilm. Due to the non-porous layer, almost all organic molecules (organic acids, proteins, polysaccharides, etc.) can be retained on the membrane surface during the filtration process. The adhered microbes may utilize these organic compounds as a source of nutrients to multiply further and form more microbial colonies.

13.2.2   Role of Extracellular Polymeric Substances (EPS)

EPS are the metabolites generated during the cell growth process, consisting mainly of polysaccharides, proteins, lipids, humic substances, and DNA [21]. It is reported that EPS accounts for 50–90% of the organic compounds in a biofilm [22]. The EPS encases cells into its polymeric structure and changes the physical-chemical properties (hydrophilicity, zeta potential, surface energy, roughness, etc.) of a membrane surface, which in turn may cause more settlement and deposition of organic contaminants. Accumulated bacteria may be further released from the colony and relocate onto other parts of the membrane surface, starting a new bacterial colony and further spreading the biofilm.

The gradual growth of bacterial colonies within the EPS polymer eventually forms an intact and stable bio-layer across the membrane surface. EPS not only enhances the adhesion of the biofilm but also shields the microorganism from the biocidal components of the cleaning process [23]. Long-term growth of the biofilm can degrade membrane materials and cause irreversible fouling on the RO membrane [13]. It has also been observed that the commonly used disinfectant sodium hypochlorite is only effective against free bacterial cells and exhibited only slight inactivation ability against biofilm capsuled cells [23]. Biofouling not only affects the membrane's lifespan but also adds an energy burden which consequently impedes the widespread application of RO technique [18].

13.2.3   Crucial Factors

The biofouling of reverse osmosis membranes results in the performance degradation of the RO plants. The structure and composition of the biofilm on the membrane
surface have considerable effect on the RO desalting system performance. The crucial factors that need to be considered to better understand the biofilm formations
are: the membrane surface type, the microbial driving force, interactions between
surfaces and microorganisms, and the factors affecting microbial adhesion.

(a) Surface type

The microbial adhesion can occur on two different surfaces through two different
mechanisms, either the pristine membrane surfaces (macroscopic adhesion) or a
surface covered with a conditioned film like protein layer covering the membrane
surface (microscopic adhesion) [24]. Macroscopic adhesion is due to the macroscopic properties of the pristine membranes, such as surface charge, hydrophilicity,
etc., that governs the microbial adhesion on the pristine membrane surface [25, 26].
Microscopic interaction controls the interaction between the conditioning film and
microorganisms [24]. This specific interaction is known as "ligand-receptor bond"
where the receptor is the protein molecules present on the conditioning film of the
membrane [27] and ligand is the substance that binds with the receptor. The interaction of microorganisms with membrane properties, such as surface charge becoming altered by the presence of conditioning films on the membrane surface [28],
often can result in enhanced cell attachment to the surfaces [29].

(b) Driving force

Different types of driving forces (Fig. 13.4), control the microbial adhesion on
the membrane surfaces:

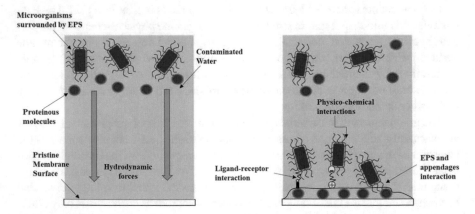

Fig. 13.4 Driving forces that control the microbial adhesion on the membrane surface with (1)
Hydrodynamic force (2) Physico-chemical interactions between the microbes and membrane surface, (3) Ligand-receptor interactions, and (4) Adhesive interaction; Reproduced with permission
from [30]; Copyright 2012 @ Elsevier

- Hydrodynamic force
- Physicochemical interactions between the microbes and membrane surface
- Ligand-receptor interactions
- Adhesive interaction

The hydrodynamic forces of convection and diffusion transports the microbes from the wastewater towards the membrane surface. Once the microorganisms reach the vicinity of the membrane, the physicochemical interactions play a crucial role in microbial adhesion on the membrane surface [25]. Initial adhesion and secondary adhesion are the two steps of microbial adhesion on the RO membrane surface. The initial adhesion occurs due to the physicochemical interactions between the membrane surface and the microbes, whereas the secondary adhesion results from the interaction between the adhered (on the membrane surface) microbes and the suspended (in solution) microbes [31]. The conditioning film layers have binding sites where the ligand-receptor interaction between the receptors of the cell membranes and the binding sites (e.g., polarized bonds, charged groups or OH groups) occurs [32]. Moreover, the adhesive nature of the EPS and the appendages of some microbes cause the adhesive interactions which can facilitate the cell attachment on the membrane surface [32]. The interaction of microbial adhesion on the membrane surface occurs when there is an attraction between microorganisms and the membrane surface (negative total free energy of the interaction exists) [31].

Different physical and chemical factors affect the transport and attachment of microorganisms, such as the mass transport condition, pH and ionic strength of the solution, surface charges, surface hydrophobicity/hydrophilicity, surface roughness, nutrient and EPS concentrations, the amount of the microorganisms, etc.

Mass transport condition affects microorganism growth and build-up on the membrane surface as well as the shear force generation. Enhanced shear force hinders microbe adhesion and restrains microbial growth on the membrane surface, thus reducing biofouling [30]. The electrostatic double layer interaction between the membrane and microorganisms is influenced by solution pH, which has significant impact on the colloids' charge [31]. In addition to solution pH, ionic strength of the solution is another key parameter affecting the electrostatic double layer interaction between the membrane and the microorganisms. As substantial amount of microbes contains negative charges which would repel the microbes from the negatively charged membrane [30]. Hydrophobicity, hydrophilicity, and surface roughness has a profound impact on biofilm formation as they influence the interaction of microbes with the membrane surfaces. In general, hydrophilic membranes interact more with water whereas hydrophobic membranes interact more with microbial matter. Rough surfaces of the membranes contain a greater quantity of sites as well as more surface area available for microbial attachment and adhesion. Moreover, the rough surfaces lead towards a reduction of the van der Waals and electrostatic double layer interactions of the membrane [31]. As the nutrients facilitate the growth of microorganisms, the decreasing nutrient concentrations in the feed stream/influent will hinder biofouling development. Studies have found that enhanced carbon concentrations in the feed cause lower microbial mass as it decreases the time of the initial growth of the microorganisms [31]. Increased amounts of microorganisms play a crucial role in biofouling as the probability of adhesion and growth of microbes is higher along with enhanced EPS concentration [31]. The development of biofilm is also dependent on the redox potential and carbon, nitrogen, and phosphorous (C:N:P) ratio. In addition, the growth rate of microorganisms depends on the following factors [13, 33]:

(1) feed water quality
(2) temperature
(3) pH
(4) dissolved oxygen content
(5) the presence of organic and inorganic nutrients
(6) pollution
(7) depth and location of the intake

13.3 Biofouling Impact on RO Membranes Performance

Biofouling will have many negative impacts on RO membrane system. Biofilm formed on an RO membrane surface can act as an extra thin layer on the membrane that increases the concentration polarization on the membrane and reduces the efficiency of the conventional transport processes (Fig. 13.5) [12, 34].

The adverse consequences of the biofilm formed on membrane surface are as follows [12–14]:

(a) Permeability declines on the RO membrane due to the formation of gel-like biofilm on RO membrane surface.
(b) Reduces the salt rejection and quality of water production due to the accumulation of dissolved ions on RO membrane surface.
(c) Degrades the RO membrane materials and causes irreversible fouling on the RO membrane, and,
(d) Increases energy consumption due to the higher-pressure requirement after the formation of biofilm.

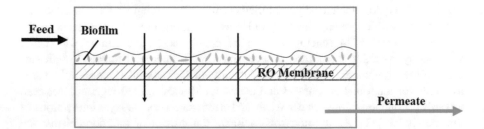

Fig. 13.5 Schematic representation of the fouled RO membrane; Reproduced with permission from [34]; Copyright 1997 @ Elsevier

Although biofouling has many adverse consequences on the RO membranes performance, the main concerns are the permeability decline, reduced salt rejection, and increased energy consumption.

13.3.1 Permeability Decline

Permeability decline is attributed to the formation of biofilm layer, which increases the hydraulic resistance and transmembrane osmotic pressure of the fouled membrane [35]. The decline rate depends on the physicochemical properties of the biofilm and microbiological properties of the feed water [36].

In most cases, permeability decline tends to exhibit two phases. Sharp permeability declines at the initial RO membrane separation stage followed by a smooth decline. The initial sharp permeability decline is attributed to the early deposited bacterial cells, which leads to increased trans-membrane osmotic pressure [35]. The biofilm layer is formed gradually at this initial stage. After that, the formation of biofilm and EPS production will reach a balanced state, which means there is an equilibrium in the loss of and growth of biofilm and EPS at the feed solution-membrane interface. In general, increased pressure is applied to compensate for the permeability decline and maintain constant water production.

In order to elucidate the mechanisms governing the decline in RO membrane performance caused by cell deposition and biofilm growth, a bench-scale investigation of RO biofouling with *Pseudomonas aeruginosa* PA01 was conducted by Herzberg [35]. The contribution of bacterial cells and EPS that impacts the permeability decline of RO membrane was evaluated by comparing the permeability decline of dead cell deposition in different solution mediums to the growth of biofilm on RO membrane (Fig. 13.6).

The decrease in flux (production of clean water) is higher for dead cells in the wastewater medium (ionic strength of 14.6 mM and pH 7.4) when compared with dead cells in deionized water with 0.01 mM $LaCl_3$ at pH 5.8. This rapid decline was attributed to the high ionic strength of the wastewater medium. The sharp permeability declines of PA01 biofilm in the wastewater medium should be attributed to the growth of biofilm and production of EPS. The proposed mechanism of permeability decline caused by growth of biofilm and EPS was further confirmed by Scanning Electron Microscopy (SEM) images of the fouling layer formed from cells and EPS layer (Fig. 13.7b) produced from PA01 cells can be easily observed by comparing with dead cells (Fig. 13.7a).

13.3.2 Salt Rejection Decline

Desalination using the RO membrane process is a pressure-driven transport of water through a membrane medium, which will lead to accumulation of solutes retained by the membrane on the feed side. Biofilm formed during the separation processes

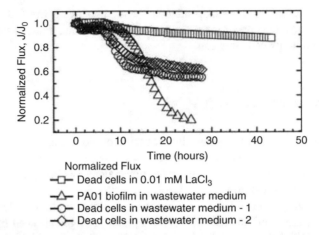

Fig. 13.6 Normalized permeability decline upon deposition of formaldehyde fixed PA01 dead cells, PA01 biofilm growth (initial concentration of 10^7 cells/mL) on the RO membrane in waste-water medium (ionic strength of 14.6 mM and pH 7.4), and PA01 dead cells (initial concentration of 10^9 cells/mL) in wastewater medium (-1 and -2 represent two replicates of fouling experiments conducted with the same synthetic wastewater used in the previous runs); Adapted with permission Ref. [35]; Copyright 2007 @ Elsevier

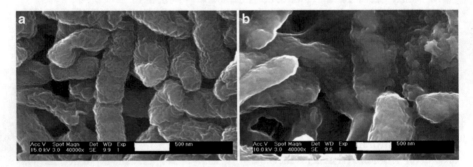

Fig. 13.7 SEM images of PA01 biofouling layers: (**a**) Dead cells fixed in formaldehyde and deposited on the RO membrane in DI water supplemented with 0.01 mM LaCl$_3$ after 38 h of deposition. (**b**) Live cells with their PES (biofilm) growth for 19 h on the RO membrane in a synthetic wastewater medium; Reproduced Adapted with permission from Ref. [35]; Copyright 2007 @ Elsevier

will inevitably increase the trans-membrane pressure (TMP) as well as the concentration polarization (CP) [37]. TMP is the pressure that is needed for the transport of water through the membrane. CP refers to the concentration gradient of solutes at the membrane surface resulted from the accumulation of solutes retained by the membrane, which is one of the most important factors influencing the performance of RO membrane separation processes. Upon the formation of a secondary biofilm membrane on RO membrane surface, the back diffusion of salt ions from membrane

Fig. 13.8 Comparison of
concentration polarization
(CP) in RO membrane
separation process. (**a**)
After membrane fouling;
(**b**) Before membrane
fouling

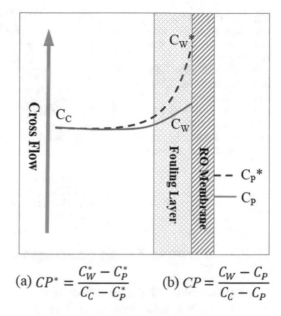

$$(a)\ CP^* = \frac{C_W^* - C_P^*}{C_C - C_P^*} \qquad (b)\ CP = \frac{C_W - C_P}{C_C - C_P}$$

to feed solution is hindered. The concentration of solutes (C_w) at the membrane surface can be significantly elevated (C_w^*) due to the development of biofilm, where $CP_* > CP$ as illustrated in Fig. 13.8 [38]. Therefore, with an increase in CP caused by the formation of fouling layer on the membrane surface, the salt passage through the RO membrane can be significantly increased.

The biofilm consisting of an EPS matrix and bacterial cells increases the TMP, which leads to a decreased salt rejection ability. Their roles in decreased salt rejection were also investigated by Herzberg [35]. A drastic increase in salt passage was observed for two experiments with the deposition of dead cells on the membrane in wastewater medium 1 and 2. This increase indicates that not only the EPS matrix, but also the deposition of dead bacterial cells on the RO membrane surface can decrease salt rejection (Fig. 13.9) [35]. Salt rejection by biofilm is due to the increased CP, whereas dead bacterial cells cause salt rejection due to causing decreased back diffusion of salt ions [27].

Another possible mechanism for salt rejection decline is the biodeterioration from the growth of biofilms. Biofilms formed on RO membrane surface can attack membranes by excreting acids and/or exoenzymes that attack the membrane materials. This process is called "biodeterioration" [39, 40]. Some reports show that cellulose acetate RO membranes can be biodegraded by microorganisms [41, 42]. Reports indicate that common membrane materials, such as polyamide and polyethersulfone, appear to not be attacked by microorganisms [34].

Fig. 13.9 Percent salt passage. PA01 biofilm in wastewater medium (initial cell concentration: 10^7 cells/mL); PA01 dead cells in wastewater medium 1 and 2 (initial cell concentration: 10^9 cells/mL); Adapted with permission from Ref. [35]; Copyright 2007 @ Elsevier

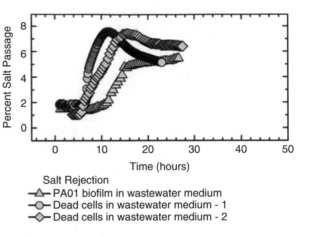

13.3.3 Increased Energy Consumption

Desalination using RO membrane system is a pressure-driven process. This method inevitably leads to the accumulation of bacterial cells, salt ions, and other materials on or within the RO membrane. Once fouling occurs, the membrane permeability decreases. To combat this, an increased applied pressure is required in order to off-set the loss of water production from the development of biofilm [43].

Theoretically, 0.7 kWh/m^3 is the minimum energy required for seawater desalination [44]. In reality, the energy consumption of seawater desalination ranges from 2 to 5 kWh/m^3 with modern materials, modules, and technologies [45, 46]. With the formation of secondary biofilm membrane on RO membrane surface, about 150% of the initial operating pressure (200 psi) is required to compensate the flux loss [34, 44].

13.4 Conclusions and Outlook

RO technique plays an irreplaceable role in seawater desalination industries. However, biofouling caused by bacterial adhesion and propagation on the RO membrane surface hinders the widespread application of RO. Biofouling in RO desalination plant is inevitable as seawater contains substantial amount of organics, nutrients, and microorganisms, especially bacteria, fungi, and yeasts. The contact between the membrane surface and contaminant containing seawater in the RO desalination plants causes the biofouling through adsorption, transport, attachment, growth, multiplication, and detachment of microorganisms that eventually lead towards biofilm formation. Continuous production of EPS facilitates the biofouling of the membrane through enhanced adhesion of the biofilm along with shielding the microbes from the cleaning agents. Biofouling increases the concentration polarization as well as the transmembrane pressure due to the formation of biofilm on the RO membrane surface. Enhanced hydraulic resistance and reduced permeability of RO membranes due to deposition of dead cells, the growth of biofilm, and production of EPS can significantly affect the RO desalination plant performance. Increased concentration polarization results in decreased salt rejection while energy consumption increases due to increased applied pressure to offset the loss of water production. The RO plant performance, as well as the efficiency, degrades due to biofouling which eventually affects the plant expenditures by requiring frequent membrane cleaning and membrane replacement. Advancement in research has adapted different strategies for biofouling mitigation through minimizing microbial concentration, e.g., feed pretreatment, biocide application, etc. and preventing microbial adhesion and/or inactivation of bacteria adhered to membrane surface through development of antibiofouling membrane through surface modification.

Although biofouling of the RO membrane is a huge challenge, RO membrane separation technology is still a promising way for desalination. With the development of diverse fouling-resistance materials and new cleaning procedures, fouling-resistance performance of RO membranes has been improved significantly. A better understanding of the fundamental of the biofouling of RO membrane will contribute towards efficient biofouling management, and hence, will enhance the RO desalination application combating the global water crisis.

References

1. L.F. Greenlee, D.F. Lawler, B.D. Freeman, B. Marrot, P. Moulin, Reverse osmosis desalination: Water sources, technology, and today's challenges. Water Res. **43**, 2317–2348 (2009)
2. P.K. Cornejo, M.V.E. Santana, D.R. Hokanson, J.R. Mihelcic, Q. Zhang, Carbon footprint of water reuse and desalination: A review of greenhouse gas emissions and estimation tools. J. Water Reuse Desalin. **4**, 238–251 (2014)
3. Almar Water Solutions, (2016) Desalination technologies and economics: CAPEX, OPEX & technological game changers to come. Mediterranean Regional Technical Meeting Marseille CMI, December 12–14
4. C. Fritzmann, J. Löwenberg, T. Wintgens, T. Melin, State-of-the-art of reverse osmosis desalination. Desalination **216**, 1–76 (2007)
5. R. Perry, D. Green, J. Maloney, *Perry's Chemical Engineers' Handbook* (McGraw-Hill Companies Inc., New York, 1997)
6. H. Lonsdale, U. Merten, R. Riley, Transport properties of cellulose acetate osmotic membranes. J. Appl. Polym. Sci. **9**, 341–1362 (1965)
7. D.R. Paul, Reformulation of the solution-diffusion theory of reverse osmosis. J. Membr. Sci. **241**, 371–386 (2004)
8. K.P. Lee, T.C. Arnot, D. Mattia, A review of reverse osmosis membrane materials for desalination – development to date and future potential. J. Membr. Sci. **370**, 1–22 (2011)
9. M.E. Mattson, M. Lew, Recent advances in reverse osmosis and electrodialysis membrane desalting technology. Desalination **41**, 1–24 (1982)
10. M. Elimelech, W.A. Phillip, The future of seawater desalination: Energy, technology, and the environment. Science **333**, 712–717 (2011)

11. J. Johnson, Membrane cleaning fundamentals: Cleaning criteria and normalization of reverse osmosis systems. WaterWorld, January (2006), https://www.waterworld.com/municipal/technologies/article/16211727/membrane-cleaning-fundamentals-cleaning-criteria-and-normalization-of-reverse-osmosis-systems

12. A. Matin, Z. Khan, S. Zaidi, M. Boyce, Biofouling in reverse osmosis membranes for seawater desalination: Phenomena and prevention. Desalination **281**, 1–16 (2011)

13. H. Maddah, A. Chogle, Biofouling in reverse osmosis: Phenomena, monitoring, controlling and remediation. Appl. Water Sci. **7**, 2637–2651 (2017)

14. T. Nguyen, F. Roddick, L. Fan, Biofouling of water treatment membranes: A review of the underlying causes, monitoring techniques and control measures. Membranes **2**, 804–840 (2012)

15. S.R. Pandey, V. Jegatheesan, K. Baskaran, L. Shu, Fouling in reverse osmosis (RO) membrane in water recovery from secondary effluent: A review. Rev. Environ. Sci. Bio. **11**, 125–145 (2012)

16. B.A. Qureshi, S.M. Zubair, A.K. Sheikh, A. Bhujle, S. Dubowsky, Design and performance evaluation of reverse osmosis desalination systems: An emphasis on fouling modeling. Appl. Therm. Eng.Appl. Therm. Eng. **60**, 208–217 (2013)

17. J. Baker, L. Dudley, Biofouling in membrane systems: A review. Desalination **118**, 81–89 (1998)

18. O. Rendueles, J.-M. Ghigo, Multi-species biofilms: How to avoid unfriendly neighbors. FEMS Microbiol. Rev. **36**, 972–989 (2012)

19. D. Monroe, Looking for chinks in the armor of bacterial biofilms. PLoS Biol. **5**, e307 (2007)

20. H.-C. Flemming, G. Schaule, T. Griebe, J. Schmitt, A. Tamachkiarowa, Biofouling – The Achilles heel of membrane processes. Desalination **113**, 215–225 (1997)

21. M. Herzberg, S. Kang, M. Elimelech, Role of extracellular polymeric substances (EPS) in biofouling of reverse osmosis membranes. Environ. Sci. Technol. **43**, 4393–4398 (2009)

22. A. Karimi, D. Karig, A. Kumar, A. Ardekani, Interplay of physical mechanisms and biofilm processes: Review of microfluidic methods. Lab Chip **15**, 23–42 (2015)

23. J. Mansouri, S. Harrisson, V. Chen, Strategies for controlling biofouling in membrane filtration systems: Challenges and opportunities. J. Mater. Chem. **20**, 4567–4586 (2010)

24. H.J. Busscher, W. Norde, P.K. Sharma, H.C. Van der Mei, Interfacial re-arrangement in initial microbial adhesion to surfaces. Curr. Opin. Colloid Interface Sci. **15**, 510–517 (2010)

25. H.J. Busscher, A.H. Weerkamp, Specific and non-specific interactions in bacterial adhesion to solid substrata. FEMS Microbiol. Rev. **46**, 165–173 (1987)

26. M. Hermansson, The DLVO theory in microbial adhesion. Colloids Surf. B. Biointerfaces **14**, 105–119 (1999)

27. P. Cuatrecasas, Membrane receptors. Annu. Rev. Biochem. **43**, 169–214 (1974)

28. R. Neihof, G. Loeb, Dissolved organic-matter in seawater and electric charge of immersed surfaces. J. Mar. Res. **32**, 5–12 (1974)

29. A.-C. Olofsson, M. Hermansson, H. Elwing, N-acetyl-L-cysteine affects growth, extracellular polysaccharide production, and bacterial biofilm formation on solid surfaces. Appl. Environ. Microbiol. **69**, 4814–4822 (2003)

30. R.A. Al-Juboori, T. Yusaf, Biofouling in RO system: Mechanisms, monitoring and controlling. Desalination **302**, 1–23 (2012)

31. J.A. Brant, A.E. Childress, Assessing short-range membrane–colloid interactions using surface energetics. J. Membr. Sci. **203**, 257–273 (2002)

32. J.T. Staley, Growth rates of algae determined in situ using an immersed microscope. J. Phycol. **7**(1), 13–17 (1971)

33. M.O. Saeed, A. Jamaluddin, I. Tisan, D. Lawrence, M. Al-Amri, K. Chida, Biofouling in a seawater reverse osmosis plant on the Red Sea coast, Saudi Arabia. Desalination **128**, 177–190 (2000)

34. H.-C. Flemming, Reverse osmosis membrane biofouling. Exp. Thermal Fluid Sci. **14**, 382–391 (1997)

35. M. Herzberg, M. Elimelech, Biofouling of reverse osmosis membranes: Role of biofilm-enhanced osmotic pressure. J. Membr. Sci. **295**, 11–20 (2007)
36. P. Goh, W. Lau, M. Othman, A. Ismail, Membrane fouling in desalination and its mitigation strategies. Desalination **425**, 130–155 (2018)
37. S. Kim, E.M.V. Hoek, Modeling concentration polarization in reverse osmosis processes. Desalination **186**, 111–128 (2005)
38. T.H. Chong, F.S. Wong, A.G. Fane, Enhanced concentration polarization by unstirred fouling layers in reverse osmosis: Detection by sodium chloride tracer response technique. J. Membr. Sci. **287**, 198–210 (2007)
39. A.H. Rose, History and scientific basis of microbial biodeterioration of materials. Econ. Microbiol. **6**, 1–18 (1981)
40. H.-C. Flemming, G. Schaule, R. McDonogh, H.F. Ridgway, Effects and extent of biofilm accumulation in membrane systems, in *Biofouling and Biocorrosion in Industrial Water Systems*, ed. by G. G. Geesey, Z. Lewandoski, H. C. Flemming, (Lewish Publishers, CRC Press, Inc, 1994), pp. 63–89
41. A.P. Murphy, C.D. Moody, R.L. Riley, S.W. Lin, B. Murugaverl, P. Rusin, Microbiological damage of cellulose acetate RO membranes. J. Membr. Sci. **193**, 111–121 (2001)
42. W. Luo, M. Xie, F.I. Hai, W.E. Price, L.D. Nghiem, Biodegradation of cellulose triacetate and polyamide forward osmosis membranes in an activated sludge bioreactor: Observations and implications. J. Membr. Sci. **510**, 284–292 (2016)
43. L. Malaeb, G.M. Ayoub, Reverse osmosis technology for water treatment: State of the art review. Desalination **267**, 1–8 (2011)
44. M. Schiffler, Perspectives and challenges for desalination in the 21st century. Desalination **165**, 1–9 (2004)
45. S.S. Shenvi, A.M. Isloor, A. Ismail, A review on RO membrane technology: Developments and challenges. Desalination **368**, 10–26 (2015)
46. S.F. Anis, R. Hashaikeh, N. Hilal, Reverse osmosis pretreatment technologies and future trends: A comprehensive review. Desalination **452**, 159–195 (2019)

Chapter 14
Approaches Towards Scale Control in Desalination

Ashish Kapoor and Sivaraman Prabhakar

14.1 Introduction

Scaling is a process which consists in the formation of a solid layer on a solid substrate. It may result due to several factors including solution chemistry, temperature, flow conditions and surface properties. Scaling is an important phenomenon encountered in the operation of desalination plants leading to reduction in water productivity and life of the desalination systems. It depends on the type of dissolved species, their concentration, surface characteristics of the solid medium, temperature and the hydrodynamics of the fluid [1].

Formation of scales is an inherent consequence of a desalination process. The extraction of pure water increases the concentration of the reject stream which, in turn, facilitates scale formation [2]. In thermal desalination processes, the scaling, particularly on the heat transfer surfaces leads to a decrease in the heat transfer efficiency. The presence of inorganic species which are constituents of sparingly soluble salts such as calcium, magnesium, iron, and silica along with counter-ionic species such as carbonates, and sulphates lead to scaling. With thermal conductivities of calcium sulfate ($CaSO_4$), calcium carbonate ($CaCO_3$) or silica being significantly less compared to heat transfer tubing materials including ferrous and non-ferrous alloys, the heat transfer efficiencies are very much reduced [3]. In reverse osmosis membrane process, scaling leads to a reduction in permeate water flux as well as deterioration of product water quality.

Prevention of scale in totality is extremely difficult but scale control is necessary to minimize the scaling rate and ensure sustainable operation of the desalination plants [4]. Preventive methods adopted for scale control such as acid treatment, ion

A. Kapoor · S. Prabhakar (✉)
Department of Chemical Engineering, SRM Institute of Science and Technology,
Kattankulathur, TN, India
e-mail: prabhaks@srmist.edu.in

© Springer Nature Switzerland AG 2020
V. S. Saji et al. (eds.), *Corrosion and Fouling Control in Desalination Industry*,
https://doi.org/10.1007/978-3-030-34284-5_14

exchange, and addition of anti-scalants are reasonably good and requires good monitoring of feed water quality. Curative methods such as cleaning of the membrane surfaces or heat transfer tubes require significant downtime (involving production loss). In practice, a combination of both the methods is normally adopted to ensure sustainability at reasonable economics [5–8] .

This chapter describes various methods of scale control particularly with reference to reverse osmosis and thermal desalination processes including physical, chemical and physico-chemical methods. With a background on the mechanism of scale formation, the invasive methods discussed include acid pretreatment, ion exchange, feed flow reversal, membrane separation using nano-filtration (NF) and addition of anti-scalants. Other methods under investigation such as the use of ultrasonic energy and application of magnetic and electrical fields for scale control have also been discussed. Development of non-scaling membranes incorporating a variety of nanoparticles such as rutile and carbon nanotubes have been highlighted. A brief discussion on scaling phenomenon in non-conventional desalination processes, such as forward osmosis (FO) and membrane distillation (MD) is included indicating the possible methods of prevention.

14.2   Approaches towards Scale Control

Theoretically, there are many approaches for scale control including the removal of the scale forming species or by binding them with a complex. Other methods involve operating the desalination plants within the limits of scaling or use of materials which do not allow adhesion of scales on its surface.

The factors which determine the scaling potential in desalination plants are feed water composition, the characteristics of solute species such as solubility, solution characteristics (ionic strength, temperature) and surface characteristics of the membrane or heat transfer surface where scaling is likely to occur (wall temperature, boundary layer concentration, presence of crystal growth sites) and charge on the crystal seeds.

The major scales formed in thermal and membrane desalination processes differ because of the difference in their operational characteristics: reverse osmosis operates at ambient temperature while thermal desalination operates over a range of temperatures. Multi-stage flash (MSF) desalination plants operate below 90 °C but some operate at higher temperatures up to 120 °C. Multi-effect distillation (MED) plants operate at much lower temperatures around 65 °C. Other than minor components like silicate scale, $CaSO_4$ and $CaCO_3$ are the two dominant solute species responsible for scale formation [9].

Preventive approaches include all the methods by which either the scale forming species are removed (by ion exchange or chemical precipitation) or prevented by modifying the scale formation steps (by addition of anti-scalants) [10]. Pretreatment with acid to remove carbon dioxide is one of the widely adopted methods in desalination processes, in the early stages of development. Thermal desalination processes tend to operate at lower temperatures to minimize hard scale. In this context, the potential of nanofiltration (NF) as a scale prevention technique has been established [11] and that may lead to further improvements.

Scaling is essentially a consequence of concentration build-up on the surface or more precisely the boundary layer concentration. The other steps such as super saturation-seeding-crystal growth etc. follow. In general, we can estimate bulk concentration at any point in the stream, but the boundary layer or wall concentration is at best a considered guess. In thermal desalination, the scaling depends on wall temperature of the heat transfer system and related flow dynamics. In the case of reverse osmosis, it is the specific flux and rejection characteristics of the membrane coupled with the hydrodynamics of the feed which govern the concentration build up. Prevention of the scale requires attacking any of the different steps leading to scale formation, even though the ideal approach would be to remove the species responsible for scaling. The next possible approach could be to prevent the concentration build up at the surface. Preventive methods result in long service cycle of the desalination plant, but it has challenges in terms of chemistry and economics, as the undesirable species have to be removed or converted to non-scaling form.

The pragmatic approach would be to remove a component of the scale forming species either the anionic (sulphates and carbonates) or cationic (calcium) from the feed by physical or chemical methods. Chemical and 'membrane-based' treatments belong to this category. In chemical treatment, carbonates are decomposed, while in the membrane treatment using nanofiltration, cations such as calcium and magnesium are removed along with bivalent anions such as sulphates. The third approach is to engage the scale-forming species with some binding agent so that the elemental species are not available to form the scales or alternately the saturation values are enhanced by a few orders of magnitude. Thus, the preventive approaches can be broadly classified as

(i) Chemical treatment

- Acid treatment
- Chemical precipitation
- Addition of polymeric anti-scalants

(ii) Physical Methods

- Periodic pressurized air backwash
- Pellet softening
- Feed flow reversal

(iii) Physico-chemical methods

- Use of membrane systems in pretreatment
- Removal of chemical species using ion exchange

The preventive methods primarily interfere with or modify any of the following steps:

- The removal of mineral ions from the feed
- Prevent the sparingly soluble salts from reaching saturation, by preferentially binding the cation or anion

- Inhibit crystal formation
- Modification of the crystal to inhibit the growth
- Prevention of adherence of the crystal to the surface

Figure 14.1 illustrates the schematic representation of pathways of scale formation and the route that follows, after addition of anti-scalant [12].

14.2.1 Chemical Treatment

14.2.1.1 Acid Treatment

Acid treatment of seawater can be done for both thermal and membrane processes in the pretreatment section. The addition of acid decomposes the carbonates releasing CO_2 to the environment, consequently minimizing the chances of carbonate scale formation [13]. pH adjustment is also one of the strategies for the prevention of silica-scales considering that both silica solubility and its polymerization kinetics are highly dependent on pH.

In the case of thermal desalination plants after addition of acid, the feed is subjected to de-aeration using ejectors lest the CO_2 present may not only reduce the efficiency of heat transfer but also promote carbonate scaling as well. Hence, acidification is to be followed by degasification to prevent carbonate scaling. Further, after acidification, the feed requires to be neutralized to minimize corrosion.

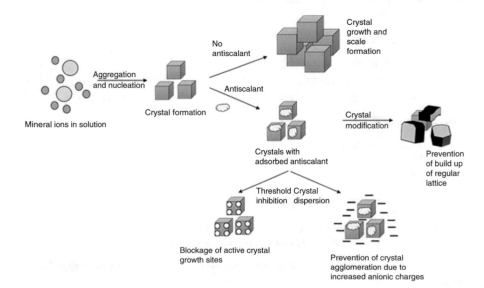

Fig. 14.1 Pathway of scaling and approaches to prevent scaling. Modified after Ref. [12]

This treatment would only help in minimizing or nearly preventing carbonate scaling in the desalination plant but is of no help in the case of magnesium hydroxide and $CaSO_4$ scaling.

In reverse osmosis process, acidification was carried out in the earlier days particularly in the beginning when cellulose-acetate based membranes were used with a two-fold purpose viz. decomposition of carbonates and maintaining a conducive pH to sustain membrane performance. No doubt the formation of carbonate scale on the membrane is prevented but the generated CO_2 would pass through the membranes making the permeate acidic, requiring pH correction before distribution.

14.2.1.2 Addition of Anti-scalants

In the last few years, the use of very good anti-scalants with acidic or non-acidic properties have encouraged designers to go ahead without acid dosing and operate the plants at lower temperatures, as it results in savings for thermal desalination plants with reference to degassing and neutralization. In the case of the reverse osmosis process, plant operation is carried on without acid addition but with anti-scalant dosing [14, 15].

The primary role of an anti-scalant is to hinder, if not to prevent the formation of scales. There are several mechanisms proposed to explain the way anti-scalants could act [12, 16]. Since calcium is identified as the main component of scaling, the anti-scalants aim to bind some how the calcium, thus preventing them from forming the scaling compound and reach saturation levels.

The prevention can be achieved either by not allowing to reach the saturation concentration or by inhibition of the formation of crystals bigger than the critical size necessary for nucleation or by the modification of the surface of the crystals. The surface modification results in distortion of crystals as they grow. The distortion slows down and often stops the growth of usually highly-ordered crystals. Many anti-scalants are available for field applications and selection of an apt anti-scalant depends on a variety of factors including their chemical nature, properties, and system under consideration.

Categories of Anti-scalants Anti-scalants can be broadly categorized into two groups: conventional anti-scalants (the majority being phosphorus-based) and unconventional greener anti-scalants. They are essentially chemical additives capable of controlling either the formation or deposition of scale or both the steps. Further, their capabilities go beyond those of acid treatment. These additives are effective not only for $CaSO_4$ scales but also for alkaline $CaCO_3$/ $Mg(OH)_2$ scales. The efficiency of scale inhibition relies on various parameters of the system of interest including ion concentration, temperature, pressure, pH along with characteristics of anti-scalants such as functional groups, molecular weight, polydispersity, molecular configuration, etc. Further, the type of anti-scalant and its dosage is also crucial. While the dosing rate less than optimal leads to scale formation, an overdose results in sludge formation and contributes to fouling [17]. Conventionally used anti-scalants come under the following three families:

(i)   Polyphosphates (hexametaphosphate (HMP), tripolyphosphate (TPP), etc.)
(ii) Organophosphates (aminotris (methylenephosphonic acid), ATMP; 1-hydroxyethane-1,1- bis(phosphonic acid), HEDP; 2-phosphonobutane 1,2,4-tricarboxylic acid (PBTC), etc.)
(iii) Organic polyelectrolytes (polyacrylates (PA); polycarboxysulfonates).

Several studies have focused on the development of novel anti-scalants and comparing their performance with commercial scale inhibitors [18]. The effectiveness of a few poly(acrylic acid)(PAA)-based scale inhibitors with hydrophobic end-groups and three different commercial scale inhibitors (Belgard EV 2030, Albrivap DSB(M) and Sokalan PM 10i were evaluated in real seawater samples [17]. PAA with mid-length hydrophobic end groups (hexyl isobutyrate-PAA, Mn = 1400, HIB-PAA; and cyclohexyl isobutyrate-PAA, Mn = 1700, CIB-PAA), low molar mass PAA with a long end group (hexadecyl isobutyrate-PAA Mn = 1700, HDIB-PAA) and PAA with mid-length hydrophobic end groups (hexyl isobutyrate-PAA, Mn = 3600) were studied to assess their anti-scaling performance. The studies indicated that low molar mass (molecular weight) PAA with end groups of moderate hydrophobicity are most effective in the role of scale inhibitor. The nature of end group is likely to play an essential role in directing selectivity of adsorption on crystals being formed. The moderately hydrophobic groups promote adsorption of PAA on the surface thus inhibiting growth of crystals. The less hydrophobic groups lack selectivity of adsorption while more hydrophobic groups tend to self-assemble.

Polyamino polyether methylene phosphonate is found to simultaneously control $CaCO_3$ and calcium sulfate scale formation and deposition at very high supersaturations. It functions by both precipitation inhibition and dispersion of precipitated material. Tripol 8510 (Trisep, USA) is a commercial anti-scalant based on polyacrylic acid and diphosphonic acid components that is shown to form precipitates with calcium/aluminum and iron ions by altering the characteristics of foulants in a hybrid coagulation-nanofiltration membrane process for purification of brackish water [19]. The polyacrylic acids and polyphosphates play a role in sequestering ions and dispersing particles. However, the excessive dosage is observed to lead to the formation of agglomerates that can deposit on membrane surfaces contributing to fouling. Gryta [20] studied the inhibition of $CaCO_3$ precipitation by polyphosphate scale inhibitors. The combined effect of high water temperature, anti-scalants concentration and duration of membrane-based process was examined. Studies have been reported on the synthesis of novel anti-scalants containing specific functional groups, such as diallylammoniopropanephosphonate-alt-(sulfur dioxide) copolymer [21], and poly amino-phosphonate [22].

One of the concerns with phosphorus-based scale-inhibitors is their limited biodegradability. The continued accumulation of phosphorus is likely to contribute to eutrophication and consequent blooming of harmful algae. In view of the growing concerns on environmental issues, there is a considerable interest in developing 'greener anti-scalants' that are biodegradable and have a minimal environmental impact [23, 24].

A relatively newer class of anti-scalants is based on chemicals such as polymale-ates (PMA), polyaspartates (PAS), and polyepoxysuccinates (PESA), as well as their various derivatives including copolymers with PA. A comparative study is reported for four phosphorus-free polymers (PAS, PESA, polyacrylic acid sodium salt (PAAS) and copolymer of maleic and acrylic acid (MA-AA) and of three phos-phonates (ATMP, HEDP and PBTC) regarding inhibition of $CaSO_4$ precipitation. Although a detailed mechanism of antiscaling behavior of the novel candidates is still to be understood, initial studies have indicated the inhibition in the formation of crystals (to allow the crystal growth) as a contributing factor to scale inhibition [25]. Polyaspartic acid (PASP), a polyamino acid and its derivatives do not contain phos-phorus and offer viable alternatives for development of biodegradable, eco-friendly anti-scalants. The functional group chemistry plays a role in scale inhibition effect of PASP. The ionization of PASP and the consequent formation of negatively charged molecular chains leading to water-soluble complexes with calcium ions appears to be the reason behind non-deposition scales on the membrane surface [26]. Phosphorus-free inhibitors, such as the sodium salt of PASP, MA/AA, PESA, and PAAS, have been shown to be effective for $CaCO_3$ scaling in alkaline conditions [27]. Tlili et al. [28] investigated the effect of different concentrations of sodium poly(acrylate) on prevention of gypsum scale formation and found it to have an impact on the precipitation rate, the texture and the morphology of gypsum.

Ketrane et al. [29] studied comparative performance of commercial anti-scalants (three polyphosphates, one polyphosphonate and one polycarboxylic acid) for $CaCO_3$ precipitation from hard water. Phosphonates were found to be superior in terms of scale inhibition when compared to polycarboxylates or polyphosphates. Zhao et al. [30] prepared environmental friendly and low-cost poly (citric acid) anti-scalant by condensation polymerization of citric acid and demonstrated its efficacy in $CaSO_4$ scaling inhibition. Poly (citric acid) interacts with active sites on surface of growing $CaSO_4$ scale crystals and distort crystal polymorphs thus checking their growth. Phosphate free polysuccinimide (PSI) derived anti-scalants were also used for $CaSO_4$ scaling [31].

A combination of more than one anti-scalants has been used to synergise the individual characteristics of the different anti-scalants. Shen et al. [32] explored the strategy of optimization of the combination of three anti-scalants for $CaCO_3$ scal-ing. The mixture of HEDP, and PAA and synthesized hydrolyzed poly maleic anhy-dride (HPMA) was optimized by using Simplex Lattice of Design-Expert software through $CaCO_3$ precipitation method. HEDP is known to modify the structure of $CaCO_3$ by incorporating into the crystals and thus inhibiting the scale formation. HPMA is shown to influence the growth of $CaCO_3$ crystals resulting from carboxyl-ate ions adsorption on the nuclei of $CaCO_3$. PAA is widely studied regarding its role in distortion and inhibition of crystal growth. The statistical optimization yielded the optimum mass ratio of HEDP, PAA and synthesized HPMA to be 10/10/80, that showed excellent $CaCO_3$ deposition inhibiting performance.

Further, in the quest for eco-friendly anti-scalants, a significant amount of interest has been seen in the development of scale inhibitors based on natural sources and their modified or derived forms [23, 33]. The compounds obtained from petrochemical origin, natural organic molecules, plant extracts, and modified natural molecules are being studied to examine their effectiveness in scale inhibition performance. Carboxymethyl inulin branded under the name *Carboxyline CMI* is an eco-friendly anti-scalant that performs three essential functions, namely complexing of metal ions, crystal growth inhibition, and dispersancy [34]. *Herniaria glabra* aqueous extract, used in the treatment of urolithiasis, is evaluated as a potential anti-scalant by studying inhibiting effects towards $CaCO_3$ scaling using by chronoamperometry and fast controlled precipitation methods [35]. Abd-El-Khalek et al. demonstrated use of palm leaves extract (*Phoenix dactylifer L*) for $CaCO_3$ scaling in cooling water [36]. Palm leaves extract were shown to decrease the rate of scale formation by chemical bond formation of cations from neutral bioextract with active components resulting in formation of soluble complex or by improving dispersion of suspended solid matter via sorption. Maher et al. [12] synthesized and demonstrated an eco-friendly anti-scalant chitosan biguanidine hydrochloride, for inhibition of the precipitation of calcium sulfate and carbonate onto the membrane surface. Chitosan biguanidine hydrochloride showed a good performance as scale inhibitor for $CaSO_4$ and $CaCO_3$ at about 10 and 15 mg/L. More details on anti-scalants can be found in Chap. 15.

Use of anti-scalants reduces scaling risk by various mechanisms as discussed. However, often anti-scalants used are costly and become less effective at high concentrations. Further, they are themselves found to contribute to fouling. Hence, novel scaling prevention approaches are being explored.

14.2.1.3 Chemical Precipitation

Chemical precipitation is cost intensive unless the precipitate is a value product. However, some studies have reported the removal of sulphate ions using barium chloride by precipitating as barium sulphate. Cob et al. [37] studied the removal of silica by chemical precipitation. Among the several methods examined including precipitation of silica with $Fe(OH)_3$, $Al(OH)_3$ and silica gel, $Al(OH)_3$ was found to be the most effective precipitant for silica removal. Considering the efforts and cost involved, chemical precipitation for the removal scale forming ionic species with reference to desalination plants is only of academic interest.

14.2.2   Physical Methods of Scale Prevention

14.2.2.1   Periodic Pressurized Air Backwash

Periodic air backwash refers to the process by which compressed air is passed through the membrane for dislodging the scales. The backwash air pressure required depends on the membrane and its pore-size. Julian et al. [38] investigated pressurized air backwash as a means of shear force enhancement on the membrane surface for the retardation of scaling in submerged vacuum membrane distillation and crystallization (VMDC) process for inland brine water treatment. Several parameters such as pressure for air-backwash, its duration, frequency, and the addition of scouring were examined to observe their influence on productivity. However, it is more difficult to practice with reference to reverse osmosis membranes and may lead to damage of the membrane.

14.2.2.2   Pellet Softening (PS)

Pellet softening aims at the removal of calcium salts as carbonates. It is carried out in a small reactor filled with fine grains of sand/calcite and conditioned to basic pH. The feed is allowed to contact from the bottom of the reactor to enable fluidization. $CaCO_3$ crystals formed under these conditions adhere to the surface of the sand/calcite crystals and grow in size and mass. Beyond a particular mass, the crystals settle down, and are removed and replaced with fresh particles of sand/calcite. The discarded materials can be utilized in cement industries. Al-Ghamdi [39] has studied the feasibility of calcite particles as a control method to avoid scaling of $CaCO_3$ in seawater desalination. It was found to be effective in reducing the scaling potential and supersaturation level of seawater. The method is not suitable for $CaSO_4$ removal at least under ambient temperature conditions because of its relatively higher solubility and requirement of large induction time.

14.2.2.3   Feed Flow Reversal (FFR)

Calcium sulphate is a major scaling component in seawater desalination processes. The induction time required for the precipitation of calcium sulphate is significantly high providing time for the removal of salts before deposition. Accordingly, in this technique demonstrated by Pomerantz et al. [40], the feed flow direction is reversed before the induction time required for $CaSO_4$, allowing for the replacement of supersaturated brine solution at the exit with the unsaturated feed, effectively 'zeroing' the nucleation time. The flow reversal was achieved by altering the location of entry and exit of the pressurized feed to the RO system. Gu et al. [41] examined the FFR with a spiral-wound RO plant in a cyclic mode and demonstrated

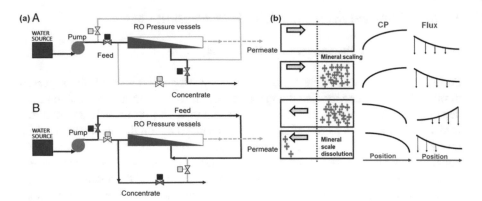

Fig. 14.2 Feed flow reversal approach for RO scale control (**a**) shows schematic of operation in (A) normal feed flow mode, and (B) feed flow reversal. (**b**) illustrates concept of feed flow reversal operation. FFR gets triggered via mineral scale detection using an external membrane monitor. Reproduced with permission from Ref. [41]; Copyright 2013 @ Elsevier

scale-free operation for brackish water desalination under conditions of high mineral scaling ($CaSO_4$) in an automated system (Fig. 14.2). Section A indicates the service cycle where water flux decreases consequent to the increase of concentration polarization. Section B illustrates the reversal wherein feed enters from the concentrate side and emerges out of the entry point thus flushing out the saturated solution.

14.2.3 Physico-Chemical Methods of Scale Prevention

14.2.3.1 Removal of Chemical Species Using Ion Exchange

Ion Exchange is another mechanism to prevent scale formation [42]. One can use either softeners to replace the calcium and magnesium ions with sodium or anionic resins to remove sulphates. Cation exchange resins are not employed in the hydrogen form as it would make the water acidic and acid would be required for regeneration of the resins. Use of ion exchange or softeners in the pretreatment step is a possibility whereby the calcium and magnesium ions are replaced by sodium ions in the solution. Since one of the scale-forming species is nearly removed without pH change, the desalination plants may not have to struggle with the scaling threat.

Zhu et al. [43] employed a weak-base anion exchange resin (Relite MG 1/P) to preferentially remove sulfate from seawater thus inhibiting formation of $CaSO_4$ scales. The resin showed high removal of sulfate and could be regenerated for successive cycles of usage. Strongly basic anion exchange resins have been used for the removal of silica [37, 44]. It is impractical on most brackish waters, because of their high level of hardness and requirement of relatively large quantities of good quality

NaCl for regeneration [45, 46]. The technical feasibility of using synthetic resins directly for seawater applications is challenging because of the presence of a large number of ionic species (contains on an average about 1200–1600 ppm of magnesium and about 400 ppm of calcium besides about 2800-3400 ppm of sulphate with a total dissolved solids content of around 35,000 -45,000 ppm), which would result in large quantities of regenerant waste. Besides, the resins may undergo plasmolysis.

Interestingly, Pless et al. [47] have studied the desalination of brackish water using hydrotalcite (HTC) as anion-exchange material and amorphous aluminosilicate as cation-exchange material. These inorganic materials may not suffer plasmolysis at higher concentration of solute but they do not preferentially remove bivalent species. Sasan et al. [25], have studied the preferential removal of silica from industrial water using two different forms (calcined and un-calcined) from aqueous solutions (Fig. 14.3) [48].

14.2.3.2   Membrane Pretreatment

Micro-filtration (MF) and ultra-filtration (UF) membranes have been extensively used for the pre-treatment of the feed for desalination through RO. Even though they are effective in the reduction of fouling, they are not effective in minimizing scaling threat as they cannot remove the scaling species, which are initially in the dissolved state.

NF with its capability for removing multivalent species preferentially appears to have a promising potential. The investigations initially started at Saline Water Conversion Corporation (SWCC) for the deployment of NF as a pretreatment for RO. Hassan et al. [49] demonstrated that NF based pretreatment step for seawater feed stream helped in the removal of very fine turbidity, residual bacteria, scale forming hardness ions and lowered the total dissolved solid (TDS) content. Initial studies reported production of high-quality water at higher fluxes thereby indicating the absence or at least minimization of scaling challenges. Later, Sofi et al. [50] reported the use of NF for water softening as a feed pretreatment step prior to sea water reverse osmosis (SWRO) as well MSF. Llenas et al. [11] investigated six

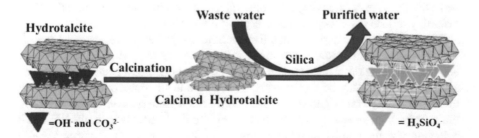

Fig. 14.3 Use of inorganic anion exchange for removal of dissolved silica. Reproduced with permission from Ref. [48]; Copyright 2017@ Elsevier

different NF membranes and analyzed their efficacy in rejecting ionic species responsible for forming scales. The advantage of using NF before RO is two-fold: namely reduction in the operating pressure of RO system and the improvement in the recovery as NF is not only effective in removing significant amounts of bivalents such as calcium, magnesium, sulphates etc., but also some amounts of NaCl etc. Unlike RO membranes, NF membranes have marginal surface charge, which hinders scale formation. Since NF rejects monovalent species as well, permeate of NF will have less concentration of solute species than what could be expected only by the removal of multi-valent species. Consequently, the down-stream RO can operate at low pressures and high recoveries. With these twin advantages, NF as a pretreatment appears to be quite attractive [51, 52] for reverse osmosis. Studies on the performance of seawater desalination using a series-combination of NF and RO membranes have been found to be encouraging with reference to recovery, purity and life [53, 54]. The dual NF-SWRO desalination process in Umm Lujj plant, Saudi Arabia exhibited increased permeate flow significantly from 91.8 to 130 m^3/h compared to a single SWRO desalination process [55]. Since then, many studies have reported the operational advantages of NF as a pretreatment to SWRO [56–58].

The main objective of the membrane-based scale prevention treatment in thermal desalination plants is the potential to operate at higher top brine temperature (TBT) to improve thermal efficiency and to reduce the overall water production cost. The feed requirement is governed more by hydro-dynamic conditions (in tune with heat transfer design requirement) and cooling water. Hence, very high recoveries in thermal plants are difficult to achieve, unlike RO. Besides, additional energy may have to be expended for NF treatment. Thus, NF as a pretreatment for a thermal desalination plant has its limitations. Neither it contributes significantly towards operation in terms of energy consumption nor has influence on product quality. Unlike reverse osmosis, in thermal desalination plants, the product quality is not dependent on feed salinity. Significant advantage can accrue on account of long service time with less maintenance due to reduced scaling threat. However, the additional investment on capital and the operational cost due to increased power consumption for NF may not be attractive to warrant adoption even though it is technically very advantageous. A few studies on use of NF as a pretreatment for thermal desalination plants were reported claiming that high percentage of distillate could be recovered based on simulated laboratory experiments [59]. Unfortunately, NF in thermal processes serve the purpose of removing only multivalent species but cannot contribute much towards productivity except reducing scaling related losses.

The experience so far indicates that technically NF as a pretreatment step is operationally advantageous; however, it is necessary to look at the economics. In NF-RO combination, a second step pressurization is necessary and the extent of energy recovered may be less as both the membrane streams operate at relatively low pressures and fairly good recoveries. Since the boundary layer concentration is likely to be high and the gestation period required for $CaSO_4$ is high, the life-cycle behavior of scaling on NF membrane surface needs to be assessed. Perhaps it would warrant use of anti-scalants. With the development of anti-fouling membranes and good anti-scalants, the economic challenges need to be convincingly addressed.

14.3   Recent Approaches

Recent developments of scale control are essentially preventive approaches. New membranes have been developed with surfaces which do not allow adhesion of the solute species. Alternately, the incoming feed is subjected to physical treatments with the application of electrochemical or magnetic energy so that ionic species are not allowed to form the salt species a step prior to crystallization.

The field of nanotechnology offers unique solutions that can overcome the limitations of conventional scaling control methods. Inclusion of nanomaterials in thin film composite (TFC) membranes has emerged as an alternative technique for improving anti-scaling characteristics of membranes. Nanomaterials such as zeolite, silica, multiwalled carbon nanotubes (MWCNTs) and titanium dioxide (TiO_2) nanoparticles have been effectively demonstrated for enhancement of anti-scaling properties of polyamide (PA) membranes [60–62]. The incorporated nanoparticles provide an unchecked pathway for transport of water species and alter the membrane structural configuration resulting in enhanced water permeability and anti-scaling characteristics. In addition to TiO_2, TFC membranes incorporated with zeolite nanoparticles have also been explored by several researchers owing to unique properties of zeolites such as molecular sieving and competitive adsorption and diffusion. Use of zeolite membranes is an attractive option for desalination due to cation exchange behavior in which the dissolved cations can be easily removed from water by exchanging with cations on the exchangeable sites of the zeolite membranes [63]. Graphene oxide incorporated membranes and nanocomposite membranes have shown promise to exhibit better flux and rejection and a bit longer life [64, 65]. Their scale prevention properties over a longer time period are under assessment. Developing scale resistant membranes is one direction that could yield some relief to scaling phenomenon in reverse osmosis.

Duan et.al [66] have reported an electrochemical method for the prevention and removal of mineral scales such as $CaSO_4$ and $CaCO_3$ using an electrically conducting CNT-PA RO membrane. They further inferred that a continuous application of electrical potential (2.5 V) to the membrane surface resulted in pushing $CaSO_4$ crystal formation away from the membrane surface, allowing the dispersal of the formed crystals. The schematic of the principle is illustrated in Fig. 14.4. In the case of near neutral membrane, consequent to preferential separation of water, the solute species both calcium and sulphate ions accumulate on the surface enabling the precipitation. On the other hand, when the membrane is electrically charged, a layer of counter ions (SO_4^{--}) accumulate near the surface while the Ca^{2+} are away from the surface. Any possibility of formation of $CaSO_4$ can occur only away from the surface.

In thermal desalination processes, modification of the scaling surface is challenging without affecting the heat transfer characteristics. Formation of scales can be minimized or prevented only by the retraction of ionic species from the surface. *Electrocoagulation* and *magnetic treatment*-based methodologies [67, 68] have been proposed for scale prevention. Masoudi et al. [69] proposed *magnetic slippery*

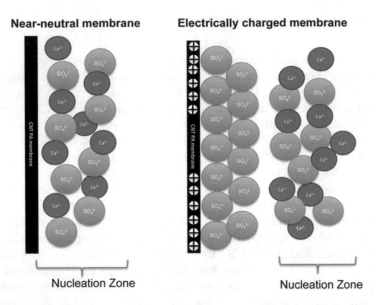

Fig. 14.4 Shifting of nucleation zone away from membrane surface on electrically charged membranes. Reproduced with permission from [66]. Copyright 2014 @ Royal Society of Chemistry

surface in two forms of Newtonian fluid (MAGSS) and gel structure (Gel-MAGSS). These surfaces provide a liquid-liquid interface to elevate the energy barrier for scale nucleation and minimize the adhesion strength of the formed scale on the surface.

Magnetic and electro chemical methods have been tried but have many challenges to overcome, even though they are useful in water transport and distribution system. Its applicability in thermal desalination is promising. Incorporating a small surface charge of the membranes as in NF membranes can aid in minimization of scaling.

14.4 Scaling in Non-conventional Desalination Systems

Forward osmosis (FO) and membrane distillation (MD) are slowly getting into prominence as small-scale desalination processes characterized by lower flux and recoveries. The type of scaling is different compared to conventional desalination processes as the membrane used in MD is hydrophobic in nature, while FO is a passive process with much lower specific fluxes.

14.4.1   Forward Osmosis (FO)

The scaling in FO can be attributed to bidirectional diffusive flow of solutes between feed and draw solute. Silica is a major contributor to scaling in FO systems, resulting in severe water flux reduction. Both colloidal silica and reactive silica have been found to be involved in scaling phenomenon. Detailed studies have indicated that the scaling mechanisms vary with types of membranes used [70]. Mono-silicic acid agglomeration, followed by deposition from bulk solution onto membrane surface appears to be the mechanism with asymmetric cellulose triacetate (CTA) membrane, while it is the interaction of mono-silicic acid with membrane surface followed by silica polymerization on the surface in polyamide TFC membranes. Similar effects of membrane surface properties were observed in gypsum scaling as well [71, 72]. In another study, Gwak and Hong [73] used anti-scalants blended draw solution in FO and concluded that the reverse diffusion of draw solute contributed to scale inhibition (Fig. 14.5). Addition of anti-scalants to the draw solute binds the scaling species (Ca^{++}) reducing the scale formation besides enhancement of the size, which in turn reduces the loss draw solute. It can be observed that rate of reduction of normalized water flux is much slower with PAspNa5 with reference incremental change in permeate volume.

14.4.2   Membrane Distillation (MD)

MD makes use of temperature gradient across a relatively hydrophobic membrane to drive transport of water vapour across the membrane pores and hence do not have scaling threats with reference to calcium salts. However, when operating with

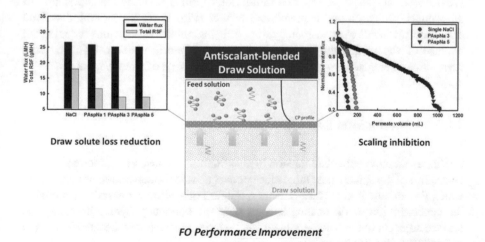

Fig. 14.5 Performance of anti-scalant blended in draw solution (a mixture of NaCl and poly aspartic acid sodium salt) (PAspNa) for scale inhibition. RSF denotes reverse solute flux. Reproduced with permission from [73]; Copyright 2017 @ Elsevier

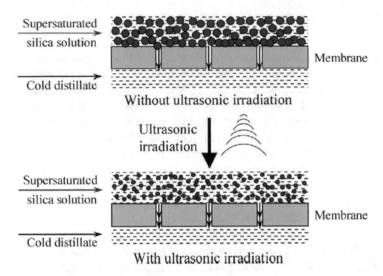

Fig. 14.6 Mitigation of silica scale through ultrasound irradiation into direct contact membrane distillation. Reproduced with permission from [79]; Copyright 2016 @ Elsevier

hypersaline brines and process concentrates, the MD process has shown vulnerability to scaling phenomenon. Majority of reported data [74, 75] deals with scaling caused by salts and minerals especially, NaCl, carbonates, and sulfates. A few studies [76–78] also deal with silica scaling phenomenon in MD systems and its mitigation. pH adjustment to acidic or highly basic range has been suggested as a pretreatment in such instances with fairly high silica concentrations up to 600 mg/L [78]. Hou et al. [79] have reported mitigation of silica scale by the application of ultrasound irradiation on the membrane surface (Fig. 14.6). Silica, both dissolved and insoluble, tends to deposit on hydrophobic membrane surface thus reducing the passage of vapour through the membrane pores. Upon irradiation with ultrasonic rays, silica aggregates are loosened allowing the passage of water vapour.

14.5   Conclusions and Outlook

Scaling is an inevitable phenomenon in desalination because of the increasing concentration of the solutes near the surface consequent to the separation of nearly pure water. Prevention or at least significant reduction of scaling is possible by reducing the concentration of the scaling species near the boundary layer. Alternately the surface adjacent to the boundary layer can be modified such that the adhesion of the precipitates does not take place.

The potential of various techniques such as ion exchange, feed flow reversal, nano-filtration and addition of anti-scalants were assessed. Ion exchange does not appear to be a feasible solution as it would involve frequent regeneration leading to high consumption of chemicals and generation of large volumes of wastewater. Reversal of feed flow applicable only to reverse osmosis desalination has apparent potential to reduce the scaling threat because of the long gestation time required for $CaSO_4$ precipitation but the sustainability of the method is yet to be established with reference to long term performance and logistics involved in the frequent flow reversal.

Use of nano-filtration as a pretreatment step has high potential both for thermal and membrane processes. The additional energy which NF requires has economic implications particularly for thermal desalination processes. In reverse osmosis, higher water recovery could be achieved by combining with NF. Further, the plant can be operated at lower pressures, thereby saving some cost on the materials of construction of the plant. The most popular method, however, is the addition of anti-scalants to prevent the scale-formation.

More developments have taken place for scale inhibition in reverse osmosis with emphasis shifting to the surface modification. Accordingly, nanocomposite membranes incorporating zeolites and multi-walled carbon nanotubes have been developed. Novel green biodegradable anti-scalants based on natural materials are under various stages of development to make the desalination processes more eco-friendly. Application of external forces including magnetic, electromagnetic and electrochemical forces are being assessed for scale mitigation, particularly with reference to developing membrane based desalination processes.

Not much of developments have been reported with reference to thermal desalination excepting for modification of anti-scalants and optimization of operating parameters including top brine temperature. Low temperature thermal desalination appears, in this context as a potential alternative, particularly for remote areas as scaling and corrosion problems are minimal.

Under normal operating conditions where the hydrodynamics of fluid flow is well maintained scale formation is fairly hindered. However, when the plant is shut down the scaling becomes a serious phenomenon as the supersaturated solution will have enough time for adhesion on the surface and crystal growth. Flushing the system immediately after shut down (after stopping the heat source) would definitely reduce the scaling in thermal processes. However, the economics of such an operation particularly in terms of the fresh water usage requires assessment.

In the case of reverse osmosis provision of suck back tanks would enable dislodging the accumulated solids by osmosis and this method could be very useful for the brackish water RO plants which do not operate continuously.

Developments and changes are the prime-drivers of technology and the search for scale mitigation and sustained long term membrane performance may lead to a better membranes in reverse osmosis. Low temperature thermal processes may become popular due to less scaling probability and the availability of waste heat in the power plants.

References

1. S. Patel, M.A. Finan, New antifoulants for deposit control in MSF and MED plants. Desalination **124**, 63–74 (1999)
2. Y. Magara, A. Tabata, M. Kohki, M. Kawasaki, M. Hirose, Development of boron reduction system for sea water desalination. Desalination **118**, 25–33 (1998)
3. K. Al-Anezi, N. Hilal, Scale formation in desalination plants: Effect of carbon dioxide solubility. Desalination **204**, 385–402 (2007)
4. A. Matin, F. Rahman, H.Z. Shafi, S.M. Zubair, Scaling of reverse osmosis membranes used in water desalination: Phenomena, impact, and control; future directions. Desalination **455**, 135–157 (2019)
5. A.E. Al-Rawajfeh, H.E.S. Fath, A.A. Mabrouk, Integrated salts precipitation and nano-filtration as pretreatment of multistage flash desalination system. Heat Transf. Eng. **33**, 272–279 (2012)
6. L. Henthorne, B. Boysen, State-of-the-art of reverse osmosis desalination pretreatment. Desalination **356**, 129–139 (2015)
7. D. Hasson, R. Semiat, Scale control in saline and wastewater desalination. Isr. J. Chem. **46**, 97–104 (2006)
8. M. Khayet, Fouling and scaling in desalination. Desalination **393**, 1 (2016)
9. R. Sheikholeslami, H.W.K. Ong, Kinetics and thermodynamics of calcium carbonate and calcium sulfate at salinities up to 1.5 M. Desalination **157**, 217–234 (2003)
10. E.E.A. Ghafour, Enhancing RO system performance utilizing anti-scalants. Desalination **153**, 149–153 (2003)
11. L. Llenas, X. Martínez-Lladó, A. Yaroshchuk, M. Rovira, J. de Pablo, Nanofiltration as pretreatment for scale prevention in seawater reverse osmosis desalination. Desalin. Water Treat. **36**, 310–318 (2011)
12. Y.A. Maher, M.E.A. Ali, H.E. Salama, M.W. Sabaa, Preparation, characterization and evaluation of chitosan biguanidine hydrochloride as a novel anti-scalant during membrane desalination process. Arab. J. Chem. (2018). https://doi.org/10.1016/j.arabjc.2018.08.006
13. H. Zidouri, Desalination in Morocco and presentation of design and operation of the Laayoune seawater reverse osmosis plant. Desalination **131**, 137–145 (2000)
14. S.F. Anis, R. Hashaikeh, N. Hilal, Reverse osmosis pretreatment technologies and future trends: A comprehensive review. Desalination **452**, 159–195 (2019)
15. R.Y. Ning, J.P. Netwig, Complete elimination of acid injection in reverse osmosis plants. Desalination **143**, 29–34 (2002)
16. M. Safari, A. Golsefatan, M. Jamialahmadi, Inhibition of scale formation using silica nanoparticle. J. Dispers. Sci. Technol. **35**, 1502–1510 (2014)
17. A.A. Al-Hamzah, C.M. Fellows, A comparative study of novel scale inhibitors with commercial scale inhibitors used in seawater desalination. Desalination **359**, 22–25 (2015)
18. Y. Bao, M. Li, Y. Zhang, Research on the synthesis and scale inhibition performance of a new terpolymer scale inhibitor. Water Sci. Technol. **73**, 1619–1627 (2016)
19. W.L. Ang, A.W. Mohammad, A. Benamor, N. Hilal, C.P. Leo, Hybrid coagulation–NF membrane process for brackish water treatment: Effect of anti-scalant on water characteristics and membrane fouling. Desalination **393**, 144–150 (2016)
20. M. Gryta, Polyphosphates used for membrane scaling inhibition during water desalination by membrane distillation. Desalination **285**, 170–176 (2012)
21. S.A. Ali, I.W. Kazi, F. Rahman, Synthesis of a diallylammonio propanephosphonate-alt-(sulfur dioxide) copolymer and its evaluation as an anti-scalant in desalination plants. Polym. Int. **63**, 616–625 (2014)
22. I.W. Kazi, F. Rahman, S.A. Ali, Synthesis of a polyaminophosphonate and its evaluation as an anti-scalant in desalination plant. Polym. Eng. Sci. **54**, 166–174 (2014)
23. A. Pervov, A. Andrianov, G. Rudakova, K. Popov, A comparative study of some novel "green" and traditional anti-scalants efficiency for the reverse osmotic Black Sea water desalination. Desalin. Water Treat. **73**, 11–21 (2017)

24. D. Hasson, H. Shemer, A. Sher, State of the art of friendly "green" scale control inhibitors: A review article. Ind. Eng. Chem. Res. **50**, 7601–7607 (2011)
25. K. Popov, G. Rudakova, V. Larchenko, M. Tusheva, S. Kamagurov, J. Dikareva, N. Kovaleva, A comparative performance evaluation of some novel "green" and traditional anti-scalants in calcium sulfate scaling. Adv. Mater. Sci. Eng. **2016**, 1–10 (2016)
26. B.K. Pramanik, Y. Gao, L. Fan, F.A. Roddick, Z. Liu, Antiscaling effect of polyaspartic acid and its derivative for RO membranes used for saline wastewater and brackish water desalination. Desalination **404**, 224–229 (2017)
27. A.G. Pervov, A.P. Andrianov, Assessment of the effectiveness of new "green" scale inhibitors used in reverse-osmosis seawater desalination. Pet. Chem. **57**, 139–152 (2017)
28. M.M. Tlili, A. Korchef, M. Ben Amor, Effect of scalant and anti-scalant concentrations on fouling in a solar desalination unit. Chem. Eng. Process. Process Intensif. **46**, 1243–1250 (2007)
29. R. Ketrane, B. Saidani, O. Gil, L. Leleyter, F. Baraud, Efficiency of five scale inhibitors on calcium carbonate precipitation from hard water: Effect of temperature and concentration. Desalination **249**, 1397–1404 (2009)
30. Y. Zhao, L. Jia, K. Liu, P. Gao, H. Ge, L. Fu, Inhibition of calcium sulfate scale by poly (citric acid). Desalination **392**, 1–7 (2016)
31. S.A. Ali, I.W. Kazi, F. Rahman, Synthesis and evaluation of phosphate-free anti-scalants to control $CaSO_4 \cdot 2H_2O$ scale formation in reverse osmosis desalination plants. Desalination **357**, 36–44 (2015)
32. Z. Shen, J. Shi, S. Zhang, J. Fan, J. Li, Effect of optimized three-component anti-scalant mixture on calcium carbonate scale deposition. Water Sci. Technol. **75**, 255–262 (2017)
33. M. Chaussemier, E. Pourmohtasham, D. Gelus, N. Pécoul, H. Perrot, J. Lédion, H. Cheap-Charpentier, O. Horner, State of art of natural inhibitors of calcium carbonate scaling. A review article. Desalination **356**, 47–55 (2015)
34. G. van Engelen, R. Nolles, A sustainable anti-scalant for RO processes. Desalin. Water Treat. **51**, 921–923 (2013)
35. O. Horner, H. Cheap-Charpentier, X. Cachet, H. Perrot, J. Lédion, D. Gelus, N. Pécoul, M. Litaudon, F. Roussi, Anti-scalant properties of Herniaria glabra aqueous solution. Desalination **409**, 157–162 (2017)
36. D.E. Abd-El-Khalek, B.A. Abd-El-Nabey, M.A. Abdel-kawi, S.R. Ramadan, Investigation of a novel environmentally friendly inhibitor for calcium carbonate scaling in cooling water. Desalin. Water Treat. **57**, 2870–2876 (2016)
37. S. Salvador Cob, B. Hofs, C. Maffezzoni, J. Adamus, W.G. Siegers, E.R. Cornelissen, F.E. Genceli Güner, G.J. Witkamp, Silica removal to prevent silica scaling in reverse osmosis membranes. Desalination **344**, 137–143 (2014)
38. H. Julian, Y. Ye, H. Li, V. Chen, Scaling mitigation in submerged vacuum membrane distillation and crystallization (VMDC) with periodic air-backwash. J. Memb. Sci. **547**, 19–33 (2018)
39. A. Al-Ghamdi, W. Omar, Application of pellet softening (PS) as a scale prevention method in seawater desalination plants. Desalin. Water Treat. **51**, 5509–5515 (2013)
40. N. Pomerantz, Y. Ladizhansky, E. Korin, M. Waisman, N. Daltrophe, J. Gilron, Prevention of scaling of reverse osmosis membranes by "zeroing" the elapsed nucleation time. Part I. Calcium sulfate. Ind. Eng. Chem. Res. **45**, 2008–2016 (2006)
41. H. Gu, A.R. Bartman, M. Uchymiak, P.D. Christofides, Y. Cohen, Self-adaptive feed flow reversal operation of reverse osmosis desalination. Desalination **308**, 63–72 (2013)
42. T. Vermeulen, B.W. Tleimat, G. Klein, Ion-exchange pretreatment for scale prevention in desalting systems. Desalination **47**, 149–159 (1983)
43. L. Zhu, C.B. Granda, M.T. Holtzapple, Prevention of calcium sulfate formation in seawater desalination by ion exchange. Desalin. Water Treat. **36**, 57–64 (2011)
44. M.B. Sik Ali, B. Hamrouni, S. Bouguecha, M. Dhahbi, Silica removal using ion-exchange resins. Desalination **167**, 273–279 (2004)
45. I. Bremere, M. Kennedy, S. Mhyio, A. Jaljuli, G.J. Witkamp, J. Schippers, Prevention of silica scale in membrane systems: Removal of monomer and polymer silica. Desalination **132**, 89–100 (2000)

46. J.A. Redondo, I. Lomax, Y2K generation FILMTEC RO membranes combined with new pretreatment techniques to treat raw water with high fouling potential: Summary of experience. Desalination **136**, 287–306 (2001)
47. J.D. Pless, M.L.F. Philips, J.A. Voigt, D. Moore, M. Axness, J.L. Krumhansl, T.M. Nenoff, Desalination of brackish waters using ion-exchange media. Ind. Eng. Chem. Res. **45**, 4752–4756 (2006)
48. K. Sasan, P.V. Brady, J.L. Krumhansl, T.M. Nenoff, Removal of dissolved silica from industrial waters using inorganic ion exchangers. J. Water Process Eng. **17**, 117–123 (2017)
49. A.M. Hassan, M.A.K. Al-Sofi, A.S. Al-Amoudi, A.T.M. Jamaluddin, A.M. Farooque, A. Rowaili, A.G.I. Dalvi, N.M. Kither, G.M. Mustafa, I.A.R. Al-Tisan, A new approach to membrane and thermal seawater desalination processes using nanofiltration membranes (Part 1). Desalination **118**, 35–51 (1998)
50. M.A.K. Al-Sofi, A.M. Hassan, G.M. Mustafa, A.G.I. Dalvi, M.N.M. Kither, Nanofiltration as a means of achieving higher TBT of $\geq 120°C$ in MSF. Desalination **118**, 123–129 (1998)
51. O. Labban, T.H. Chong, J.H. Lienhard, Design and modeling of novel low-pressure nanofiltration hollow fiber modules for water softening and desalination pretreatment. Desalination **439**, 58–72 (2018)
52. C. Kaya, G. Sert, N. Kabay, M. Arda, M. Yüksel, Ö. Egemen, Pre-treatment with nanofiltration (NF) in seawater desalination-preliminary integrated membrane tests in Urla, Turkey. Desalination **369**, 10–17 (2015)
53. M.A.K. Al-Sofi, A.M. Hassan, O.A. Hamed, A.G.I. Dalvi, M.N.M. Kither, G.M. Mustafa, K. Bamardouf, Optimization of hybridized seawater desalination process. Desalination **131**, 147–156 (2000)
54. A.M. Hassan, A.M. Farooque, N.M. Kither, A. Rowaili, A demonstration plant based on the new NF – SWRO process. Desalination **131**, 157–171 (2000)
55. A.S. Al-Amoudi, A.M. Farooque, Performance restoration and autopsy of NF membranes used in seawater pretreatment. Desalination **178**, 261–271 (2005)
56. O.A. Hamed, A.M. Hassan, K. Al-Shail, M.A. Farooque, Performance analysis of a trihybrid NF/RO/MSF desalination plant. Desalin. Water Treat. **1**, 215–222 (2009)
57. Y. Song, X. Gao, C. Gao, Evaluation of scaling potential in a pilot-scale NF-SWRO integrated seawater desalination system. J. Memb. Sci. **443**, 201–209 (2013)
58. Y. Song, X. Gao, T. Li, C. Gao, J. Zhou, Improvement of overall water recovery by increasing RNF with recirculation in a NF-RO integrated membrane process for seawater desalination. Desalination **361**, 95–104 (2015)
59. D. Zhou, L. Zhu, Y. Fu, M. Zhu, L. Xue, Development of lower cost seawater desalination processes using nanofiltration technologies – A review. Desalination **376**, 109–116 (2015)
60. M. Fathizadeh, A. Aroujalian, A. Raisi, Effect of added NaX nano-zeolite into polyamide as a top thin layer of membrane on water flux and salt rejection in a reverse osmosis process. J. Memb. Sci. **375**, 88–95 (2011)
61. B. Rajaeian, A. Rahimpour, M.O. Tade, S. Liu, Fabrication and characterization of polyamide thin film nanocomposite (TFN) nanofiltration membrane impregnated with TiO_2 nanoparticles. Desalination **313**, 176–188 (2013)
62. T.A. Saleh, V.K. Gupta, Synthesis and characterization of alumina nano-particles polyamide membrane with enhanced flux rejection performance. Sep. Purif. Technol. **89**, 245–251 (2012)
63. P. Swenson, B. Tanchuk, A. Gupta, W. An, S.M. Kuznicki, Pervaporative desalination of water using natural zeolite membranes. Desalination **285**, 68–72 (2012)
64. Y. Jiang, P. Biswas, J.D. Fortner, A review of recent developments in graphene-enabled membranes for water treatment. Environ. Sci. Water Res. Technol. **2**, 915–922 (2016)
65. B.H. Jeong, E.M.V. Hoek, Y. Yan, A. Subramani, X. Huang, G. Hurwitz, A.K. Ghosh, A. Jawor, Interfacial polymerization of thin film nanocomposites: A new concept for reverse osmosis membranes. J. Memb. Sci. **294**, 1–7 (2007)
66. W. Duan, A. Dudchenko, E. Mende, C. Flyer, X. Zhu, D. Jassby, Electrochemical mineral scale prevention and removal on electrically conducting carbon nanotube-polyamide reverse osmosis membranes. Environ. Sci. Process. Impacts **16**, 1300–1308 (2014)

67. D. Ghernaout, M.W. Naceur, B. Ghernaout, A review of electrocoagulation as a promising coagulation process for improved organic and inorganic matters removal by electrophoresis and electroflotation. Desalin. Water Treat. **28**, 287–320 (2011)
68. V. Kozic, A. Hamler, I. Ban, L.C. Lipus, Magnetic water treatment for scale control in heating and alkaline conditions. Desalin. Water Treat. **22**, 65–71 (2010)
69. A. Masoudi, P. Irajizad, N. Farokhnia, V. Kashyap, H. Ghasemi, Anti-scaling magnetic slippery surfaces. ACS Appl. Mater. Interfaces **9**, 21025–21033 (2017)
70. M. Xie, S.R. Gray, Silica scaling in forward osmosis: From solution to membrane interface. Water Res. **108**, 232–239 (2017)
71. M.I. Baoxia, M. Elimelech, Gypsum scaling and cleaning in forward osmosis: Measurements and mechanisms. Environ. Sci. Technol. **44**, 2022–2028 (2010)
72. M. Xie, S.R. Gray, Gypsum scaling in forward osmosis: Role of membrane surface chemistry. J. Memb. Sci. **513**, 250–259 (2016)
73. G. Gwak, S. Hong, New approach for scaling control in forward osmosis (FO) by using an anti-scalant-blended draw solution. J. Memb. Sci. **530**, 95–103 (2017)
74. L.D. Tijing, Y.C. Woo, J.S. Choi, S. Lee, S.H. Kim, H.K. Shon, Fouling and its control in membrane distillation-A review. J. Memb. Sci. **475**, 215–244 (2015)
75. D.M. Warsinger, J. Swaminathan, E. Guillen-Burrieza, H.A. Arafat, V.J.H. Lienhard, Scaling and fouling in membrane distillation for desalination applications: A review. Desalination **356**, 294–313 (2015)
76. J. Gilron, Y. Ladizansky, E. Korin, Silica fouling in direct contact membrane distillation. Ind. Eng. Chem. Res. **52**, 10521–10529 (2013)
77. P. Zhang, A. Pabstmann, S. Gray, M. Duke, Silica fouling during direct contact membrane distillation of coal seam gas brine with high sodium bicarbonate and low hardness. Desalination **444**, 107–117 (2018)
78. J.A. Bush, J. Vanneste, E.M. Gustafson, C.A. Waechter, D. Jassby, C.S. Turchi, T.Y. Cath, Prevention and management of silica scaling in membrane distillation using pH adjustment. J. Memb. Sci. **554**, 366–377 (2018)
79. D. Hou, L. Zhang, C. zhao, H. Fan, J. Wang, H. Huang, Ultrasonic irradiation control of silica fouling during membrane distillation process. Desalination **386**, 48–57 (2016)

Chapter 15
Chemical Methods for Scaling Control

Argyro Spinthaki and Konstantinos D. Demadis

15.1 Introduction

"Scale" is defined as a solid mass separating critical equipment surfaces from a working fluid (commonly water) [1]. It can be inorganic or organic (or a mixture), depending on the respective formation process, or the process fluid [2]. Scale creates several operational problems in industrial systems, such as inefficient heat transfer, under-deposit corrosion, and increased back pressure [3]. Industrial operations, such as cooling systems, make extensive use of either surface water or seawater, due to its cost-effectiveness, high heat capacity and abundance in nature [4]. Such systems often suffer from scale growth and, thus, the need for scale inhibition and control is urgent. One of the most effective strategies for scale/deposit mitigation is the utilization of chemical additives that act as scale inhibitors [5].

15.2 Inorganic Scales: Formation and Characterization

15.2.1 Calcium Carbonate

Calcium is the third element in Group IIA of the Periodic Table. Its cation (Ca^{2+}) is abundant in surface or ground waters [6]. On the other hand, carbonate (as HCO_3^-) is one of the most common anions found in either potable or industrial waters, due to the dissolution of atmospheric CO_2. Hence, their combination often leads to the formation of calcium carbonate ($CaCO_3$), one of the most abundant minerals in

A. Spinthaki · K. D. Demadis (✉)
Crystal Engineering, Growth and Design Laboratory, Department of Chemistry,
University of Crete, Crete, Greece
e-mail: demadis@uoc.gr

© Springer Nature Switzerland AG 2020
V. S. Saji et al. (eds.), *Corrosion and Fouling Control in Desalination Industry*,
https://doi.org/10.1007/978-3-030-34284-5_15

nature. $CaCO_3$ occurs in three main polymorphs: calcite, aragonite and vaterite [7, 8]. $CaCO_3$ poses a potential threat when precipitating in bulk water, or depositing on industrial piping or heat exchangers [9]. Calcium-containing mineral salts frequently encountered in geothermal waters are $CaCO_3$ and calcium silicate. Since the nineteenth century, it was known that even a small amount of CO_2 in the atmosphere is sufficient to greatly increase the solubility of $CaCO_3$ in water solutions, most likely due to the lowering of solution pH. In 1952, Miller calculated the solubility of $CaCO_3$ in temperatures ranging from 0 °C up to 105 °C and pressures of 1–100 bars. It was concluded that higher temperatures result in lower solubility, thus making $CaCO_3$ an inverse-solubility salt [10].

$CaCO_3$ forms in aqueous systems in three principal forms: calcite, aragonite and vaterite. The formation reaction is the following.

$$Ca^{2+}(aq) + CO_3^{2-}(aq) \Leftrightarrow CaCO_3(s) \qquad (15.1)$$

Figure 15.1 shows views of the crystal structures of calcite, aragonite and vaterite, and their respective X-ray diffraction (XRD) diagrams. The structure consists of trigonal planar carbonate groups coordinated to Ca^{2+} ions in distorted octahedra. The carbonates lie in layers parallel to [001]. The carbonate groups within a single layer have the same orientation. In the neighbouring layers the carbonate groups are rotated by 60° about the [001] plane.

Aragonite is denser than calcite, and is generally believed to be stable only at higher pressures. However, it readily crystallizes at ambient conditions and persists metastably for millions of years. The structure is nearly hexagonal with c as the unique axis. The pseudo-hexagonality of the structure is responsible for the pseudo twinning by rotation of 60° about c.

In vaterite, the Ca^{2+} ions form a primitive-hexagonal array, and sit at the corners of trigonal prismatic interstices which lie in layers parallel to (002). The carbonate groups occupy half of the Ca_6 interstices in one layer and the other half in the next layer, so as to surround each Ca^{2+} ion in an approximate octahedron, while the arrangement of carbon groups is approximately hexagonal close-packed.

15.2.2 Calcium Sulfate

Calcium sulfate ($CaSO_4$) appears in several crystalline forms, such as calcium sulfate dihydrate ($CaSO_4 \cdot 2H_2O$), widely known as gypsum, calcium sulfate anhydrous ($CaSO_4$), and calcium sulfate hemihydrate ($CaSO_4 \cdot 1/2H_2O$). In natural deposits, the main form is the dihydrate one; however, small quantities of anhydrite mineral are also present in most areas, to a lesser nevertheless extent [11]. Since the beginning of the twentieth century, studies tried to elucidate the scale formation mechanisms for $CaSO_4$. The liquid-solid equilibrium between Ca^{2+} and SO_4^{2-} ions and solid $CaSO_4 \cdot 2H_2O$ can be described by the following equation:

$$Ca^{2+}(aq) + SO_4^{2-}(aq) + 2H_2O \Leftrightarrow CaSO_4 \cdot 2H_2O(s) \qquad (15.2)$$

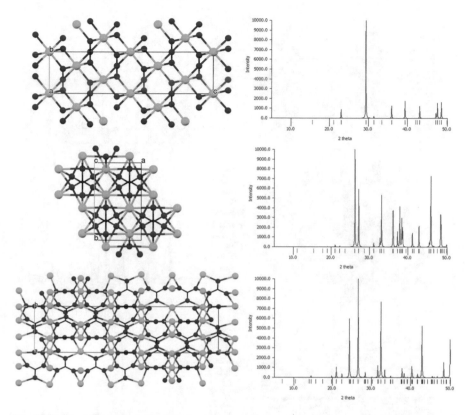

Fig. 15.1 Crystal structures of calcite (upper, viewed along the a axis), aragonite (middle, viewed along the c axis) and vaterite (lower, viewed along the a axis), and their respective XRD diagrams. Color codes: Ca green, O red, C black

In 1970, Hasson and Zahavi proposed the transient nucleation mechanism for $CaSO_4$ formation [12]. Still, there is no in-depth analysis about the exact stages of the formation process of calcium sulfate scale in heat exchange systems. In 1997 Linnikov studied the kinetics and mechanism of nucleation of the initial period of sulfate scale formation, and determined the energy required for nucleation activation. He also studied the effect of temperature in a solution and correlated metal surface roughness and crystal growth [13]. Figure 15.2 shows views of the crystal structures of the three common forms of calcium sulfate, gypsum, hemihydrate and anhydrite, and their respective XRD diagrams. Powder XRD diagrams are useful tools for identifying the presence of the three forms of calcium sulfate in precipitates and deposits for the diagnosis of the scales in membrane systems.

The crystal structure of gypsum consists of CaO_8 and SO_4 polyhedra stacked along the b-axis to form layers. Water molecules are located between the layers. The water molecule shares an oxygen atom with the Ca-polyhedron and hydrogen atoms form (at ambient pressure) weak hydrogen bonds with the oxygen atoms belonging to SO_4 and CaO_8 polyhedra. The structure of calcium sulfate hemihydrate is very

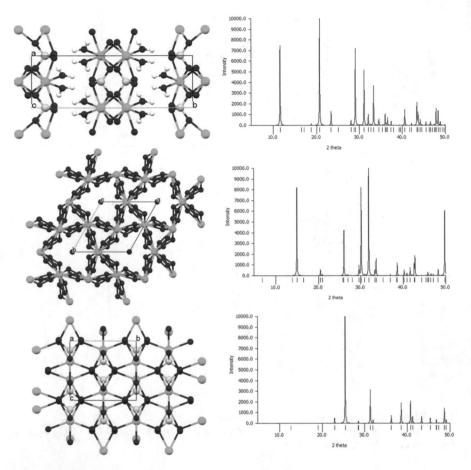

Fig. 15.2 Crystal structures of gypsum (upper, viewed along the c axis), calcium sulfate hemihydrate (middle, viewed along the c axis) and anhydrite (lower, viewed along the a axis), and their respective XRD diagrams. Color codes: Ca green, O red, S yellow, H white

different. The Ca atoms are coordinated by six O atoms, all originating from different sulfate anions. The Ca-sulfate coordination creates triangles within which the water molecule is situated, thus creating a "channel" running parallel to the a axis. In contrast to gypsym, the water is not coordinated to a Ca center, but can be better described as lattice water.

The anhydrite form of calcium sulfate contains no water (either bound, or in the lattice). The coordination of Ca by the oxygens of the sufate anion creates a dense, three-dimensional structure. The Ca atoms are coordinated by seven oxygens, originating from different sulfate anions.

15.2.3 *Amorphous Silica*

Amorphous or colloidal silica is a form of silicon dioxide (SiO_2) that is encountered as a precipitate or deposit in industrial waters. It is perhaps one of the most "peculiar" water-formed deposit because it is not a mineral salt, but a covalent solid composed of Si-O bonds. As it will be demonstrated later, its chemical control/inhibition is unconventional. In contrast to the majority of "traditional" scales, it still is one of the persistent unresolved issues in deposit control, and it has been coined as the "Gordian knot" of water treatment [14]. Its formation is based on a peculiar polycondensation mechanism [15] and that usually takes place in the pH regime from 7.0 to 8.5, provided that the soluble silica levels (i.e. mono-, and di-silicic acid, the former being the major species) are >180 ppm. It is worth-while to take a closer look at some important details of silica condensation chemistry.

In dilute aqueous solutions, soluble silica is found as monosilicic acid, $Si(OH)_4$. This monomer exists in two forms in about-neutral pH environment: protonated (major, $Si(OH)_4$) and mono-deprotonated (minor, $Si(OH)_3O^-$). As pH increases, the concentration of the deprotonated form ($Si(OH)_3O^-$) also tends to increase [16]. By increasing the silicate concentration and adjusting the pH value to ~ 7, the two co-existing forms of soluble silica start to react with each other, leading to condensation of these two monomers, with simultaneous loss of a hydroxide anion and formation of disilicic acid. This S_{N2} – type reaction can be described by the following equation [17]:

$$Si(OH)_4 + Si(OH)_3 O^- \rightarrow (HO)_3 Si\text{-}O\text{-}Si(OH)_3 + OH^- \qquad (15.3)$$

Disilicic acid continues to incorporate an additional monomer, forming the trimer. Continuation of this process leads to the formation of oligomers and finally, silica nanoparticles (see Fig. 15.3) [18, 19]. These can further agglomerate and create large silica particles and (if a suitable surface is provided) silica deposits.

The aforementioned dimerization step is the most crucial one as far as the kinetics of the polycondensation is concerned and is the rate-determining step in the complex silica polycondensation process [20]. The next steps involve the formation of trimers, tetramers and then higher oligomeric species until 1–2 nm amorphous silica nanoparticles are formed [21]. It is worth mentioning that the pKa of the polysilicic acids at these stages of the polymerization is about 6.5 and therefore negatively charged silicate species become predominant above pH 7–8 [22]. In these conditions, silicic acid on the surface of colloidal species exists in equilibrium with dissolved silicic acid molecules and, thus, further particle growth occurs via the Ostwald ripening process, leading to stable sols. In contrast, below pH 7, particles are only slightly charged, so no electrostatic repulsion prevents them from aggregation, leading to gel formation [23].

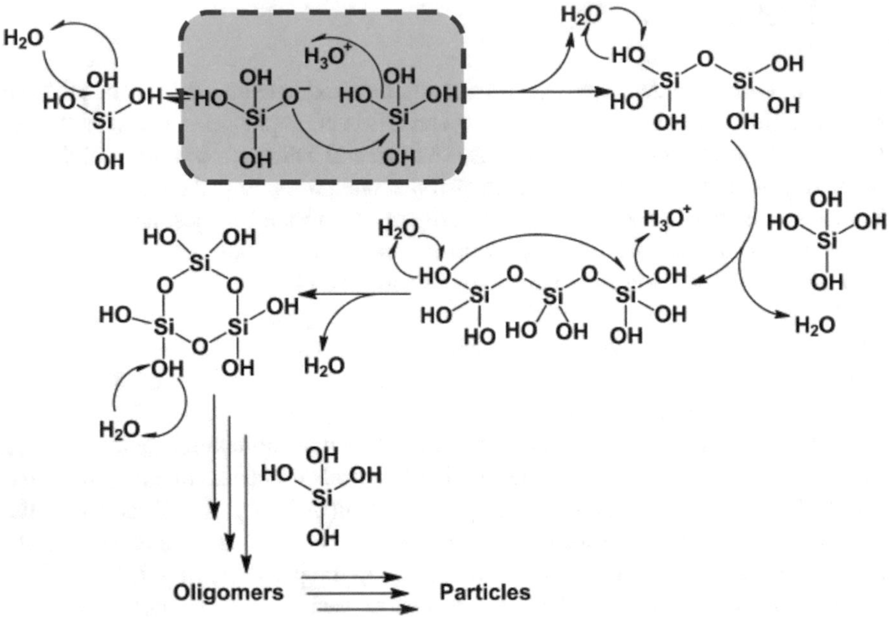

Fig. 15.3 S_{N2} – like mechanism of silicic acid polycondensation and silica particles formation

15.2.4 Metal Silicates

15.2.4.1 Magnesium Silicate

Magnesium silicates constitute a family of geological entities that embrace several types of minerals consisting of both magnesium (in its Mg^{2+} state) and silicon (in its SiO_4^{4-} state). They exist in a variety of different Mg:Si stoichiometries and hydration states. In the field of geology, these structures are well-defined, but when it comes to inorganic precipitation/deposition in the water treatment industry, their true identity is elusive, and, occasionally, controversial. Here, magnesium silicate is any water-formed solid that contains both Mg and Si in no specified stoichiometry. The molecular formula of such compounds may be expressed as $MgSiO_3 \cdot xH_2O$ [24, 25]. The powder XRD diagram represents an amorphous material, as expected. The Fourier-transform infrared (FT-IR) spectrum (see Fig. 15.4) resembles that of amorphous silica, as the vibrational band at 1016 cm^{-1}, is characteristic of the Si–O–Si symmetrical stretching vibration [26, 27].

Such magnesium silicate deposits (with the aforementioned definition) have been observed in cooling towers [28], heat exchangers [29] and geothermal systems [30]. Formation, precipitation, and deposition of magnesium silicates have been at the epicenter of intense interest [31]. Such scaling, like any other, poses a severe threat to the smooth process of industrial systems, but it requires harsh conditions, and especially high temperature to turn into a geologically recognized material.

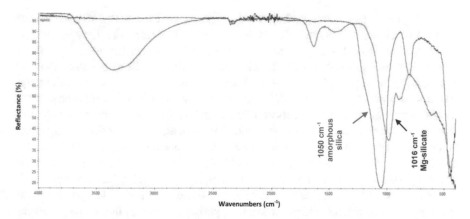

Fig. 15.4 FT-IR spectrum of amorphous silica (blue) and magnesium silicate (red)

Whatever the case might be, there are some important facts regarding its formation in waters rich in both Mg and Si. Just like the (colloidal/amorphous) silica system, the magnesium silicate system is highly pH–dependent. At pH below 7 there is practically no magnesium silicate precipitation because "silica" exists as silicic acid monomer [Si(OH)$_4$], in a non–ionized, and thus, unreactive form. Above pH 9 (close to the pKa$_1$ of silicic acid ~ 9.3 [32]), silica exists in the form of reactive silicate ions [mono-deprotonated Si(OH)$_3$O$^-$], which favor the attachment of the cationic Mg^{2+}. Another crucial factor for magnesium silicate formation is temperature. It has been suggested that precipitation begins at a lower pH if the temperature is sufficiently high [33]. It is also important to mention that silica precipitates at a lower temperature compared to magnesium silicate. It has also been proposed that magnesium silicate can be formed by co–precipitation of magnesium hydroxide and silicic acid, producing non–stoichiometric forms of the scale. Another proposal is that magnesium hydroxide occurs first and then reacts with colloidal silica. The fact is that in any case the scale is amorphous, it is not recognized as one of the geological minerals and it does require a well-thought strategy for its effective mitigation [34].

15.2.4.2   Aluminium Silicate

Among the first approaches of the scientific community was the clarification of the effects of several aluminum-containing silicates in soils [35]. Typically, Al^{3+} appears either as 6- or 4- coordinated in its compounds. This second case is very usual among minerals, because Al^{3+} can substitute Si^{4+}, using an extra cation to balance the charge difference [36]. These scales are governed by low solubility constants, in fact, there is such high affinity between the two species that silicic acid is able to compete with strong ligands, such as catechol, for aluminum cations [37]. The exact identity of aluminum silicate, as it is formed in water systems, remains elusive, regardless of the intense research during the past several decades [38]. When it comes to geothermal brines, scaling of such type can be a severe threat to the smooth process of industrial systems, due to deposition. Aluminum silicate typically precipitates at pH from 4 to 10, although the rate of precipitation increases at pH above 9 and decreases at pH 5–8 [39]. In addition, Al (III) is said to be "notorious" in complexing with phosphonates [40]. Water chemistries such as the Salton Sea example are proof that even if Al^{3+} appears in low soluble concentrations, it can cause precipitation and it is concentrated in scales [41]. As is probably by now understood, the complex speciation of aluminium in aqueous solutions is the major reason that the scales that industries have to tackle are not always in agreement to the structure of the geological minerals [42, 43]. In the precipitates obtained from several geothermal plants, the Al:Si ratios vary, from 1:8, 1:10, 1:20 and even 1:40. The easiest conclusion probably to be drawn is that aluminium is incorporated in the amorphous silica matrix, and in some cases, perhaps catalyzes the whole procedure [44]. A reasonable conclusion is that aluminum hydroxide adsorbs high concentrations of silicic acid, perhaps catalyzing the whole procedure, an assumption also made when iron (III) is present in the brine [45]. The Fourier-transform infrared (FT-IR) spectrum (see Fig. 15.5) resembles that of amorphous silica, as the vibrational band at ~ 1050 cm^{-1}, is characteristic of the Si–O–Si symmetrical stretching vibration [30], but shifted to lower frequencies, just like in the magnesium silicate case.

15.2.5 Metal Sulfides (Zinc Sulfide)

In contrast to the metal silicates mentioned above, zinc sulfide is a well–characterized mineral salt and has a solubility constant (K_{sp}) of about 2×10^{-25} [46]. The formation of zinc sulfide in aqueous solutions is not simple, because of zinc [47] and sulfide speciation [48]. Some crucial observations are warranted. An increase in

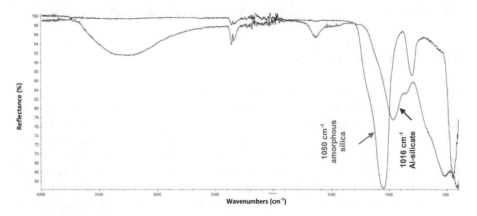

Fig. 15.5 FT-IR spectrum of amorphous silica (blue) and aluminum silicate (red)

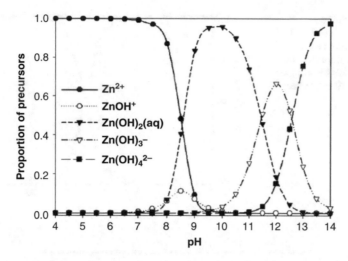

Fig. 15.6 Zinc speciation in a wide pH range. Reproduced with permission from Ref. 47; Copyright 2013 © Royal Society of Chemistry

pH reduces the actual concentration of Zn^{2+}, due to zinc hydroxide precipitation, a competitive reaction to the formation of zinc sulfide (see Fig. 15.6). In addition, S^{2-} is the dominant species in pH values greater than 11, while at pH values of about neutral, HS^- plays the key role (see Fig. 15.7). The extremely low solubility constant of ZnS, allows the fast deprotonation of HS^-, which shifts the equilibrium and reduces the pH.

The precipitates occurring from the *in situ* formation of zinc sulfide at ambient conditions are mainly zinc blend or sphalerite, as indicated by the powder XRD diagrams (see Fig. 15.8) [49].

15.2.6 Calcium Phosphate(s)

Calcium orthophosphates are a vast category of sparingly soluble salts that have attracted the interest of many interdisciplinary fields of science, such as geology, chemistry, biology and medicine. The reader can refer to several excellent reviews on the topic [50–52]. The initial attempts to establish their chemical composition were performed by Berzelius in the middle of the nineteenth century [53]. Approximately 70 years afterward, the idea on the existence of different crystal phases of calcium orthophosphates was introduced [54]. Mixtures of calcium orthophosphates had been called apatites until then. By definition, the calcium orthophosphate minerals consist of calcium, phosphorus and oxygen. These three chemical elements are present in abundance on the surface of our planet: oxygen is the most widespread chemical element of the earth's surface, calcium occupies the fifth place, and phosphorus is among the first twenty most abundant chemical

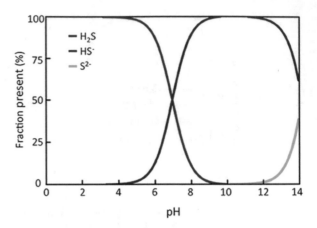

Fig. 15.7 Sulfide speciation in a wide pH range. Reproduced with permission from Ref. 48. Copyright 2014 © Holmer and Hasler-Sheetal under CC BY

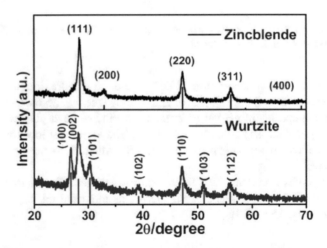

Fig. 15.8 Powder X-ray diagram of two polymorphs of zinc sulfide. Reproduced with permission from Ref. 49. Copyright 2012 © Royal Society of Chemistry

elements [55]. The calcium phosphate mineral family consists of a great number of compounds of different stoichiometries and hydration states. The chemical composition of many calcium orthophosphates includes hydrogen, either as an acidic orthophosphate anion, or as incorporated water. Diverse combinations of oxides of calcium and phosphorus provide a large variety of calcium phosphates, which are distinguished by the type of the phosphate anion: ortho- (PO_4^{3-}), meta- (PO_3^-), pyro- $(P_2O_7^{4-})$, and poly- $[(PO_3)_n^{m-}]$. In the case of multi-charged anions (orthophosphates and pyrophosphates), calcium phosphates are also differentiated by the number of hydrogen ions attached to the anion. Examples include mono- $(Ca(H_2PO_4)_2)$, di- $(CaHPO_4)$, and tri- $(Ca_3(PO_4)_2)$ calcium phosphates, and pyrophosphates $(Ca_2P_2O_7)$ [56–58]. These structures are well-defined and are of great interest to the scientific community, not only for their major role in biominerization [59, 60], but

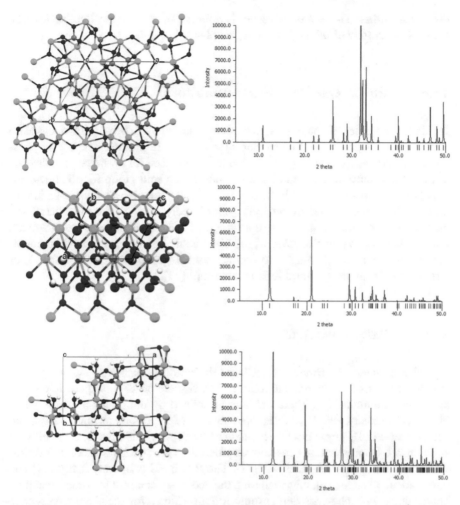

Fig. 15.9 Crystal structures of hydroxyapatite (upper, viewed along the c axis), brushite (middle, viewed along the b axis) and calcium pyrophosphate (lower, viewed along the c axis), and their respective XRD diagrams. Color codes: Ca green, O red, P orange, H white

also for their notorious role as sparingly soluble scales in several industrial systems [61–64].

Figure 15.9 shows portions of the structures of hydroxyapatite, brushite and calcium pyrophosphate, as well as the corresponding powder XRD diagrams. Powder XRD is a useful tools for identifying the presence of the three forms of calcium phosphate in precipitates and deposits for the diagnosis of the scales in membrane systems.

Crystalline mineral calcium phosphates do not appear as precipitates or deposits in water systems. Usually amorphous calcium phosphates are noted, but these cannot be characterized by diffraction techniques. Hence, this topic will not be discussed further. The interested reader is refereed to several excellent reviews on the crystallography of mineral calcium phosphates [50–52].

15.3 Main Categories of Scale Inhibitors

In this section, selected major classes of scale inhibitors will be presented, based on their functional groups and molecular structure. An important category is phosphonic acids (phosphonates) [65]. Phosphonates are used for a variety of scales due to the high affinity of the (deprotonated) phosphonic acid group for alkaline-earth ions [66] and crystal surfaces [67]. Carboxylic acids are also used as scale inhibitors, but to a much lesser extent [68]. Anionic polymers, on the other hand, are integral parts of most water treatment programs [69]. Most anionic polymers are polyacrylate-based polyelectrolytes [70]. They could be polyacrylic acid homopolyers, copolymers (containing a second functional group, such as sulfonate), or terpolymers (containing a third functional group) [71].

15.3.1 Phosphonic Acids

Phosphonic acids are widely utilized for different applications, including drug delivery systems [72], corrosion control [73], dispersion [74], sequestration [75] and scale inhibition [76]. Their structure is defined by the presence of the phosphonic acid moiety, $-PO_3H_2$, exhibiting a number of important attributes, discussed below. The $-PO_3H_2$ group is a di-acid and can be deprotonated twice, depending on solution pH. Deprotonation constants vary based on the specific structural features, with pKa1 being <2 and pKa1 ~ 7 [77]. The P-C bond is hydrolytically [78] and enzymatically stable [79], thus rendering the necessary stability for water treatment applications. Phosphonates demonstrate a high affinity for metal ions present in industrial water streams, as well as a tendency to attach to mineral surfaces. The phosphonate molecule can bear other functional groups (such as carboxylate, amino, sulfono), which can positively impact scale inhibition performance [80].

More specifically, our group has previously reported the influence of five phosphonate-based chemical additives on the precipitate formation in the presence of Mg^{2+} ions and soluble silica (silicic acid) [81]. The phosphonate additives tested were 2-phosphonobutane-1,2,4-tricarboxylic acid (PBTC) [82], hydroxyethylidene-1,1-diphosphonic acid (HEDP) [83], amino-tris(methylenephosphonic acid) (AMP) [84], hexamethylenediamine-tetrakis(methylenephosphonic acid) (HDTMP) [85], and bis diethylenetriamine-penta(methylenephosphonic acid), DETPMP [86], and their schematic structures are shown in Fig. 15.10.

The addition of phosphonate additives to a supersaturated solution retards the formation of calcium-containing scales, most commonly calcium carbonate [87]. Phosphonate additives also inhibit calcium sulfate dihydrate (gypsum) scale

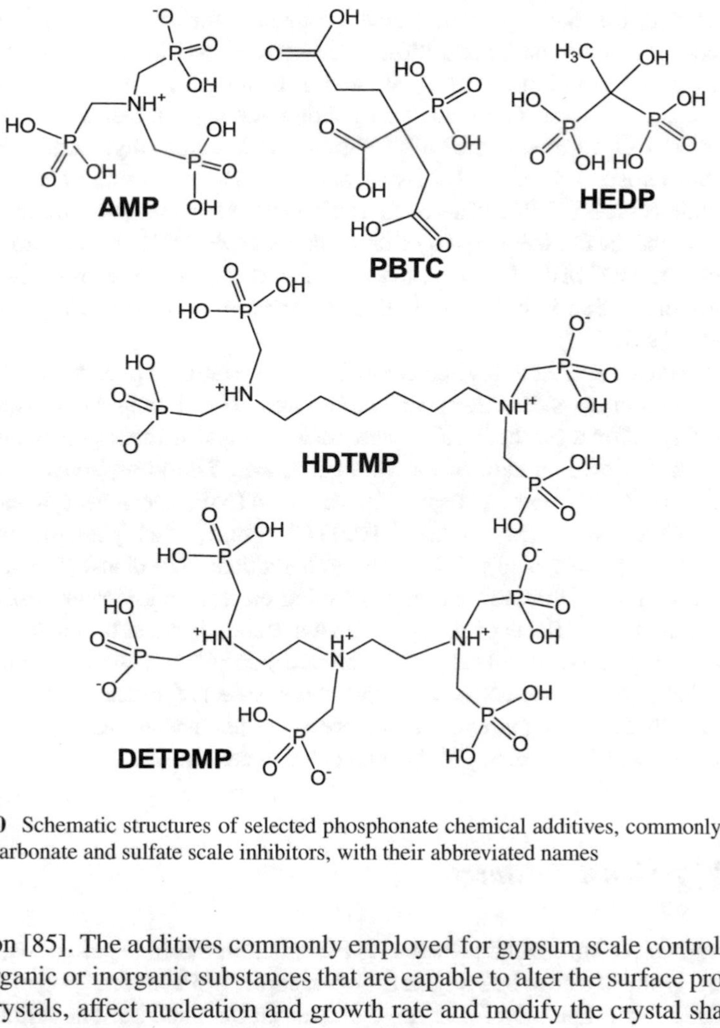

Fig. 15.10 Schematic structures of selected phosphonate chemical additives, commonly used as calcium carbonate and sulfate scale inhibitors, with their abbreviated names

formation [85]. The additives commonly employed for gypsum scale control can be either organic or inorganic substances that are capable to alter the surface properties of the crystals, affect nucleation and growth rate and modify the crystal shape and agglomeration behavior [88].

The effects of three phosphonic acids [ethylenediamine-tetra(methylenephos-phonic acid), EDTMP; hexamethylenediamine-tetra(methylenephosphonic acid), HDTMP; and diethylenetriamine-penta(methylenephosphonic acid), DETPMP] were evaluated on the growth of $CaCO_3$ [89]. The result revealed the following inhibitor effectiveness: DETPMP > EDTMP > HDTMP. Molecular dynamic simu-lations on the interaction of these phosphonic acids with the calcite (104) surface indicated that strong electrostatic interactions between the oxygen atoms in the phosphonate functional groups and the Ca^{2+} centers of the calcite (104) face played a dominant role in their adsorption. The weakest inhibitor of $CaCO_3$ was found to be HDTMP because of only one phosphonate group was found to effectively inter-act with the $CaCO_3$ surface.

Sousa and Bertran reported a new procedure for the evaluation of scale inhibitors for calcium carbonate [90]. It is based on continuous measurement of particle size distribution by laser diffraction techniques with simultaneous pH monitoring. Data were obtained during homogeneous nucleation and particle growth of $CaCO_3$ formed in the bulk for the phosphonates ethylenediamine-tetra-methylene phosphonic acid, EDTMP; and diethylenetriamine-pentamethylene phosphonic acid, DETPMP. The comparative bulk crystallization inhibition efficiency for the evaluated inhibitors showed that DETPMP is a more efficient inhibitor than EDTMP. Moreover, it was postulated that the main mechanism of the inhibiting action is based on both nucleation inhibition and crystal growth retardation [91].

Antiscalants are used in reverse osmosis (RO) systems to prevent salt precipitation but might affect side-stream concentrate treatment. Precipitation experiments were performed on a synthetic RO concentrate with and without phosphonate antiscalants for the precipitation of calcium carbonate. The phosphonates evaluated were aminotri(methylene phosphonic acid), ATMP; hexamethylenediamine-tetra(methylenephosphonic acid), HDTMP; and diethylenetriamine-penta (methylenephosphonic acid), DETPMP [92]. Particle size distributions, calcium precipitation, microfiltration flux, and scanning electron microscopy (SEM) were used to evaluate the effects of phosphonate antiscalant type, antiscalant concentration, and precipitation pH on calcium carbonate precipitation and filtration. Results showed that phosphonate antiscalants could decrease precipitate particle size and change the shape of the particles. The presence of antiscalant during precipitation can also decrease the mass of precipitated calcium carbonate.

15.3.2 Anionic Polymers

The charge of the antiscalant polymers is important in the end application, depending on the type of scale to be mitigated. For example, while anionic polyelectrolytes are suitable for mitigation of calcium carbonate/sulfate and barium sulfate scales, they are inactive for silica scales. Hence, a reasonable approach would be to study them based on their charge.

Anionic polymers are widely known as efficient scale inhibitors. Most of these are polyacrylate-based polyelectrolytes. Polymers of acrylic or maleic acids have been used as inhibitors for calcium carbonate [93] or calcium sulfate [94], but commonly their derivatives (either co-polymers, or ter-polymers) are also preferred. Based on literature data, usually ter-polymers are more efficient additives, however, the manufacturing cost is higher (Fig. 15.11).

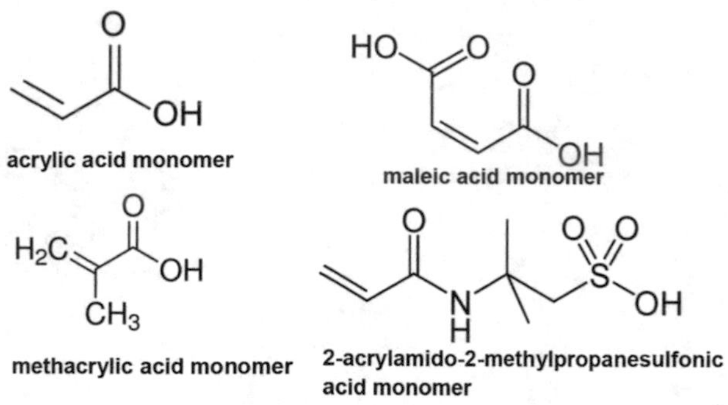

acrylic acid monomer

maleic acid monomer

methacrylic acid monomer

2-acrylamido-2-methylpropanesulfonic acid monomer

Fig. 15.11 Schematic structures of monomers used for the synthesis of homo-, co-, and ter-polymers as scale inhibitors and dispersants

15.3.3 Cationic Polymers

Cationic polymers were found to retard the condensation of silicic acid when added at low dosage (up to around 100 ppm) to supersaturated solutions of silicic acid (500 ppm, ~ 8 mM). Examples of such compounds are poly(1-vinylimidazole) [95], amine-terminated polyaminoamide dendrimers (PAMAMs) [96–98], poly(acrylamide-*co*-diallyldimethylammonium chloride) [99], polyethyleneimine (branched and linear) [100, 101], and cationically-modified inulin [102]. Also, cationically-grafted polyethylene glycol (PEG) oligomers with phosphonium groups were shown to be efficient inhibitors for silica [103]. Chemical structures of selected additives are shown in Fig. 15.12.

Cationic polymers are specific inhibitors for silica scale. As mentioned before, colloidal silica is a very peculiar scale because it is not a mineral (as it is formed in water systems), but a random polymer composed of Si-O bonds. Hence, a polymeric silica inhibitor must have the ability to stabilize the silica scale precursor, i.e. silicic acid $Si(OH)_4$ or its deprotonated form silicate $Si(OH)_3O^-$. The positive charge on the polymer is assumed to stabilize silicate through ionic interactions at pH regions from 7.0 to 8.5.

The additives shown in Fig. 15.12 have been used for the stabilization of silicic acid in aqueous solutions, in order to control its polycondensation. Some comparative results are shown in Fig. 15.13. The most efficient inhibitors are the dendrimers PAMAM-1 and PAMAM-2, PPEI, and PEGP+-4000.

15.3.4 Neutral Polymers

As for cationic polymers, neutral polymers (Fig. 15.14) are also used to combat silica scale, by stabilizing neutral silicic acid $Si(OH)_4$. Here, the dominant stabilization mechanism is different. The polymeric scale inhibitor must possess moieties

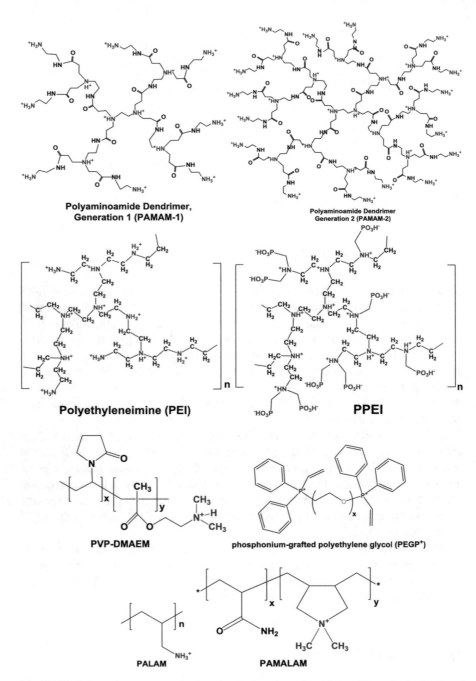

Fig. 15.12 Schematic structures of selected cationic polymeric additives. The cationic functional groups are colored red

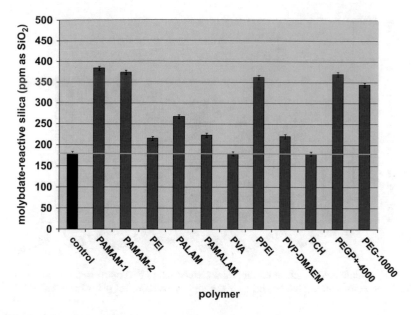

Fig. 15.13 Silicic acid stabilization (at initial concentration 500 ppm) in the presence of various additives (at 40 ppm concentration) at pH = 7.0 and after 24 h. The line corresponds to the "control" and has been added to aid the reader

Fig. 15.14 Schematic structures of selected neutral polymeric additives

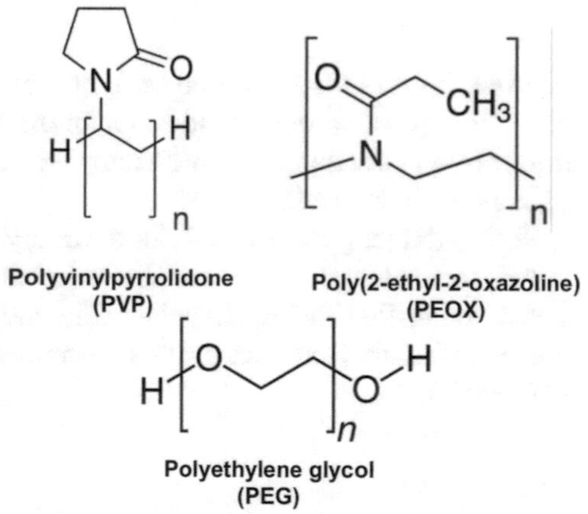

Polyvinylpyrrolidone
(PVP)

Poly(2-ethyl-2-oxazoline)
(PEOX)

Polyethylene glycol
(PEG)

with electronegative atoms (e.g., O or N) that are capable of forming hydrogen bonds with silicic acid. Some of the previously studied approaches included polyvinylpyrrolidone [104] and various molecular weights of PEG [105]. These efforts resulted in the important outcome that silicic acid can be stabilized, not only via electrostatic interactions (as found for the cationic polymers), as was widely

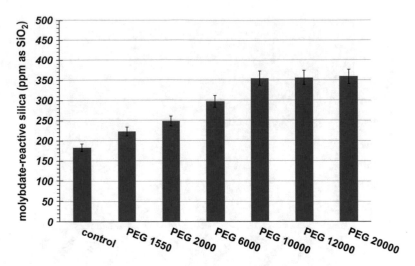

Fig. 15.15 Silicic acid stabilization (at initial concentration 500 ppm) in the presence of PEG additives of various molecular weights (at 100 ppm concentration) at pH = 7.0 and after 8 h

Fig. 15.16 Schematic structure of the polypeptide P5S3

H$_2$N-GSYS**RR**AS*LG[**KSKKL RR**AS*LG]$_3$**KL RR**AS*L-COOH

believed, but via the formation of hydrogen bonds that include either the neutral silanol group as hydrogen bond donor, or the respective mono de-protonated moiety as the H-bond acceptor. Another example of neutral polymers regarding silica scaling is polyethyloxazoline [106].

PEG polymers are available in a variety of molecular weights. The effect of polymer chain length on the stabilization of silicic acid has been studied (Fig. 15.15). It was found that the stabilization efficiency of the PEG polymers increases as molecular weight increases, but this phenomenon levels off at a molecular weight of ~ 10,000 Daltons.

15.3.5 "Green" Inhibitors

Increasing environmental concerns and discharge limitations have imposed additional challenges in treating process waters. Therefore, the discovery and successful application of chemical additives that have a mild environmental impact have been the focus of several researchers [107]. The need for not only efficient but environmentally friendly inhibitors is slowly but surely leading a great part of the scientific research away from inhibitors with unclear environmental impact and towards "green" and thus universally accepted antiscaling molecules [108].

A recent example is the simultaneous effective silicic acid polycondensation inhibition and biological activity of the P5S3 polypeptide [109]. The polypeptide is shown in Fig. 15.16 in an extended polyproline II conformation, which is the most prominent secondary structure motif in P5S3, and in amino acid single letter code representation. In the structural model, atoms are shown in grey (carbon), red (oxygen) and blue (nitrogen). In the single-letter amino acid representation, the cationic amino acids lysine (K) and arginine (R) are shown underlined and bold; the amino- and carboxy-termini are indicated by "H$_2$N-" and "-COOH", respectively. The central sequence (squared brackets) is repetitive and occurs three times in P5S3. Serines in the protein kinase A target site are marked by an asterisk. The P5S3 polypeptide has been shown to stabilize at pH ~ 7, ~ 350 ppm silicic acid at a concentration of 100 ppm. At circumneutral pH the polypeptide P5S3 is positively charged due to its high isoelectric point at 12.64. Therefore, stabilization of silicic acid occurs through an interaction of the oppositely charged, deprotonated silanol groups and lysine (pKa = 9.8) plus arginine (pKa = 12.5) side chains.

Working on a different scale, carboxylmethylpullulan (CMP) polymer is currently tested against calcium carbonate scaling and its efficiency is proving worthmentioning. CMP is a close analogue of carboxylmethylinulin (CMI), which is already proven to be a promising calcium oxalate inhibitor.

Inhibitory effects of phosphonated chitosan (PCH, synthesized from chitosan (CHS) by a Mannich-type reaction, see Fig. 15.17) were studied on the formation of amorphous silica [110]. Specifically, the ability of PCH to retard silicic acid condensation in aqueous supersaturated solutions at circumneutral pH is studied. It was discovered that when PCH is added in dosages up to 150 ppm, it can inhibit silicic acid condensation, thereby maintaining soluble silicic acid up to 300 ppm (for 8 h, from a 500 ppm initial stock solution). PCH was also found to affect colloidal silica particle morphology.

In addition, PCH shows synergy with either purely cationic (polyethyleneimine, PEI) or purely anionic (carboxymethyl inulin, CMI) [111]. It was found that the action of inhibitor blends is not cumulative. PCH/PEI blends stabilize the same level of silicic acid as PCH alone in both short-term (8 h) and long-term (72 h) experiments. PCH/CMI combinations, on the other hand, can only achieve short-term inhibition of silicic acid polymerization, but fail to extend this over the first 8 h. PCH and its combinations with PEI or CMI affect silica particle morphology,

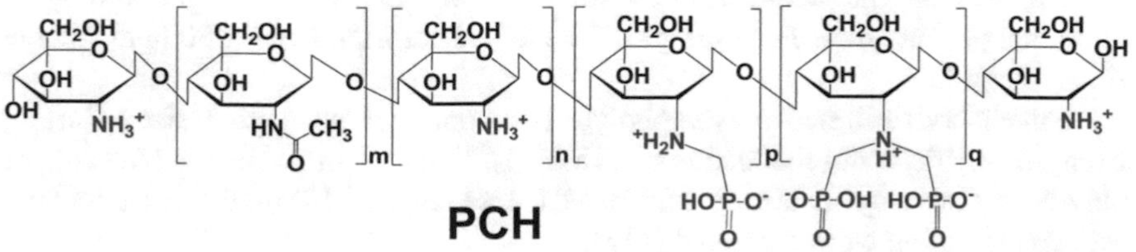

Fig. 15.17 Schematic structure of the zwitterionic phosphonomethylated chitosan backbone (m = 0.16, n = 0.37, p = 0.24, q = 0.14)

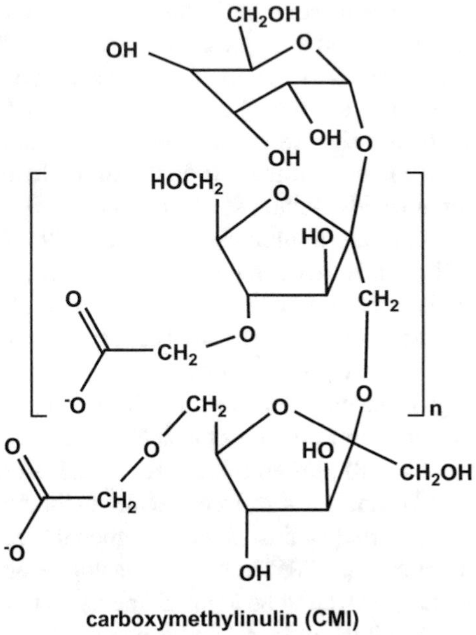

Fig. 15.18 Schematic structure of the CMI polymer

carboxymethylinulin (CMI)

studied by SEM. Spherical particles and their aggregates, irregularly shaped particles and porous structures are obtained depending on additive or additive blend. It was demonstrated by FT-IR that PCH is trapped in the colloidal silica matrix.

The effect of a biodegradable, environmentally friendly polysaccharide-based polycarboxylate, carboxymethyl inulin (CMI, see Fig. 15.18), on the crystal growth kinetics of calcium oxalate was studied at 37 °C [112, 113]. CMI is produced by carboxymethylation of inulin, the latter extracted from chicory roots. The spontaneous crystallization method was utilized to investigate the crystallization kinetics of calcium oxalate (CaC_2O_4, CaO_x). The experimental results show that the retardation in mass transport in the growth process is controlled by the carboxylation degree of CMI and also its concentration. The studies also indicated that CMI was effective in directing calcium oxalate crystallization from calcium oxalate monohydrate (COM) to calcium oxalate dihydrate (COD).

Six PEG polymers were used as silica scale inhibitors [114]. Their molecular weights range from 1550 to 20,000. There was a profound dependence of inhibitory performance on the additive molecular weight. PEG polymers showed good inhibition performance. For example, PEG 20000 can stabilize ~ 350 ppm soluble silica after 8 h.

A methacrylate-based polyampholyte homopolymer was synthesized starting from N-methacryloyl-L-histidine (MHist). The inhibitory effects of poly-N-methacryloyl-L-histidine (poly-MHist, see Fig. 15.19) on the in vitro silicic acid condensation were evaluated [115].

In particular, the ability of poly-MHist to retard silicic acid condensation in aqueous supersaturated solutions at three pH values, 5.5, 7.0, and 8.5, was studied.

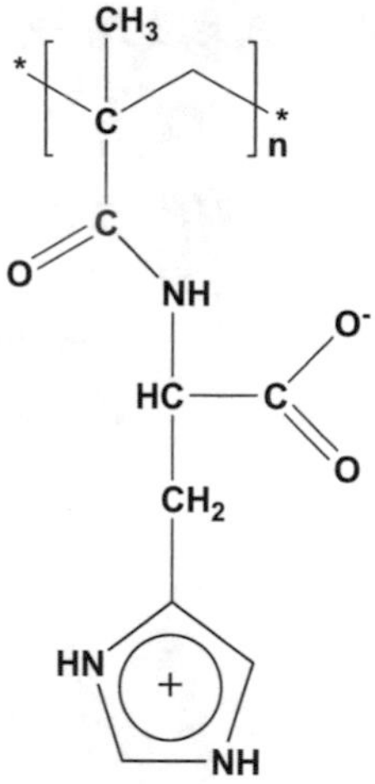

Fig. 15.19 Schematic
structure of the poly-MHist
polymer

The direct role of the imidazole ring was confirmed by substantial changes in silicic acid stabilization efficiency based on the following observations: (a) the protonation degree of the imidazole ring affects stabilization. At a relatively low pH of 5.5, the imidazole is protonated and the entire polymer acquires a zwitterionic character (−COO− is also present). Inhibitory activity increases considerably. In contrast, at a high pH of 8.5, the imidazole ring is neutral and the polymer backbone is anionic due to the presence of −COO− moieties. This results in total inactivity with respect to silicic acid stabilization. (b) The use of a similar, but pH-insensitive derivative poly(N-acryloyl-L-phenylalanine) (poly-PHE), which contains a phenylalanine instead of histidine, results in total loss of silicic acid stabilization activity. Finally, poly-MHist also shows effects on colloidal silica particle morphology. Increasing the poly-MHist concentration resulted in a reduced size of silica particles.

Further information on green polymeric inhibitors for silica scale can be found in selected literature reviews [18, 19, 116, 117].

15.3.6 Tagged Scale Inhibitors

One of the recently recognized problematic issues of scale inhibitors is their inability to sustain long-term inhibitory activity. Several causes are responsible: (a) inhibitor degradation (commonly by oxidizing biocides [84], [118]), (b) loss of inhibitor due to attachment onto mineral surfaces of the precipitating scales [119], (c) loss of inhibitor due to precipitation as metal-inhibitor "complexes" [120].

Hence, it is imperative that the precise concentration of the inhibitor in the water system is known to ensure inhibitory performance. One of the recent approaches in quantifying the fate of the inhibiting molecule is to "tag" the additive (either small molecule or polymer) with a fluorescent moiety. Such an approach can also facilitate

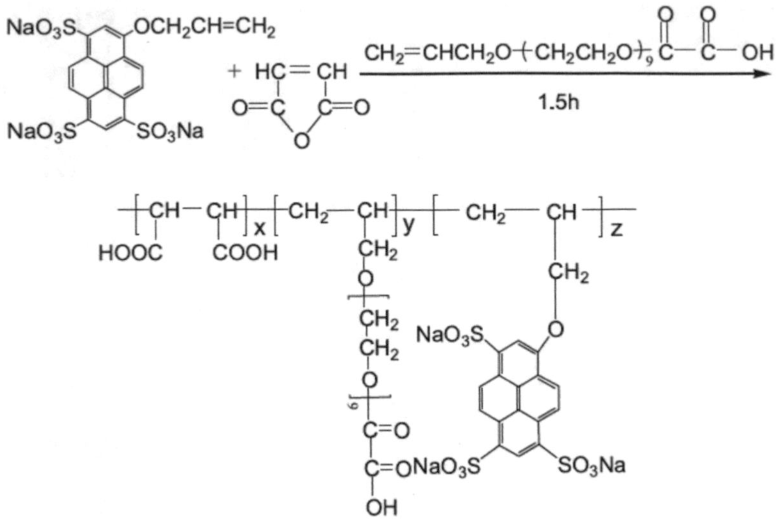

Fig. 15.20 Synthetic scheme for the preparation of the tagged polymer MA–APEM–APTA. Reproduced with permission from Ref. 123; Copyright 2014 © Elsevier

the on-line monitoring of the inhibition program [121]. An excellent review of the subject of tagged inhibitors has been published [122].

A novel fluorescent-tagged scale inhibitor, maleic anhydride–oxalic acid–allypolyethoxy carboxylate–8-hydroxy-1,3,6-pyrene trisulfonic acid trisodium salt (pyranine) (MA–APEM–APTA) was synthesized from maleic anhydride (MA), oxalic acid–allypolyethoxy carboxylate (APEM) and 8-hydroxy-1,3,6-pyrene trisulfonic acid trisodium salt (APTA), see Fig. 15.20 [123]. The polymer MA–APEM–APTA plays an important role on $CaCO_3$ inhibition, with higher inhibitory activity than polyacrylate. The effect on the formation of $CaCO_3$ was investigated with a combination of techniques, such as SEM, transmission electron microscopy (TEM), XRD and FT-IR analysis. The MA–APEM–APTA polymer was used to accurately measure polymer consumption on line.

In an important publication, Popov reported the synthesis of a tagged hydroxyethylidene-1,1-diphosphonic acid molecule (HEDP-F) [124], see Fig. 15.21.

HEDP-F was synthesized and used for fluorescent microscope visualization of gypsum crystal formation in supersaturated aqueous solutions. The visualization of HEDP-F location on gypsum crystals has demonstrated that the bisphosphonate molecules do not act as they are expected to do according to the current scale inhibition theory. At ambient temperature, the gypsum macrocrystals are found to form, and then to grow without visible sorption of bisphosphonate on the crystal edges or any other gypsum crystal growth centers.

Generally, the fluorescent markers (tags) should meet the following requirements: (i) synthetic availability of the dyes capable for polymerization; (ii) dye

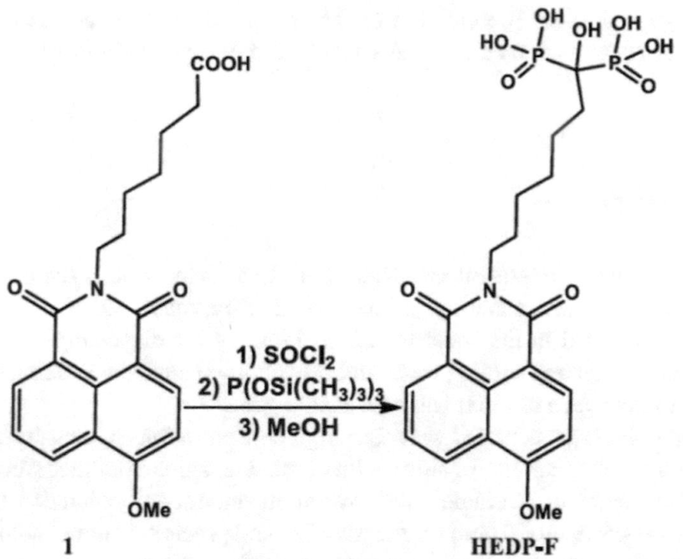

Fig. 15.21 Synthetic scheme for the preparation of tagged hydroxyethylidene-1,1-diphosphonic acid molecule (HEDP-F). Reproduced with permission from Ref. 124. Copyright 2019 © Wiley

chemical stability during polymerization; (iii) minimal influence of a dye on the polymer structure and on its molecular weight; (iv) the polymer structure should not affect the optical properties of a marker.

15.4 Inhibition Strategies

15.4.1 Threshold Inhibition

The threshold effect is explained by the adsorption of the inhibitor onto the crystal growth sites of the submicroscopic crystallites that are initially produced in the supersaturated solution, interfering with crystal growth and morphology [13]. This process can prevent crystal growth or at least delay it for prolonged periods of time [125]. Therefore, scale inhibition by threshold inhibitors is based on kinetic and not thermodynamic effects [126]. They are called threshold inhibitors and describe the mechanism of scale inhibition at sub-stoichiometric ratios.

In 1949, the first theory of threshold scale inhibition was proposed by Raistrick [127]. He noticed that the molecular dimensions of a polyphosphate chain corresponded very nearly to those of a calcite lattice. He proposed that, as embryo crystals of calcium carbonate started to form from a supersaturated solution, polyphosphate molecules became adsorbed onto their surfaces because of the close similarity of structure.

Although a detailed presentation of the threshold phenomenon is outside the scope of the present chapter, the reader is referred to several reviews on the subject [128–130].

15.4.2 Dispersion

In the field of water treatment and scale inhibition, dispersion is the avoidance of small scale particles adhering onto critical equipment surfaces. Dispersion approaches are used in the water treatment industry for dispersing scale-forming salts in boiler water and cooling water and preventing them from sticking to parts of the water system such as pipes and heat exchangers.

For example, supersaturated water has high concentrations of dissolved calcium, magnesium or other cations, together with carbonate, sulfate or other anions. These cause the formation of insoluble salts in water, the identity of which depends on the specific water chemistry. These salts crystallize and precipitate out of solution, and, eventually adhere to metal surfaces, agglomerating to form tenacious deposits. Dispersion chemistries and technologies can combat this phenomenon. Commonly, dispersing additives are polyelectrolytes, normally anionic. The polymer absorbs onto the crystal nucleus of the forming scale and changes its surface properties, such as surface charge. These modified particles demonstrate a much lower tendency to agglomerate due to electrostatic repulsion forces. The end result is that they remain in solution and do not adhere to the surfaces, for example a RO membrane. Some notable examples follow.

Amjad [131] has evaluated the dispersing effect of several synthetic and natural polymers on ferric oxide dispersion, such as starch, alginic acid, tannic acid, fulvic acid, lignosulfonate, poly(acrylic acid), poly(maleic acid), poly(acrylamide), poly(acrylic acid:2-acryalamido-2-methylpropane sulfonic acid), poly(acrylic acid:2-acryalamido-2-methylpropane sulfonic acid:sulfonated styrene), carboxymethyl inulin, and poly(diallyldimethyl ammonium chloride) [131]. It was found that the natural polymers displayed minor effects. CMI showed increased dispersing activity as the carboxylation degree increased. Finally, the ter-polymers were the most effective.

A double hydrophilic PEG-based block polymer, PEG double-ester of maleic anhydride-acrylic acid (PEGDMA-AA) was synthesized and evaluated as an Fe(III) dispersant [132]. The inhibition mechanism toward iron scales was supposed to be the formation of PEGDMA-AA-Fe^{3+}, while the PEG segments enhance the solubility of the Fe-polymer complex.

The influence of multifunctional polyelectrolytes on the dispersion of aluminum hydroxide particles was studied [133]. Three multifunctional polyelectrolytes were compared, a terpolymer of acrylic acid (AA), 2-acrylamide-2-methyl propane sulfonic acid (AMPS), and N-vinylpyrrolidone (NVP) (P(AA/SA/NVP)), an acrylic acid homopolymer (P(AA)), and a copolymer of AA and AMPS (P(AA/SA)). The influence of monomer units acting as functional groups, with respect to particle size and zeta potential was evaluated. The most effective dispersant was P(AA/SA/NVP), which prevented further coagulation among the initial particles and shifted the zeta potential to the most negative value. From the results, the authors concluded that the prominent dispersing capability of P(AA/SA/NVP) was due to its preferred extended conformation on the particle surface due to a subtle balance between the moderate affinity of NVP and the relatively higher affinities of AA and AMPS for aluminum hydroxide in an aqueous solution and the hydrophobicity of the amide groups of AMPS.

The dispersion performance of polyepoxysuccinate (PESA) on inorganic scales was evaluated with regards to ferric oxide [134]. The experimental results demonstrated that PESA functioned as an excellent dispersant for ferric oxide.

Polyaspartic acid–melamine grafted copolymer (PASPM) was synthesized by using polysuccinimide and melamine as the starting materials [135]. The grafted copolymer product was evaluated for its dispersion capacity for ferric oxide. It was found that PASPM was able to efficiently disperse Fe_2O_3.

The stability of alumina slurry in high ionic concentrations was evaluated using the settling rate of alumina abrasives in various slurries [136]. The ζ-potential was used to interpret the effect of electrostatic interaction. Various commercially available phosphonate dispersants (e.g., AMP) were evaluated in terms of the settling rate of alumina abrasives in the slurry. The stabilization behavior of those dispersants was not well correlated to the ζ-potential because the ζ-potentials were small (less than -10 mV). It may be related to the adsorption strength on the alumina surface relative to citric acid on the alumina particle surface.

The effectiveness of a mixed phosphonate/carboxylate additive, PBTC as dispersant for concentrated alpha-alumina ultrafine powder suspensions was studied [138]. Three grades of alpha alumina powders with diameters in the range of 100–300 nm were tested. The study was based on direct measurements of electrokinetic properties, adsorption isotherms and rheology and showed that PBTC was a good candidate for dispersion. In addition, PBTC was adsorbed as a monolayer through an inner sphere complex. The key contribution of the phosphonate group of the PBTC molecule in the adsorption mechanism was demonstrated and the influence of alumina surface chemistry, especially hydroxylation, on adsorption was emphaszed. Finally, it was shown that the concentrated suspensions could be used to implement a shaping process by direct coagulation casting (DCC).

15.4.3   Chelation

Several anionic additives, mainly phosphonates, are strong chelants for metal ions. In principle, scale inhibition could be achieved if all scaling cations (eg. Ca^{2+}, Mg^{2+}, Ba^{2+}, etc.) could be "inactivated" by a chelation process. However, such an approach would not be economically feasible, as large quantities of the chelating molecule would be required. Nevertheless, in certain applications, chelation is a viable strategy. For example, in water systems where both Mg^{2+} and silica exist at high concentrations, the risk of magnesium silicate precipitation and deposition is high. A successful approach was based on the use of the widely used ethylene-tetraacetic acid (EDTA) to chelate Mg^{2+} ions, and thus inhibit the formation of magnesium silicate [138].

Chelating agents can also be used for dissolution of existing scale deposits. For example, DETPA was used for the dissolution of barite [139]. Atomic Force Microscopy (AFM) was used to observe the dissolution of (001) cleavage surfaces of barite ($BaSO_4$). The etch pit geometry suggested that the active phosphonate groups of DETPA molecule formed bonds with the Ba^{2+} ions along with the [110] directions.

Several chelants were used for the dissolution of amorphous silica scale [140, 141]. These included ascorbic acid (vitamin C, ASC), citric acid (CITR), carboxy-methyl inulin (CMI), 3,4-dihydroxybenzoic acid (catechuic acid, DHBA), 3,4,5-trihydroxybenzoic acid (gallic acid, GA), dopamine hydrochloride (DOPA), iminodiacetic acid (IDA), histidine (HIST), phenylalanine (PHALA), and malic acid (MAL). It was found that all studied molecules showed variable dissolution efficiency, with MAL, CMI, HIST, and PHALA being the slowest/least effective dissolvers, and the catechol-containing DHBA, GA, and DOPA being the most effective ones. IDA and CITR have intermediate efficiency.

A thorough review on various chelants has appeared in the literature [142].

15.5 Problems Associated with the Application of Scale Inhibitors

15.5.1 Calcium Tolerance (Ca²⁺ Stress)

Since most mineral scale inhibiting additives are anionic in nature, they tend to strongly interact with dissolved metal cations present in the water system. Because Ca^{2+} is the most abundant metal ion, one of the most desirable properties of scale inhibitors is calcium tolerance. It could be defined as the ability of a certain scale inhibitor to be present in a Ca-rich solution, without precipitating out as an insoluble Ca-inhibitor "complex".

The precipitation of Ca-phosphonate salts can: (a) cause fouling of heat exchanger and reverse osmosis membrane surfaces and (b) decrease the phosphonate concentration in the system to the extent that severe calcium carbonate scaling can occur. It should be noted that Ca-inhibitor scales usually have the same inverse solubility features commonly observed with scales and they also impede heat transfer.

Three of the most common $CaCO_3$ scale inhibitors were tested for their calcium tolerance ability [143], AMP, HEDP, and PBTC. This quantified as the concentration of the inhibitor (in ppm) to withstand a 1000 ppm Ca-containing solution at pH 9. The ranking was found to be PBTC (185 ppm) > AMP (12 ppm) > HEDP (8 ppm).

In a separate study, Amjad found the same ranking under slightly different conditions (presence of 250 mg/L Ca, pH 9.50, 25 °C): PBTC (98 ppm) > AMP (26 ppm) > HEDP (12 ppm) [144]. The adverse effects of low calcium tolerance can be alleviated by the addition of cationic electrolytes, such as organoammonium-based polymers [145].

It should be noted that in some applications, precipitation of metal-inhibitor "complexes" is desirable. Such insoluble metal-inhibitor precipitates are used for controlled release of the active phosphonate inhibitor [146–148].

15.5.2   Sensitivity of Scale Inhibitors to Oxidizing Biocides

Biocides are additives that combat microbiological growth in process waters [149]. A drawback of certain scale inhibitors, such as HEDP, and certain aminomethylene phosphonates, such as AMP, HDTMP and DETMP, is their sensitivity to oxidizing biocides, such as chlorine or bromine-based biocides (necessary to control microbiological growth) [150, 151]. Orthophosphate (PO_4^{3-}), one of the degradation products, can cause calcium phosphate scale deposition in high hardness process waters. Knowledge of this susceptibility to oxidizers may help water system operators on decisions regarding which phosphonate additive to apply, at what dosage level and for how long.

The degradation of AMP in the presence of a hypobromite-based biocide at two temperatures 25 °C and 43 °C was studied [84, 118]. The principal conclusions of the studies revealed that AMP was susceptible to the biocide tested. In addition, AMP was not degraded completely, but only to a ~ 20% level (at 25 °C) and up to a ~ 25% level (at 43 °C).

Polymer performance for calcium phosphate inhibition is a critical factor. Amjad has studied the effect of oxidizing (chlorine-based) and non-oxidizing (isothiazoline, gluteraldehyde, quaternary ammonium compounds) biocides on acrylate-based dispersant polymers [152]. The incorporation of sulfonic groups in the polymer increases the capability to inhibit calcium phosphate at low dosages. The use of polymers containing acrylic acid, sulfonic acid, and sulfonated styrene (AA:SA:SS) perform better than copolymers tested. Overall, the oxidizing biocides had no impact on the performance of calcium phosphate inhibiting polymers. The non-oxidizing biocides that contained no charged groups had minimal effect on calcium phosphate inhibition.

The highest risk for water treatment systems is the degradation of phosphonates by oxidizing biocides. For example, in cooling water systems, both components must be present for different reasons. Phosphonates control scale formation and biocides mitigate microbiological growth. Degradation of phosphonate scale inhibitors can cause uncontrolled scale formation (since the scale inhibitor has been depleted). In addition, since orthophosphate is the major inorganic by-product of this decomposition, the risk of calcium phosphate scale formation is higher. In addition, the biocide itself is consumed when degrading the phosphonate inhibitor, thus increasing the risk of further microbiological growth.

15.5.3 Scale Inhibitor Entrapment and Inactivation

One of the most crucial issues to be tackled with when applying chemical inhibitors in industrial waters is their tendency of getting trapped in the forming scale matrix. This entrapment leads, in most cases, to inhibitor loss from solution. Practically, this means that a much lower critical inhibitor concentration is present in the water system, leading to ineffective scale control.

A notable example is silica scale. As mentioned before, either cationic or neutral polymers are used for silica scale mitigation. Colloidal silica that forms in the system is negatively charged and can strongly interact with the cationic polymers. The most often-encountered case is the one driven from oppositely charged moieties. This effect was observed with polyaminoamide dendrimers [99] and zwitterionic chitosan-based macromolecules [111, 112]. Similar phenomena have also been observed with the neutral silica inhibitors, PEG, [106] and polyvinylpyrrolidone [105]. These events can be explained by the tendency of such polymers to form hydrogen bonds with the hydroxylated silica surface.

Scale inhibitors, in particular phosphonates, can also be deactivated by precipitation with hardness ions, present in the system. However, it should be noted that, in certain cases, such precipitation, if appropriately controlled, can lead to effective corrosion control. For example, combination of Zn^{2+} ions and AMP generate a Zn-AMP crystalline precipitate on mild steel surfaces [153]. The composition of this protective layer is a Zn-AMP material based on spectroscopic comparisons and was found to be the same as an authentically prepared sample of Zn-AMP with the composition $\{Zn[HN(CH_2PO_3H)_3(H_2O)_3]\}_x$. Based on mass loss measurements the corrosion rate for the "control" sample is 2.5 mm/year (at pH 3), whereas for the Zn-AMP protected sample 0.9 mm/year, a 270% reduction in corrosion rate. The filming material is collected and subjected to FT-IR, XRF and energy dispersive spectroscopy (EDS) studies.

Additionally, the protective material acting as a corrosion barrier was studied in a system including Sr^{2+} and Ba^{2+} with HDTMP [154]. The effectiveness of the synergistic combinations of M^{2+} (M = Sr or Ba) and HDTMP, in a 1:1 ratio is dramatically pH-dependent. At "harsh" pH regions (pH 2.2), mass loss from the steel specimens was profound, resulting in high corrosion rates. At pH 7.0 corrosion rates were appreciably suppressed in the presence of combinations of Sr^{2+} or Ba^{2+} and HDTMP. Corrosion rates are concentration-dependent for Sr-HDTMP, whereas they are insensitive to Ba-HDTMP levels. The reader is referred to a concise review of the topic [155].

15.6 Conclusions and Outlook

In this chapter, a general overview of chemical control strategies and challenges was presented. There is vast literature dealing with a variety of scale deposits. Presentation of the plethora of cases in this chapter is beyond its scope, but the reader is referred to relevant literature that summarizes the topic. On the other hand, crystallization and deposit formation can be mitigated by chemical additives that are supplied purposely to the process water. These are called scale inhibitors and they could be either polymeric in nature, or "small molecules" (most commonly phosphonates). There are several kinds of such additives, depending on the nature of the scale to be inhibited.

The quest for the ideal scale/deposition inhibitor is still on-going. Unfortunately, a "universal" scale inhibitor is not available. Each mineral scale has its own idiosyncrasies, requiring special treatment. Water system operators must take into account several system variables for the proper selection of the scale inhibitor. These include system pH, supersaturation, presence (or absence) of oxidizing biocides, inhibitor tolerance to hardness, temperature, and potential of composite scale formation.

Future challenges in chemical scale control must face the increasingly stricter environmental regulations for less toxic chemical additives and lower discharge limitations. Also, the design and successful large scale synthesis of novel scale inhibitors are important challenges for researchers and industries working in the water treatment field.

References

1. Z. Amjad, K.D. Demadis, *Mineral Scales and Deposits: Scientific and Technological Approaches* (Elsevier, Amsterdam, 2015). ISBN: 9780444632289
2. K.D. Demadis, *Water Treatment Processes* (Nova Science Publishers, Inc, New York, 2012). ISBN: 9781621003670
3. T.R. Bott, *Fouling of Heat Exchangers* (Elsevier Science, 1995). ISBN: 9780444821867
4. F.N. Kemmer, *The Nalco Handbook* (McGraw-Hill, New York, 1998). ISBN: 0070458723
5. K. Sangwal, *Additives and Crystallization Processes: From Fundamentals to Applications* (Wiley, Chichester, 2007). ISBN: 9780470061534
6. J.C. Cowan, D.J. Weintritt, *Water-Formed Scale Deposits* (Gulf Publishing Co., Houston, 1976). ISBN: 0872018962
7. W.D. Carlson, Reviews in mineralogy, in *Mineralogical Society of America*, ed. by R. J. Reeder, vol. 11, (New York, 1975), p. 191
8. L.N. Plummer, E. Busenberg, Dissolution of aragonite-strontianite solid solutions in non-stoichiometric $Sr(HCO_3)_2-Ca(HCO_3)_2-CO_2-H_2O$ solutions. Geochim. Cosmochim. Acta **46**, 1011–1040 (1982)
9. S.N. Kazi, Fouling and fouling mitigation of calcium compounds on heat exchangers by novel colloids and surface modifications. Rev. Chem. Eng. (2019). https://doi.org/10.1515/revce-2017-0076
10. J.P. Miller, A portion of the system calcium carbonate-carbon dioxide-water, with geological implications. Am. J. Sci. **250**, 161–203 (1952)
11. A. Lancia, D. Musmarra, M. Prisciandaro, Calcium sulphate, in *Kirk-Othmer Encyclopedia of Chemical Technology*, (Wiley, New York, 2011), pp. 1–22
12. D. Hasson, J. Zahavi, Mechanism of calcium sulfate scale deposition on heat transfer surfaces. Ind. Eng. Chem. Fundam. **9**, 1–10 (1970)
13. O. Linnikov, Investigation of the initial period of sulphate scale formation Part 1. Kinetics and mechanism of calcium sulphate surface nucleation at its crystallization on a heat-exchange surface. Desalination **122**, 1–14 (1999)

14. K.D. Demadis, Inhibition and control of colloidal silica: Can chemical additives untie the "Gordian knot" of scale formation?, Corrosion 2007, NACE-07058, 62nd Annual Conference Expo., Nashville, 11–15 March 2007
15. R.K. Iler, *The Chemistry of Silica: Solubility, Polymerization, Colloid and Surface Properties, and Biochemistry* (Wiley, New York, 1979). ISBN: 978-0-471-02404-0
16. T. Coradin, J. Livage, Effect of some amino acids and peptides on silicic acid polymerization. Colloids Surf. B. Biointerfaces **21**, 329–336 (2001)
17. S. Binauld, M.H. Stenzel, Acid-degradable polymers for drug delivery: A decade of innovation. Chem. Commun. **49**, 2082–2102 (2013)
18. K.D. Demadis, M. Preari, I. Antonakaki, Naturally-derived and synthetic polymers as biomimetic enhancers of silicic acid solubility in (bio)silicification processes. Pure Appl. Chem. **86**, 1663–1674 (2014)
19. A. Spinthaki, G. Skordalou, A. Stathoulopoulou, K.D. Demadis, Modified macromolecules in the prevention of silica scale. Pure Appl. Chem. **88**, 1037–1047 (2016)
20. K.S. Soppimath, T.M. Aminabhavi, A.R. Kulkarni, W.E. Rudzinski, Biodegradable polymeric nanoparticles as drug delivery devices. J. Control. Release **7**, 1–20 (2001)
21. Z.-G. Zhao, Adsorption of phenylalanine from aqueous solution onto active carbon and silica gel. Chin. J. Chem. **10**, 325–330 (1992)
22. J.H. Kennedy, HPLC purification of pergolide using silica gel. Org. Process. Res. Dev. **1**, 68–71 (1997)
23. H. Ehrlich, K.D. Demadis, P.G. Koutsoukos, O. Pokrovsky, Modern views on desilicification: Biosilica and abiotic silica dissolution in natural and artificial environments. Chem. Rev. **110**, 4656–4689 (2010)
24. O.O. Taspinar, S. Ozgul-Yucel, Lipid adsorption capacities of magnesium silicate and activated carbon prepared from the same rice hull. Eur. J. Lipid Sci. Technol. **110**, 742–746 (2008)
25. A. Spinthaki, G. Petratos, J. Matheis, W. Hater, K.D. Demadis, The precipitation of "magnesium silicate" under geothermal stresses: Formation and characterization. Geothermics **74**, 172–180 (2018)
26. I.M. Ali, Y.H. Kotp, I.M. El-Naggar, Thermal stability, structural modifications and ion exchange properties of magnesium silicate. Desalination **259**, 228–234 (2010)
27. I. Rashid, N. Daraghmeh, M. Al-Remawi, S.A. Leharne, B.Z. Chowdhry, A. Badwan, Characterization of chitin-metal silicates as binding superdisintegrants. J. Pharm. Sci. **98**, 4887–4901 (2009)
28. P.R. Young, *Magnesium Silicate Precipitation*. Paper 466, Corrosion 93, NACE International: Houston, 1993
29. K.D. Demadis, Water treatment's "Gordian Knot". Chem. Process. **66**, 29–32 (2003)
30. H. Kristmanndóttir, M. Ólafsson, S. Thórhallsson, Magnesium silicate scaling in district heating systems in Iceland. Geothermics **18**, 191–198 (1989)
31. M. Brooke, *Magnesium Silicate Scale in Circulating Cooling Systems*. Paper 327, Corrosion 84, NACE International: Houston, 1984
32. T.M. Seward, Determination of the first ionization constant of silicic acid from quartz solubility in borate buffer solutions to 350°C. Geochim. Cosmochim. Acta **38**, 1651–1664 (1974)
33. M. Morita, Y. Goto, S. Motoda, T. Fujino, Thermodynamic analysis of silica-based scale precipitation induced by magnesium ion. J. Geotherm. Res. Soc. Japan **39**, 191–201 (2017)
34. K.D. Demadis, Recent developments in controlling silica and magnesium silicate in industrial water systems, in *Science and Technology of Industrial Water Treatment*, (CRC Press, London, 2010), pp. 179–203
35. L.H.P. Jones, K.A. Handreck, The effect of iron and aluminium oxides on silica in solution in soils. Nature **198**, 852–853 (1963)
36. C.J. Gabelich, W.R. Chen, T.I. Yun, B.M. Coffey, I.H. Suffet, The role of dissolved aluminum in silica chemistry for membrane processes. Desalination **180**, 307–319 (2005)

37. J. Liu, S. Bi, L. Yang, X. Gu, P. Ma, N. Gan, X. Wang, X. Long, F. Zhang, Speciation analysis of aluminium(III) in natural waters and biological fluids by complexing with various catechols followed by differential pulse voltammetry detection. Analyst **127**, 1657–1665 (2002)
38. T. Yokoyama, Y. Sato, Y. Maeda, T. Tarutani, R. Itoi, Siliceous deposits formed from geothermal water. I. The major constituents and the existing states of iron and aluminum. Geochem. J. **27**, 375–384 (1993)
39. D.L. Gallup, Aluminum silicate scale formation and inhibition: Scale characterization and laboratory experiments. Geothermics **26**, 483–499 (1997)
40. S. Lacour, V. Deluchat, J.-C. Bollinger, B. Serpaud, Complexation of trivalent cations (Al(III), Cr(III), Fe(III)) with two phosphonic acids in the pH range of fresh waters. Talanta **46**, 999–100 (1998)
41. D.L. Gallup, Aluminium silicate scale formation and inhibition (2): Scale solubilities and laboratory and field inhibition tests. Geothermics **27**, 485–501 (1998)
42. P. Canizares, F. Martınez, C. Jimenez, J. Lobato, M.A. Rodrigo, Comparison of the aluminum speciation in chemical and electrochemical dosing processes. Ind. Eng. Chem. Res. **45**, 8749–8756 (2006)
43. C.C. Perry, K.L. Shafran, The systematic study of aluminium speciation in medium concentrated aqueous solutions. J. Inorg. Biochem. **87**, 115–124 (2001)
44. B.A. Browne, Soluble aluminium silicates: Stochoiometry, stability and implications for environmental geochemistry. Science **256**, 1667–1670 (1992)
45. T. Yokoyama, Y. Sato, M. Nakai, K. Sunahara, R. Itoi, Siliceous deposits formed from geothermal water in Kyushu, Japan: II. Distribution and state of aluminum along the growth direction of the deposits. Geochemical J **33**, 13–18 (1999)
46. D. Carolina Figueroa Murcia, P.L. Fosbøl, K. Thomsen, E.H. Stenby, Determination of zinc sulfide solubility to high temperatures. J. Solut. Chem. **46**, 1805–1817 (2017)
47. C.-H. Choi, Y.-W. Su, C.-H. Chang, Effects of fluid flow on the growth and assembly of ZnO nanocrystals in a continuous flow microreactor. CrystEngComm **15**, 3326–3333 (2013)
48. M. Holmer, H. Harald Hasler-Sheetal, Sulfide intrusion in sea grasses assessed by stable sulfur isotopes – a synthesis of current results. Front. Marine Sci. **1**, 64 (2014). https://doi.org/10.3389/fmars.2014.00064
49. X. Liang, P. Guo, G. Wang, R. Deng, D. Pan, X. Wei, Dilute magnetic semiconductor Cu_2MnSnS_4 nanocrystals with a novel zincblende and wurtzite structure. RSC Adv. **2**, 5044–5046 (2012)
50. S.V. Dorozhkin, Calcium orthophosphates. Occurrence, properties, biomineralization, pathological calcification and biomimetic applications. Biomatter. **1**, 121–164 (2011)
51. S.V. Dorozhkin, Calcium orthophosphates ($CaPO_4$): Occurrence and properties. Prog. Biomater. **5**, 9–70 (2016)
52. S.V. Dorozhkin, Amorphous calcium orthophosphates: Nature, chemistry and biomedical applications. Int. J. Mater. Chem. **2**, 19–46 (2012)
53. J. Berzelius, Ueber basische phosphorsaure kalkerde. Ann. Chem. Pharmac. **53**, 286–288 (1845)
54. S.V. Dorozhkin, Calcium orthophosphates and human beings: A historical perspective from the 1770s until 1940. Biomatter. **1**, 53–70 (2012)
55. D.R. Lide, *The CRC Handbook of Chemistry and Physics*, 86th edn. (CRC Press, Boca Raton, 2005). ISBN: 0849304865
56. R.Z. Legeros, *Calcium Phosphates in Oral Biology and Medicine* (Karger, Basel, 1991). ISBN: 978-3-8055-5236-3
57. J.C. Elliot, *Structure and Chemistry of the Apatite and Other Calcium Orthophosphates* (Elsevier, Amsterdam, 1994). ISBN: 9780444815828
58. Z. Amjad, *Calcium Phosphates in Biological and Industrial Systems* (Kluwer Academic Publishers, Norwell, 1998). ISBN: 9781461375210
59. H.A. Lowenstam, S. Weiner, Transformation of amorphous calcium phosphate to crystalline dahillite in the radular teeth of chitons. Science **227**, 51–53 (1985)

60. J.D. Termine, A.S. Posner, Infrared analysis of rat bone: Age dependency of amorphous and crystalline mineral fractions. Science **153**, 1523–1525 (1966)
61. S. Weiner, I. Sagi, L. Addadi, Choosing the crystallization path less travelled. Science **309**, 1027–1028 (2005)
62. Y.T. Cheng, W.L. Johnson, Disordered materials: A survey of amorphous solids. Science **235**, 997–1002 (1987)
63. E.D. Eanes, I.H. Gillessen, A.S. Posner, Intermediate states in the precipitation of hydroxy-apatite. Nature **208**, 365–367 (1965)
64. P. Gras, C. Rey, G. André, C. Charvillat, S. Sarda, C. Combes, Crystal structure of mono-clinic calcium pyrophosphate dihydrate (m-CPPD) involved in inflammatory reactions and osteoarthritis. Acta Cryst. B **72**, 96–101 (2016)
65. C.M. Sevrain, M. Berchel, H. Couthon, P.-A. Jaffrès, Phosphonic acid: Preparation and applications. Beilstein J. Org. Chem. **13**, 2186–2213 (2017)
66. V. Deluchat, B. Serpaud, E. Alves, C. Caullet, J.-C. Bollinger, Protonation and complexation constants of phosphonic acids with cations of environmental interest. Phosphorus Sulfur **109**, 209–212 (1996)
67. T. Jain, E. Sanchez, E. Owens-Bennett, R. Trussell, S. Walker, H. Liu, Impacts of antiscalants on the formation of calcium solids: Implication on scaling potential of desalination concentrate. Environ. Sci. Water Res. Technol. **5**, 1285–1294 (2019)
68. N. Wada, K. Yamashita, T. Umegaki, Effects of carboxylic acids on calcite formation in the presence of Mg^{2+} ions. J. Colloid Interface Sci. **212**, 357–364 (1999)
69. Y. Chen, Y. Zhou, Q. Yao, Y. Bu, H. Wang, W. Wu, W. Sun, Preparation of a low-phosphorous terpolymer as a scale, corrosion inhibitor, and dispersant for ferric oxide. J. Appl. Polym. Sci. **132**, 41447 (2015)
70. S.B. Ahmed, M.M. Tlili, M.B. Amor, Influence of a polyacrylate antiscalant on gypsum nucleation and growth. Cryst. Res. Technol. **43**, 935–942 (2008)
71. S.P. Carvalho, L.C.M. Palermo, L. Boak, K. Sorbie, E.F. Lucas, The influence of terpolymer based on amide, carboxylic and sulfonic groups on the barium sulphate inhibition. Energy Fuels **31**, 10648–10654 (2017)
72. K.E. Papathanasiou, A. Moschona, A. Spinthaki, M. Vassaki, K.D. Demadis, Silica-based polymeric gels as platforms for delivery of phosphonate pharmaceutics, in *Polymer Gels: Synthesis and Characterization*, ed. by M. K. Thakur, (Springer, 2018), pp. 127–140
73. A. Moschona, N. Plesu, G. Mezei, A.G. Thomas, K.D. Demadis, Corrosion protection of carbon steel by tetraphosphonates of systematically different molecular size. Corros. Sci. **145**, 135–150 (2018)
74. Z. Amjad, Investigations on the influence of phosphonates in dispersing iron oxide (rust) by polymeric additives for industrial water applications. Int. J. Corros. Scale Inhib. **3**, 89–100 (2014)
75. E.N. Rizkalla, M.T.M. Zaki, I.M. Ismail, Metal chelates of phosphonate-containing ligands-V Stability of some 1-hydroxyethane-1,1-diphosphonic acid metal chelates. Talanta **27**, 715–719 (1980)
76. M.M. Reddy, G.H. Nancollas, Calcite crystal growth inhibition by phosphonates. Desalination **12**, 61–73 (1973)
77. R.M. Cigala, M. Cordaro, F. Crea, C. De Stefano, V. Fracassetti, M. Marchesi, D. Milea, S. Sammartano, Acid−base properties and alkali and alkaline earth metal complex formation in aqueous solution of diethylenetriamine -N,N,N′,N″,N″-pentakis(methylenephosphonic acid) obtained by an efficient synthetic procedure. Ind. Eng. Chem. Res. **53**, 9544––9553 (2014)
78. D. Drzyzga, J. Lipok, Analytical insight into degradation processes of aminopolyphosphonates as potential factors that induce cyanobacterial blooms. Environ. Sci. Pollut. Res. Int. **24**, 24364–24375 (2017)
79. A. Obojska, B. Lejczak, M. Kubrak, Degradation of phosphonates by streptomycete isolates. Appl. Microbiol. Biotechnol. **51**, 872–876 (1999)

80. D. Villemin, M.A. Didi, Aminomethylenephosphonic acids syntheses and applications (A review). Orient. J. Chem. **31**, 1–12 (2015)
81. A. Spinthaki, J. Matheis, W. Hater, K.D. Demadis, Antiscalant-driven inhibition and stabilization of "magnesium silicate" under geothermal stresses: The role of magnesium–phosphonate coordination chemistry. Energy Fuels **32**, 11749–11760 (2018)
82. K.D. Demadis, P. Lykoudis, Chemistry of organophosphonate scale growth inhibitors: 3. Physicochemical aspects of 2–phosphonobutane–1,2,4–tricarboxylate (PBTC) and its effect on CaCO$_3$ crystal growth. Bioinorg. Chem. Appl. **3**, 135–149 (2005)
83. M.F. Mady, A. Bagi, M.A. Kelland, Synthesis and evaluation of new bisphosphonates as inhibitors for oilfield carbonate and sulfate scale control. Energy Fuel **30**, 9329–9338 (2016)
84. K.D. Demadis, Chemistry of organophosphonate scale growth inhibitors: 4. Stability of amino–tris–methylene phosphonate towards oxidizing biocides. Phosphorus Sulfur **181**, 167–176 (2006)
85. E. Akyol, M. Öner, E. Barouda, K.D. Demadis, Systematic structural determinants of the effects of tetraphosphonates on gypsum crystallization. Cryst. Growth Des. **9**, 5145–5154 (2009)
86. P. Zhang, C. Fan, H. Lu, A.T. Kan, M.B. Tomson, Synthesis of crystalline-phase silica-based calcium phosphonate nanomaterials and their transport in carbonate and sandstone porous media. Ind. Eng. Chem. Res. **50**, 1819–1830 (2011)
87. J. Jiang, J. Zhao, Y. Xu, Molecular simulations and critical pH studies for the interactions between 2-phosphonobutane-1,2,4-tricarboxylic acid and calcite surfaces in circular cooling water systems. Des. Wat. Treat. **57**, 2152–2158 (2016)
88. E. Shall, H. Rashad, M.M.A. Abdel-Aal, Effect of phosphonate additive on crystallization of gypsum in phosphoric and sulfuric acid medium. Crystal Res. Technol. **37**, 1264–1273 (2002)
89. M. Xia, C. Chen, Probing the inhibitory mechanism of calcite precipitation by organic phosphonates in industrial water cooling system. Int. J. Environ. Sci. Develop. **6**, 300–304 (2015)
90. M.F.B. Sousa, C.A. Bertran, New methodology based on static light scattering measurements for evaluation of inhibitors for in bulk CaCO$_3$ crystallization. J. Colloid Interface Sci. **420**, 57–64 (2014)
91. M.A. Kelland, Effect of various cations on the formation of calcium carbonate and barium sulfate scale with and without scale inhibitors. Ind. Eng. Chem. Res. **50**, 5852–5861 (2011)
92. L.F. Greenlee, F. Testa, D.F. Lawler, B.D. Freeman, P. Moulin, The effect of antiscalant addition on calcium carbonate precipitation for a simplified synthetic brackish water reverse osmosis concentrate. Wat. Res. **44**, 2957–2969 (2010)
93. Z. Amjad, P.G. Koutsoukos, Evaluation of maleic acid based polymers as scale inhibitors and dispersants for industrial water applications. Desalination **335**, 55–63 (2014)
94. B. Senthilmurugan, B. Ghosh, S. Kundu, M. Haroun, B. Kameshwari, Maleic acid based scale inhibitors for calcium sulfate scale inhibition in high temperature application. J. Petr. Sci. Eng. **75**, 189–195 (2010)
95. V.V. Annenkov, E.N. Danilovtseva, E.A. Filina, Y.V. Likhoshway, Interaction of silicic acid with poly(1-vinylimidazole). J. Polym. Sci. A **44**, 820–827 (2006)
96. E. Neofotistou, K.D. Demadis, Silica scale inhibition by polyaminoamide STARBURST® dendrimers. Colloids Surf. A Physicochem. Eng. Asp. **242**, 213–216 (2004)
97. K.D. Demadis, A structure/function study of polyaminoamide dendrimers as silica scale growth inhibitors. J. Chem. Technol. Biotechnol. **80**, 630–640 (2005)
98. K.D. Demadis, E. Neofotistou, Synergistic effects of combinations of cationic polyaminoamide dendrimers/anionic polyelectrolytes on amorphous silica formation: A bioinspired approach. Chem. Mater. **19**, 581–587 (2007)
99. A. Stathoulopoulou, K.D. Demadis, Enhancement of silicate solubility by use of "green" additives: Linking green chemistry and chemical water treatment. Desalination **224**, 223–230 (2008)

100. K.D. Demadis, A. Stathoulopoulou, Novel, multifunctional, environmentally friendly additives for effective control of inorganic foulants in industrial water and process applications. Mater. Perform. **45**, 40–44 (2006)
101. K.D. Demadis, A. Stathoulopoulou, Solubility enhancement of silicate with polyamine/polyammonium cationic macromolecules: Relevance to silica-laden process waters. Ind. Eng. Chem. Res. **45**, 4436–4440 (2006)
102. A. Ketsetzi, A. Stathoulopoulou, K.D. Demadis, Being "green" in chemical water treatment technologies: Issues, challenges and developments. Desalination **223**, 487–493 (2008)
103. K.D. Demadis, A. Tsistraki, A. Popa, G. Ilia, A. Visa, Promiscuous stabilisation behavior of silicic acid by cationic macromolecules: The case of phosphonium-grafted dicationic ethylene oxide bolaamphiphiles. RSC Adv. **2**, 631–641 (2012)
104. K. Spinde, K. Pachis, I. Antonakaki, E. Brunner, K.D. Demadis, Influence of polyamines and related macromolecules on silicic acid polycondensation: Relevance to "soluble silicon pools"? Chem. Mater. **23**, 4676–4687 (2011)
105. M. Preari, K. Spinde, J. Lazic, E. Brunner, K.D. Demadis, Bioinspired insights into silicic acid stabilization mechanisms: The dominant role of polyethylene glycol-induced hydrogen bonding. J. Am. Chem. Soc. **136**, 4236–4244 (2014)
106. E. Neofotistou, K.D. Demadis, Use of antiscalants for mitigation of silica (SiO_2) fouling and deposition: Fundamentals and applications in desalination systems. Desalination **167**, 257–272 (2004)
107. M.A. Quraishi, L.H. Farooqi, P.A. Saini, Investigation of some green compounds as corrosion and scale inhibitors for cooling systems. Corrosion **55**, 493–497 (1999)
108. S.Q.A. Mahat, I.M. Saaid, B. Lal, Green silica scale inhibitors for alkaline-surfactant-polymer flooding: A review. J. Petrol. Explor. Prod. Technol. **6**, 379–385 (2016)
109. A. Spinthaki, C. Zerfass, H. Paulsen, S. Hobe, K.D. Demadis, Pleiotropic role of recombinant silaffin-like cationic polypeptide P5S3: Peptide-induced silicic acid stabilization, silica formation and inhibition of silica dissolution. Chem. Select **2**, 6–17 (2017)
110. K.D. Demadis, A. Ketsetzi, K. Pachis, V.M. Ramos, Inhibitory effects of multicomponent, phosphonate-grafted, zwitter-ionic chitosan biomacromolecules on silicic acid condensation. Biomacromolecules **9**, 3288–3293 (2008)
111. K.D. Demadis, K. Pachis, A. Ketsetzi, A. Stathoulopoulou, Bioinspired control of colloidal silica in vitro by dual polymeric assemblies of zwitterionic phosphomethylated chitosan and polycations or polyanions. Adv. Colloid Interf. Sci. **151**, 33–48 (2009)
112. B. Akın, M. Öner, Y. Bayram, K.D. Demadis, Effects of carboxylate-modified, "green" inulin biopolymers on the crystal growth of calcium oxalate. Cryst. Growth Des. **8**, 1997–2005 (2008)
113. K.D. Demadis, I. Léonard, Carboxymethylinulin "green" polymeric additives for control of calcium oxalate in industrial water and process applications. Mater. Perform. **50**, 40–44 (2011)
114. K.D. Demadis, M. Preari, "Green" scale inhibitors in water treatment processes: The case of silica scale inhibition. Desalin. Water Treat. **55**, 749–755 (2015)
115. K.D. Demadis, S. Brückner, E. Brunner, S. Paasch, I. Antonakaki, M. Casolaro, The intimate role of imidazole in the stabilization of slicic acid by a pH-responsive, histidine-grafted poly-ampholyte. Chem. Mater. **27**, 6827–6836 (2015)
116. K.D. Demadis, M. Öner, Inhibitory effects of "green" additives on the crystal growth of sparingly soluble salts, in *Green Chemistry Research Trends*, ed. by J. T. Pearlman, (Nova Science Publishers, New York, 2009), pp. 265–287
117. A. Spinthaki, K.D. Demadis, Bioinspired "green" scale inhibitors for mitigation of silica scales, in *Industrial Water Treatment: Trends, Challenges, and Solutions*, ed. by Z. Amjad, T. Chen, (NACE Publications, 2017), pp. 71–86
118. K.D. Demadis, A. Ketsetzi, Degradation of water treatment chemical additives in the presence of oxidizing biocides: "Collateral damages" in industrial water systems. Sep. Sci. Technol. **42**, 1639–1649 (2007)

119. W. El Malti, D. Laurencin, G. Guerrero, M.E. Smith, P.H. Mutin, Surface modification of calcium carbonate with phosphonic acids. J. Mater. Chem. **22**, 1212–1218 (2012)
120. K.D. Demadis, G. Angeli, "Good scale"–"bad scale". How metal–phosphonate materials contribute to corrosion inhibition, in *Mineral Scales in Biological and Industrial Systems*, (Taylor and Francis, New York, 2013), pp. 353–370
121. B. Moriarty, On-line monitoring of water treatment chemicals, in *Mineral Scales and Deposits: Scientific and Technological Approaches*, ed. by Z. Amjad, K. D. Demadis, (Elsevier, Amsterdam, 2015), pp. 737–745
122. M. Oshchepkov, K. Popov, Fluorescent markers in water treatment, in *Desalination and Water Treatment*, ed. by M. Eyvaz, (InTech Open, 2018), pp. 311–331
123. H. Wang, Y. Zhou, Q. Yao, S. Ma, W. Wu, W. Sun, Synthesis of fluorescent-tagged scale inhibitor and evaluation of its calcium carbonate precipitation performance. Desalination **340**, 1–10 (2014)
124. M. Oshchepkov, S. Semen Kamagurov, S. Sergei Tkachenko, A. Ryabova, K. Popov, Insight into the mechanisms of scale inhibition: A case study of a task-specific fluorescent-tagged scale inhibitor location on gypsum crystals. ChemNanoMat **5**, 586–592 (2019)
125. X. Li, D. Hasson, H. Shemer, Flow conditions affecting the induction period of $CaSO_4$ scaling on RO membranes. Desalination **431**, 119–125 (2018)
126. J. Fink, *Petroleum Engineer's Guide to Oil Field Chemicals and Fluids* (Elsevier, Amsterdam, 2015). 2015, and references therein. ISBN: 9780123838445
127. B. Raistrick, The influence of foreign ions on crystal growth from solution. 1. The stabilization of the supersaturation of calcium carbonate solutions by anions possessing O-P-O-P-O chains. Discuss. Faraday Soc. **5**, 234–237 (1949)
128. T.A. Hoang, Mechanisms of scale formation, in *Mineral Scales and Deposits: Scientific and Technological Approaches*, ed. by Z. Amjad, K. D. Demadis, (Elsevier, Amsterdam, 2015), pp. 47–83
129. P. Zhang, A.T. Kan, M.B. Tomson, Oil field mineral scale control, in *Mineral Scales and Deposits: Scientific and Technological Approaches*, ed. by Z. Amjad, K. D. Demadis, (Elsevier, Amsterdam, 2015), pp. 603–617
130. S. Dobberschütz, M.R. Nielsen, K.K. Sand, R. Civioc, N. Bovet, S.L.S. Stipp, M.P. Andersson, The mechanisms of crystal growth inhibition by organic and inorganic inhibitors. Nat. Commun. **9**, 1578 (2018)
131. Z. Amjad, D. Guyton, Biopolymers and synthetic polymers as iron oxide dispersants for industrial water applications. Mater. Perform. **51**, 48–53 (2012)
132. Y. Liu, Y. Zhou, Q. Yao, W. Sun, Evaluating the performance of PEG-based scale inhibition and dispersion agent in cooling water systems. Desalin. Water Treat. **56**, 1309–1320 (2015)
133. I. Nishida, Y. Okaue, T. Yokoyama, Effects of adsorption conformation on the dispersion of aluminum hydroxide particles by multifunctional polyelectrolytes. Langmuir **26**, 1663–11669 (2010)
134. X. Zhou, Y. Sun, Y. Wang, Inhibition and dispersion of polyepoxysuccinate as a scale inhibitor. J. Environ. Sci. **23**, S159–S161 (2011)
135. Y. Xu, L.L. Zhao, L.N. Wang, S.Y. Xu, Y.C. Cui, Synthesis of polyaspartic acid–melamine grafted copolymer and evaluation of its scale inhibition performance and dispersion capacity for ferric oxide. Desalination **286**, 285–289 (2012)
136. Q. Luo, Stabilization of alumina polishing slurries using phosphonate dispersants. Ind. Eng. Chem. Res. **39**, 3249–3254 (2000)
137. A.L. Penard, F. Rossignol, H.S. Nagaraja, C. Pagnoux, T. Chartier, Dispersion of alpha-alumina ultrafine powders using 2-phosphonobutane-1,2,4-tricarboxylic acid for the implementation of a DCC process. J. Europ. Ceram. Soc. **25**, 1109–1118 (2005)
138. K.D. Demadis, A. Ketsetzi, E.-M. Sarigiannidou, Catalytic effect of magnesium ions on silicic acid polycondensation and inhibition strategies based on chelation. Ind. Eng. Chem. Res. **51**, 9032–9040 (2012)

139. A. Putnis, J.L. Junta-Rosso, M.F. Hochella, Dissolution of barite by a chelating ligand: An atomic force microscopy study. Geochim. Cosmochim. Acta **59**, 4623–4632 (1995)
140. K.D. Demadis, E. Mavredaki, Green additives to enhance silica dissolution during water treatment. Environ. Chem. Lett. **3**, 127–131 (2005)
141. K.D. Demadis, E. Mavredaki, M. Somara, Additive-driven dissolution enhancement of colloidal silica. 2. Environmentally friendly additives. Ind. Eng. Chem. Res **50**, 13866–13876 (2011)
142. T. Knepper, Synthetic chelating agents and compounds exhibiting complexing properties in the aquatic environment. Trends Anal. Chem. **22**, 708–724 (2003)
143. K.D. Demadis, B. Yang, P.R. Young, D.L. Kouznetsov, D.G. Kelley, Rational development of new cooling water chemical treatment programs for scale and microbial control, in *Advances in Crystal Growth Inhibition Technologies*, ed. by Z. Amjad, (Plenum Publishing Corporation, New York, 2000), pp. 215–234
144. Z. Amjad, R.W. Zuhl, J.F. Zibrida, Factors Influencing the Precipitation of Calcium-Inhibitor Salts in Industrial Water Systems. Association of Water Technologies, Inc. Annual Convention, Phoenix, September 17–20 2003
145. Z. Amjad, R.W. Zuhl, Influence of cationic polymers on the performance of anionic polymers as precipitation inhibitors for calcium phosphonates. Phosphorus Res. Bull. **13**, 59–65 (2002)
146. V. Tantayakom, H.S. Fogler, P. Charoensirithavorn, S. Chavadej, Kinetic study of scale inhibitor precipitation in squeeze treatment. Cryst. Growth Des. **5**, 329–335 (2005)
147. A.T. Kan, J.E. Oddo, M.B. Tomson, Formation of two calcium diethylenetriaminepentakis(m ethylenephosphonic acid) precipitates and their physical chemical properties. Langmuir **10**, 1450–1455 (1994)
148. R. Pairat, C. Sumeath, F.H. Browning, H.S. Fogler, Precipitation and dissolution of calcium-ATMP precipitates for the inhibition of scale formation in porous media. Langmuir **13**, 1791–1798 (1997)
149. K.M. Wiencek, J.S. Chapman, Water Treatment Biocides: How Do they Work and why Should you Care? Corrosion 99, NACE – 99308, NACE International, 25–30 April 1999, San Antonio
150. R.D. Bartholomew, Bromine-based biocides for cooling water systems: A literature review. International Water Conference, Paper no. 74, 1998, p. 523
151. M. Vaska, W. Go, Microbial control. Evaluation of alternatives to gaseous chlorine for cooling water, Industrial Water Treatment, March/April, 39 1993
152. Z. Amjad, R.W. Zuhl, J.F. Zibrida, The effect of biocides on deposit control polymer performance. Association of Water Technologies, Inc. 2000 Annual Convention, Oct 31 – Nov 4, 2000, Honolulu
153. K.D. Demadis, S.D. Katarachia, M. Koutmos, Crystal growth and characterization of zinc–amino–tris(methylenephosphonate) organic–inorganic hybrid networks and their inhibiting effect on metallic corrosion. Inorg. Chem. Commun. **8**, 254–258 (2005)
154. K.D. Demadis, E. Barouda, R.G. Raptis, H. Zhao, Metal tetraphosphonate "wires" and their corrosion inhibiting passive films. Inorg. Chem. **48**, 819–821 (2009)
155. M. Papadaki, K.D. Demadis, Structural mapping of hybrid metal phosphonate corrosion inhibiting thin films. Comment Inorg. Chem. **30**, 89–118 (2009)

Chapter 16
Technologies for Biofouling Control and Monitoring in Desalination

H. J. G. Polman, H. A. Jenner, and M. C. M. Bruijs

16.1 Introduction

Seawater desalination plants using a distillation process began operation in the mid-1960s. The standard technologies applied were Multiple-effect distillation (MED) and Multi-stage flash distillation (MSF). In the past decades, application of Reverse Osmosis (RO) increased across the world since its higher energy efficiency in comparison with distillation. Nowadays, RO is the preferred technology in most parts of the world, although in the Middle East the major large seawater desalination plants in operation still use the distillation process [1]. Currently, with the increasing scale of desalination facilities, also in the Middle East and North Africa (MENA country), RO is used at a larger scale due to its lower energy consumption. The increase in production capacity of RO-desalination plants is shown in Fig. 16.1 [2].

The volume of seawater is directly related to the production capacity of the plant and goes up to tens of cubes per second. With seawater, a wide variety of marine biofouling organisms enter the intake system. Intake structures are in general an ideal environment for settlement and growth of marine biofouling communities as it provides optimal conditions. A number of factors directly influence the settlement and development of biofouling communities inside intake systems. These are local environmental conditions and species composition (as determined by geographical location, such as water temperature and local hydrobiological features) and design and operation parameters of the intake system (flow, materials and intake layout).

H. J. G. Polman (✉)
H2O Biofouling Solutions B.V., Bemmel, The Netherlands
e-mail: hpolman@h2obfs.com

H. A. Jenner
Aquator, IJsselstein, The Netherlands

M. C. M. Bruijs
Pecten Aquatic, Lent, The Netherlands

© Springer Nature Switzerland AG 2020
V. S. Saji et al. (eds.), *Corrosion and Fouling Control in Desalination Industry*,
https://doi.org/10.1007/978-3-030-34284-5_16

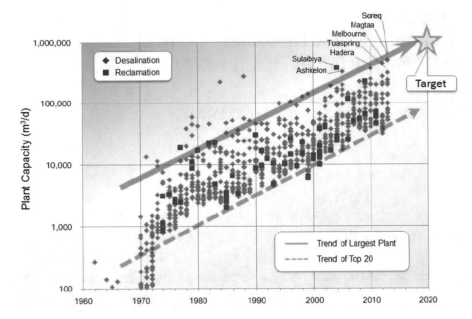

Fig. 16.1 Change in RO desalination plant size. Reproduced with permission from Ref. [2]; Copyright 2018 MDPI @ Creative Commons License

Marine biofouling results in an increased wall roughness and reduction of the inner pipe diameter, thereby resulting in a significant head loss in the intake basin and pump pit. This has a high impact on the operational reliability and could eventually result in an unplanned shutdown of the facility. Efficient mitigation of biofouling is therefore key to maintain an efficient and reliable operation of large-scale intakes. As biological and physical parameters are site-specific, so are the control strategy options. The standard industry practice for large-scale (cooling) water intakes is the application of oxidative biocides, most commonly sodium hypochlorite. For desalination facilities, the application of oxidants is, however, depending on the type of desalination process, restricted by it's potential impact on material integrity of the various components of the desalination facility and the coastal environment.

This chapter provides insight in the causes of biofouling settlement in the seawater intake pipes of desalination plants. The impact of biofouling is described and an overview of the available technologies (conventional and advanced) for biofouling control in seawater intake systems is presented.

16.2   Biofouling

Man-made solid surfaces that are in constant contact with surface water, such as conduits and pipe work of a seawater intake of a both thermal and membrane-based desalination plants, are colonized by fouling organisms in standard fouling patterns, see Fig. 16.2. Firstly, organic molecules are deposited, followed immediately by the attachment of bacteria, which in their sessile form produce 'slime' (Extracellular Polymeric Substances, EPS) as a part of their metabolic functioning to develop their microenvironment against medium stresses. After this biofilm formation, the colonization of the surfaces by other organisms becomes possible. Both the microfouling or bacterial slimes and the larger animals or macroinvertebrates constitute the overall biofouling community. Clearly, the types of fouling are dependent on the geographical location, the salinity and quality of the water and that also varies with seasonal changes. The overall biofouling community forms a specific ecological entity within the environment of an industrial water intake system, which is nearly optimal for all settled organisms due to the absence of predators, constant nutrition supply and oxygenation due to relatively high water velocities and above all, a surplus of the available substrate. The latter condition shows typically to be the regulating factor.

It is conventional to distinguish between the two types of biofouling: *microfouling*, involving bacteria, fungi and diatoms, and *macrofouling*, involving organisms such as mussels, oysters, barnacles and hydroids. Macrofouling gives rise to gross blockages of condenser tubes and pipelines, while the predominant effect of microfouling is reduction of heat transfer efficiency in condensers and heat exchangers [3]. Microfouling also results in RO membranes performance decline. The two are inextricably related, the formation of a biofilm very often being a necessary precursor to the successful settlement of larger organisms (see Fig. 16.2). Where an exposed metal becomes fouled, Microbial Influenced Corrosion (MIC) is an extra threat.

Seawater intake systems provide in general optimal conditions for macrofouling species to settle and grow. This is due to:

- Optimal seawater flow conditions (generally between 0.5–2 m/s)
- Water turbulence inside the seawater culverts that facilitates settlements of larvae

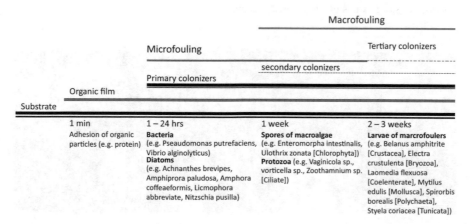

Fig. 16.2 Overview of biofouling settlement sequence from Ref. [4]; Copyright 1995 @ Inter-Research; Creative Commons License

- Continuous supply of nutrition and oxygen that stimulates growth
- Absence of predators that cannot pass the sieves or withstand the flow
- Absence of direct sunlight which most macrofouling species prefer

16.2.1 Biofouling Species

Biofouling communities exist in a wide range of sessile species of different taxa, such as mussels, oysters, hydroids, tubular worms and bryozoans. However, of all biofouling organisms' bivalves (mussels, oysters and clams) and barnacles are particularly known to cause serious fouling problems to industrial intake water systems. An example is presented in Fig. 16.3.

The macrofouling species present in coastal areas can be native or invasive species from other geographical areas, distributed through ship hull fouling and transport via ballast water. Table 16.1 provides a general overview of biofouling species of prime interest for intake systems in the Middle East.

16.2.2 Seasonality of Biofouling Settlement

The reproductive cycle of biofouling species in geographic regions with temperate climate conditions (Continental climatic zone) is characterized by distinct spawning periods, generally occurring in early spring (April – May) and late summer (September – October). In sub-tropical climate regions (Middle East, Asia, Pacific), climatic seasonal variations are often small, and spawning occurs more or less year-round. Hence, the ingress of biofouling species in these regions is a constant phenomenon and development of (severe) fouling in the absence of adequate control is a rather rapid process of only a few weeks.

Fig. 16.3 Macrofouling communities. (Source: H2O Biofouling Solutions B.V)

Table 16.1 Overview of main macrofouling species in the Middle East

Marine fouling organism	Scientific name
Oysters	– *Pinctada radiata; P. margaritifera* – *Crassostrea cucullata; C. delettrei; C. Gryphoides* – *Ostrea subucula* – *Malvufundus regula*
Mussels	– *Mytilus saidi* – *Septifer bilocularis* – *Perna picta* – *Brachidontes variabilis; B. pharaonis* – *Modiolus auriculatus; M. cf barbatus; M. ligneus* – *Chama pacifica* – *Barbatia lacerate; B. virescens* – *Limaria fragilis*
Barnacles	– *Balanus amphitrite* – *Chthamalus stellatus* – *Megabalanus sp*

The implication of this for fouling control is that maintaining a correct performance of the biofouling control measure is a daily task. Biofouling control not only concerns proper execution of control measures, but it also concerns the timely monitoring of equipment performance and efficacy of the treatment.

16.2.3   Typical Problems Due to Biofouling

Upon entering the intake system, biofouling organisms can readily colonize the available substrates, in the intake and piping system, which feeds the desalination plant. Settlement takes place on all surface types, i.e. concrete, metal, and glass-reinforced plastic (GRP) surfaces, especially where there is turbulence, such as at rubber dilatation joints, manholes and bends. The settled biofouling larvae can grow quickly into a mature community covering all available surfaces. This settlement results in an increased wall roughness of the material and reduces the pipe diameter resulting in hydraulic head loss.

Intake structures are designed to deliver seawater towards the desalination plant to produce enough potable water to meet the contract with the off taker. To provide a sufficient water supply to the production plant, the required head as to be delivered by the pumps should overcome the head losses in the pumping station of the seawater system itself. Head loss due to biofouling is mainly caused at the riser head, intake conduits (including chlorination system) and screens.

One of the typical events leading to increased fouling is the malfunctioning of the hypochlorite dosing lines towards the intake dome, most often by broken joints. Also, poor material selection of the chlorine dosing lines has been shown to result in severe fouling problems and tripping of the plant [5].

16.3   Impact of Biofouling on Desalination Plants

Biofouling is very relevant to the operational efficiency of membranes by biofilm formation on the membrane surface. The bacteria utilize organic and inorganic nutrients (organic carbon, nitrogen and phosphorous) from the seawater and produce biofilm to provide a functional barrier between the bacterial community and the seawater. This biofilm appears as a brown-orange jelly covering the surface of the membrane and collects in the membrane feed spacers, thereby restricting the available area for flow across the membrane surface. This increases the differential pressure (DP) across the membrane leading to increased energy consumption.

If the biofouling is allowed to increase then the biofouling can eventually block the flow to whole parts of the membrane surface, thereby reducing the effective membrane area and increasing flux. This causes increased feed pressure (and therefore high electricity consumption) and can cause membrane design guidelines to be exceeded in the active area of the membrane element. Blocked areas are also extremely hard to access by chemical cleaning, since the chemical is carried by feed to brine flow, resulting in what are effectively dead zones of the membrane. Finally, the DP creates a very great force on the structure of the membrane. It is common for the feed spacer to be moved along the membrane element and become extruded out the far end. This can also cause loss of effective membrane area, and finally, there is no other option but to replace the membrane.

16.3.1   Operational

Biofouling in intake systems can have a significant impact on the operational reliability of the plant as a result of hydraulic impact due to head loss. The head loss over a pipeline is determined by the flow velocities, wall roughness, and local losses [9]. The local losses are small compared to the losses by wall roughness and are therefore neglected. The flow velocities in a riser head are low, and the effect of biofouling is relatively small. The total head loss is therefore mainly determined by the friction losses in the intake pipelines.

$$\Delta H = \lambda \frac{L}{D} \frac{v^2}{2g} \text{ with } \frac{1}{\sqrt{\lambda}} = 2 \log\left(\frac{2,51}{\text{Re}\sqrt{\lambda}} + \frac{k}{3,71D} \right)$$

The head loss, ΔH, is calculated as follows [6]:
Wherein:
λ = friction factor (−);
L = length of the pipe (m);
D = diameter of the pipe (m);
v = velocity (m/s);
g = gravitational acceleration (m/s^2);
Re = Reynolds number (−);
k = roughness value (m)

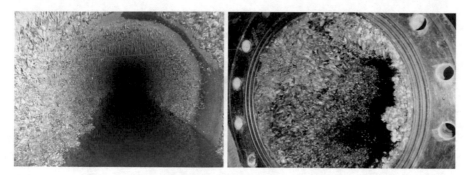

Fig. 16.4 Biofouling in sea water intake pipelines and headers. (Source: H2O Biofouling Solutions B.V.)

Both the diameter of the pipeline and the wall roughness are affected by biofouling, which can have a significant impact on hydraulics and can result in major operational problems, including unplanned plant shutdowns due to high levels of head loss or clogging of a pipe. Figure 16.4 shows severity of biofouling in seawater intake pipelines and headers.

In competition for substrate, biofouling species can grow on top of each other (cluster development) potentially forming patches of thick layers along the walls. The total biomass of biofouling settled in intake systems may potentially reach up to hundreds of tons within 2 years, depending on growth rates under seasonal conditions. Mussel species and several oyster species have byssal threads to connect to the surface. After effective mitigation, the shells are detached quickly. In contrast, the effect of marine biofouling species that cement themselves to surfaces (e.g. barnacles, tubular worms and some oyster species) is irreversible. An effective biocide dosing can kill the species, but part of the shell structure remains attached, with a remaining turbulent surface area.

16.3.2   Financial

Biofouling results in an increased head loss and decreased efficiency for the pumping station [7, 8], often resulting in an unplanned outage of the plant due to tripping pumps. The increased head loss generally results in a significant cost impact [9] due to additional required pump capacity. This will also result in increased CO_2 emissions.

Cleaning of heavily fouled seawater intake pipes is time-consuming and a very costly exercise requiring additional chemicals or physical cleaning method. For physical cleaning the seawater flow needs to be stopped entirely, requiring a total plant outage. The loss of production capacity and sometimes penalties due to non-compliance of the water production contract, most often has a significant cost impact. Other manners of cleaning the intake pipes are by using a pigging system or by hydro blasting. This, however, needs an outage of (part of) the plant and is very costly. In addition, due to water delivery commitments, a desalination plant needs to schedule an outage way ahead, and an unexpected outage needs to be prevented by all means.

The financial losses due to biofouling can arise to millions of dollars; however, the total impact is often not quantifiable due to lack of information. Therefore, it is important to prevent settlement of marine biofouling larvae in the seawater system.

16.4 Control Methods

The search for the most efficient methods to control biofouling in large-scale (cooling) water intake systems, has started in the late 20's [10], and is still a topical work field for both environmental and financial reasons. A wide variety of both mechanical and chemical-based measures have been investigated and implemented. Still, the efficacy of control measures is very site-specific and application, especially of chemical compounds, is subject to stringent permit requirements.

Many design and layout features directly, or indirectly affect the efficacy of biofouling control measures in a seawater intake system. Many antifouling technologies require special provisions to be integrated in the design of the seawater system. Examples of these are filtering devices, mechanical cleaning devices and facilities, backwash systems, provisions for recirculation of the seawater (for thermal treatment) or special dosing racks and dosing points. Provisions can also be of a simpler nature, for example, connection points for chemical and biological monitoring devices.

16.4.1 Physical Methods

To mitigate biofouling settlement in seawater systems, physical methods can be used. A number of techniques are available for reducing the effects of potential fouling in culverts or other parts of the seawater system. However, in general, physical methods do not prevent settlement but are aimed at reducing the number of biofouling larvae to enter the system (e.g. filtration), reduce favourable conditions for settlement (design, high flow velocities) or elimination of existing growth of biofouling (e.g. thermal shock, sonic methods). In addition, physical methods typically impact operation and some systems require (partly) outage.

In this chapter we, only listed the methods which are feasible for desalination plants are listed. An overview is presented in Table 16.2. In principle, physical methods are not species-specific.

Below, only the methods which are feasible for desalination plants are discussed.

Table 16.2 Overview of physical methods

Physical method	Basic principle	Applicability for desalination
Filtration	Removal of fouling larvae	Only applicable to limited water flow rates.
Water velocity/ design	Prevent/reduce settlement	Basic principle to keep fouling to a minimum, no full control.
Thermal treatment	Killing of fouling community	Requires heat source, dedicated design intake conduit lay out and impacts operation; no fouling settlement prevention.
Sonic technology	Killing of larvae	Only localized treatment, no full control.
Magnetic fields	Killing of larvae	Only localized treatment, no full control.
Ultraviolet light	Killing of larvae	Localized treatment, no large-scale experience yet.
Oxygen depletion	Killing of fouling community	Requires redundant intake conduit that can be closed; no fouling settlement prevention
Physical removal	Removing fouling community	No fouling settlement prevention, impact on operation

16.4.1.1   Filtration

Micro-filtration is a method to reduce the zoo- and phytoplankton load (including larvae) of the cooling water, which also reduces the organic load. This also potentially reduces the biocide demand. Existing methods are rotating drum filters and sand filters. Beach wells are applied at some desalination plants to retrieve seawater.

Continuously backwashed microfilters have been developed with mesh sizes of 50–100 µm, which claim to give effective protection against macrofouling. These filters are able to deal with water flows up to 4 m³/s. The youngest stages of mussels and barnacles have lengths of above 150 µm, so a microfilter with 75 µm mesh width is likely to give effective protection against the entrance of larvae of mussels and barnacles. These filters require a back-flush flow and cause a pressure loss in the seawater system.

Pressure filters are specifically designed for the filtration of large particles (mm-range) from raw water streams, such as cooling water intakes from rivers, lakes, or the sea. The raw water is filtered by means of a filter basket with a mesh size of 1 mm. At certain time intervals, the filter is automatically backflushed.

The costs for engineering and implementation of such a filter is high. Also, there is an effort needed to operate the filters, controlling the cleanliness and filtration flow to keep optimal filtration rate and prevent clogging. Also, the residual filtered material must be removed, probably as waste and possibly treated before it can be dumped. For large flow rates, filtration has limited applicability.

16.4.1.2   Water Velocity

Water velocity and the potentially complex hydrodynamics of seawater system design are important factors in the 'fouling potential' of a given seawater system. Some species are well adapted to slow-running or even stagnant water, while others require strong water currents.

Water velocity varies considerably in seawater circuits from the water intake structure to the outlet. It is low near filtering devices like traveling screens and in basins. It is high in pipes leading to seawater pre-filtrating steps for e.g. RO. The operating regime of pumps in any given circuit must also be considered as this can cause both variations in flow rate locally and even periods of stagnation.

For organisms which settle on circuit walls, the continuous seawater supply is an extremely favorable factor as it provides a source of nutrients and oxygenation. For this reason, when water velocity is not excessive, organisms show optimum growth, generally more rapid than that of the corresponding population in the natural environment. When the velocity exceeds a critical threshold, larvae are no longer able to settle, and adults are not able to feed well. They may even become detached from the substratum by shear stress. In the absence of water circulation, dissolved oxygen may become eventually a limiting factor and can cause mortality by asphyxia in 1–3 weeks, depending on the water temperature and the organisms present.

A first step in preventing biofouling is applying a proper design of the cooling water trajectory in which hydraulic conditions that favor fouling settlement are kept as low as possible [11]. When designing a seawater system, stagnant zones and sharp curves in conducts (e.g. manholes and dead-end lines) should be kept to a minimum to reduce turbulence as much as possible. In the stagnant zones, biological growth thrives as water velocities are low which favours biofouling settlement. The effect of water velocity on macro fouling settlement is clearly visible in the data provided by Ackerman [12] and Kawabe [13]. Despite the discrepancy in the different observations in the literature, it is concluded that, in distribution pipes maintained without the presence of dead legs or low flow areas, marine fouling generally do not settle in seawater where the velocity is near 3.0 m/s, and that macrofouling is already reduced at 1.8 m/s. A general rule of thumb for water intake conduits is to keep water velocities above 2.5 m/s, to reduce macrofouling settlement.

16.4.1.3   Thermal Treatment

'Thermoshock' is a well-known and generally accepted antifouling method for power stations [14], which can completely replace the use of a biocide. However, the application of thermal treatment requires a special design of the intake system that should be implemented in an early stage of plant construction. Adaptations afterwards are often technically difficult and expensive. Heat treatment consists of heating the seawater water to a temperature of 38–45 °C by means of (partial) recirculation or steam injection and maintaining this for a sufficient period of time (hrs) to guarantees elimination of existing growth. Treatment temperature and time should be attuned to the fouling type. Typically oysters require sufficiently high temperature and treatment time. In seawater systems of power stations, this should be performed 3–4 times a year, meaning settlement and growth is accepted to an extent. The crucial factor here is that shells detaching from the walls are still small enough (~ 1 cm) to pass through heat-exchanger tubes. For desalination plants, the heat source (e.g. condenser) for performing thermal treatment is often not available, nor is the design suited for recirculating. In addition, the treatment itself will impact the desalination process (unavailability of sufficient water during recirculation and high water temperatures).

16.4.1.4   Sonic Technology

The principle underlying the application of sound is that the vibration created by the energy associated with the transmission of sound will remove deposits on surfaces, by "shaking" the deposit free. Cavitation produced by the propagation of sonic waves in the continuous phase near the deposit surface, can also assist the removal process. Claudi and Mackie [15] describe how acoustic energy in the range of 39–41 kHz fragments early stages of mussels within a few seconds. It also killed attached adults within 19–24 h. The main disadvantages of the technique are the high energy costs involved and the potential harm to the integrity of the seawater system. To date there are no commercial devices available using this technique at the scale of large-scale seawater intakes, making the practical value of this technique limited.

16.4.1.5   Ultraviolet

Previously industry attempts to deploy UV for biofouling control, utilizing medium-pressure units, had not been regarded as proven and cost-effective against biofouling in industrial settings [16]. The UV systems did not provide a consistent reduction of veliger settlement, while taking up more space and utilizing more power. New development of Hydro-Optic Disinfection (HOD) UV systems has led to more cost-effective control of macrofouling in seawater, including larvae of blue mussels, barnacles, oysters, and colonial hydroids.

A study carried out at the turbine cooling water lines of Parker Dam in the US indicated that HOD UV treatment reduced maintenance due to biofouling despite lower than expected HOD UV doses [17].

16.4.1.6   Oxygen Depletion

For once-through systems, sometimes oxygen depletion can be used. There is quite a lot of literature available, however, biofouling organisms are capable to close their valves and change their metabolism from aerobic to anaerobic, which create a kind hibernation strategy by the mussels [18, 19].

For desalination facilities, mitigation by oxygen depletion can be feasible if the design of the intake is implemented in a way that the intake risers can be closed at the intake chamber. Also, the screening compartments can be separately closed (both sides) by using a motorized penstock. Furthermore, as the production plant requires availability of seawater at all times, an additional redundant intake conduit is needed to perform this procedure. This allows the system to create stagnant zones during normal plant operation. A serious problem is the cleaning of the remaining shells and debris after 100% killing of the fouling.

16.4.1.7 Physical Removal

In accessible areas, fouling can be physically removed by a variety of means, including scraping, pressure washing (by divers), or pigging. Pressures of 2000–3000 psi should remove mussels, but it may take higher pressure to remove oysters and barnacles as these cement themselves to the surface. Pigging would not be practical in pipes and conduits with lots of bends or size changes. Such physical removal requires entire outage and closure of the intake conduit. In addition, physical removal is most often labour-intensive and time-consuming which may pose problems meeting the required operational window of the plant. Once the fouling is removed, the potentially large volume of remains will have to be removed from the conduits and disposed of.

16.4.2 Chemical Methods

Chemical methods to control biofouling can be categorized into oxidizing and non-oxidizing chemicals. In the following sections, an overview of these products is provided with a practical point of view.

16.4.2.1 Oxidizing Biocides

In most cases, the industry practice is chemical treatment. The existing chemicals in use are distinguished as oxidising and non-oxidising compounds (Tables 16.3 and 16.4). Oxidising biocides include chlorine, bromine, and iodine. These chemicals act by destroying cell membranes or their extracellular enzymes, which leads to cell death. Non-oxidising biocides include numerous chemicals that act by interfering with a necessary life function such as metabolism or reproduction. In practice, application of sodium hypochlorite is most widely in use, due to extensive research and clear understanding of the mode of action, the cost-beneficial application, commercial availability and controllability to apply proven control regimes.

Of all the chemical compounds, sodium hypochlorite is the most widely applied industry oxidizing biocide. However, due to the potential environmental and public health implications of by-products associated with the application of sodium hypochlorite, many efforts have been undertaken to search for alternatives or optimize the chemical dosing.

Table 16.3 Overview of oxidizing biocides

Group	Biocide	Chemical formula	Kinetic reaction	Target organisms	Applicability to desalination plants[a]
Chlorine-based	Sodium hypochlorite	NaOCl	Fast	All	++
	Sodium dichloroisocyanurate	$C_3HCl_2N_3Na$	Fast	All	−
	Chlorine dioxide	ClO_2	Very fast	All	+
Bromine-based	Sodium hypochlorite + NaBr	NaOCl + NaBr	Fast	All	+
	1-bromo-3-chloro-5,5,dimethylhydanthoide (BCDMH)	$C_5N_2O_2H_6ClBr$ NaOCl	Fast	All	+/−
Other	Ozone	O_3	Very fast	All	+/−
	Hydrogen peroxide	H_2O_2	Fast unless stabilized	All	+/−
	Peracetic acid	$C_2H_4O_3$	Fast	All	+/−
Oxone	Monopersulphate	$2KHSO_5 \cdot KHSO_4 \cdot K_2SO_4$	Fast	All	−

[a]Expert judgement; (++) very suitable (proven practice); (+) suitable; (+/−) suitable under certain conditions; (−) not suitable

Table 16.4 Overview of non-oxidizing biocides

Group	Name of biocide	Chemical formula	Reaction	Half-life[a]	Target organisms	Applicability to desalination plants[d]
Isothiazolones	2-methyl-4-isothiazolin-3-one	$C_4H_4NO_5$	slow	long	all	+/−
	5-chloro-2-methyl-4-isothiazolin-3-one	$C_4H_4ClNO_5$	slow	long[b]	all	+/−
	1,2-benzoisothiazolin-3-one	$C_9H_6N_2S_3$	slow	long	all	−
QACs	alkyl-dimethyl-ethyl-benzyl-ammoniumchloride	$R(CH_3)2(CH_3-CH_6H_4)-NCl-R-(CH_3)2(C_8H_9)-N.Cl$	average	average	all	−
	didecyl-dimetyl-ammoniumchloride	$C_{22}NH_{48}.Cl$	average	average	all	−
	alkyl-dimethyl-benzyl-ammoniumchloride	$R-CH_3CH_3CH_2C_6H_5NCl$	average	average	all	−
	poly[oxyethylene(dimethyliminio)ethylene-(dimethyl-imino)-ethylenedichloride]	$C10H_{24}N_2O.Cl_2$	average	average	all, mostly used as algaecide	−
Other	b-bromo-b-nitrostyrene	$C_8H_6NO_2Br$	fast	short	all	−
	2,2-dithiobisbenzamide	$C1_4S_2NH_{10}$	c	−	−	−
	methylenebisthiocyanate	$CH_2(SCN)2$	fast	short	all, except algae	−
	2-bromo-2-nitropropane-1,3,-diol (BNPD)	$C_3H_6NO_4Br$	average	long	all, except algae	−
	2,2,-dibromo-3-nitrilo-propionamide (DBNPA)	$C_4N_2H_6OBr_2$	fast	short	all	−
	Glutaraldehyde	$CHO-(CH_2)3-CHO$	average	average	all, except fungi and algae	−

[a]Based on hydrolysis. The half-life of non-oxidizing biocides varies significantly with pH value and temperature. This variation can be as much as ×100. In this table a short half-life is defined as 0–10 h, average as 10 h to 7 days and long as 7 days, and longer

[b]The half-life of 5-chloro-2methyl isothiazolone is claimed to be less than the non-chlorinated isothiazolones cited

[c]Information was not available

[d]Expert judgement; (+/−) suitable under certain conditions; (−) not suitable

(a) Chlorine

Chlorine can be used in different forms. Sodium hypochlorite (NaOCl) is either purchased in bulk or generated on-site by an electrochlorination unit by electrolysis of seawater. Hypochlorite stored on-site is subject to decay, which is temperature-dependent. The commercially available hypochlorite solution (120–150 g/L) may be used as such, but more often it is diluted prior to dosing to about 500–2000 mg Cl_2/L to improve the mixing properties with the seawater. Chlorine gas (Cl_2) is still applied at sites and is the least expensive of the chlorine products, but the inherent handling and storage difficulties have restricted its use. Chlorine gas is a greenish-yellow chemical with a burning odour which is purchased as a compressed gas and is converted into hypobromite (in seawater) after injection into the intake, typically at 10 mg/L but always less than 35 mg/L to avoid excessive losses due to evaporation/outgassing.

When dissolved in natural waters, chlorine gas or a sodium hypochlorite solution produces different oxidizing compounds depending on the reaction of hydrolysis and oxidation of ammonia, leading to chloramines or bromamines. The oxidizing compounds also react with organic matter to produce chlorinated or brominated organics. As a result, water chlorination chemistry involves many molecular and ionic species or groups, whether oxidants or non-oxidants, whose terminology must be precisely defined. Chlorine is described in the literature as 'free', 'active', 'available', 'combined' or 'residual'– or by a combination of these adjectives.

Figure 16.5 summarizes the general understanding of 'chlorine' chemistry associated with coastal power stations [20, 21]. Seawater contains about 68 mg/L bromide at full salinity: when chlorine is added it oxidises the bromide ions yielding hypobromous acid (HOBr). This reaction is rapid, with 99% conversion within 10 s at full seawater salinity and within 15 s even at half seawater salinity. The free oxidants formed by the chlorination of seawater are thus predominantly composed of HOBr and the hypobromite ion OBr^-.

Total Residual Chlorine (TRC) is the sum of the free oxidants + combined oxidant (as available in chloramines, bromamines or other compounds with a N-C link). In seawater this adds an additional effect due to the bromamines which have a toxic effect on biofouling organisms, in contrast with the lower toxic effect of chloramines. This makes TRC a valid parameter to be express the effectivity of the chlorine concentrations to control biofouling in a system where seawater is used. When there is an increase in ammonia concentration the difference between Free Residual Chlorine (FRC) and TRC usually increases.

The typical formation of chlorine byproducts and environmental impact has led to research on new dosing schemes and strategies to reduce the volumes required to minimize environmental impact, whilst maintaining sufficient control.

(i) Chlorine dosing strategies

In many cases, standard dosing regime, combining a continuous dosing with regular shock dosing or solely shock dosing, has proven not to be the effective dosing under all operating conditions within the cooling water networks. Often, FRC

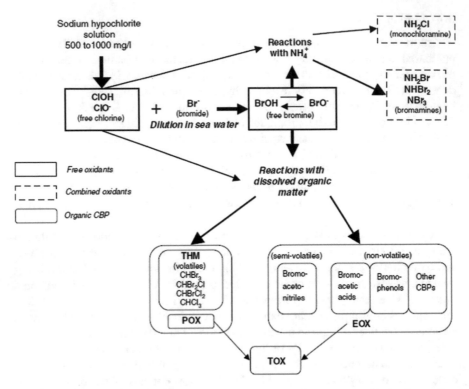

Fig. 16.5 The chemistry consequent upon sodium hypochlorite or electrochlorination product injection in natural seawater. Reproduced with permission from Ref. [21]; Copyright 2006 @ Elsevier

values are not specifically specified as initial dosing concentrations or effective concentration at a strategic point in the cooling water distribution network but are maintained as such at arbitrary sampling locations. However, it should be taken into account that it is important to maintain an effective chlorine concentration at the most critical assets of the facility, which is at the condenser inlets or the last heat exchangers in a production facility.

The FRC concentrations maintained within the network are often observed to be too low according to practical experience and literature references, which leads to increased risks of fouling, especially when technical failures in the dosing system appear or changes in operation, leading to periods of low chlorine levels or no chlorine at all. It is common practice that free residual chlorine concentrations in realtime differ from set values via Distributed Control System (DCS) and measured by inadequate online monitoring equipment. Thereby the concentrations are often too low to efficiently mitigate macrofouling, providing the specimens time to maintain normal metabolism and grow and develop larger communities. Continuously maintaining the effective chlorine level is the most important aspect to apply an effective continuous dosing regime.

At many facilities, an additional shock dosing procedure is applied in the erroneous notion that it prevents fouling species from adapting to continuous chlorination.
There is a lack of industry awareness that shock dosing as a method is by principle
not effective to control bivalve fouling species (oysters, mussels) nor other fouling
species such as barnacles and hydroids. A shock dosing is a short period, typically
one to maximum several hrs, performed daily or weekly, either applied as sole procedure or on top of a continuous dosing. Bivalves are capable of protecting themselves from the deleterious effects of chlorine by closing their shells. They have the
ability to switch over from aerobic to anaerobic metabolism for a considerable
length of time, up to weeks. After the shock dosing, when the concentration is
reduced again to lower levels as applied for the continuous dosing, and if too low,
the bivalves will restart filtration for oxygen and nutrition and will fully recover
within a day.

In view of the above, it should be considered by industries to fully cancel the
shock dosing procedure and solely apply a continuous dosing with sufficient (effective) free residual chlorine level to mitigate the fouling settlement and growth within
seawater networks. This will also lower the corrosion stress on thermal desalination
units. A well-performed continuous chlorine dosing is able to mitigate the larvae
when they enter and settle which will thus not develop into larger communities
within the cooling water system.

As a further reduction in chlorine usage a site dedicated dosing strategy can be
applied (i.e. Ecodosing™, Pulse-Chlorination®). This type of dosing strategy
replaces the continuous and/or shock dosing with an intermittent dosing strategy
which is based on the reactional behavior of local marine biofouling species. These
technologies have internationally reached the proven technology status and are
increasingly being applied worldwide [22, 23, 24] and in past years successfully
applied at seawater RO (SWRO) plants seawater intakes.

(ii) Chlorine Dosing Options for SWRO

Finding the balance of an optimum intake seawater system and the prevention of
biofouling growth on the RO membranes, presents significant operational challenges for plant owners and operators. For SWRO plants two types of biofouling
risks need to be taken into account:

- Marine biofouling organisms enter and foul the intake system
- Bacterial biofouling growth on the RO membranes

Seawater intake structures provide an ideal environment and provide optimal conditions for settlement and growth of marine biofouling organisms. Marine biofouling
results in an increased wall roughness and reduction of the inner pipe diameter which
leads to a significant head loss in the intake structure. This has a high impact on the
operational reliability of the intake system and often results in an unplanned shutdown.

Globally, the typical industrial anti-fouling practice involves continuous low-
concentration chlorination of seawater and/or periodic shock-dosing at higher concentrations. Ineffective mitigation strategy will cause settlement of biofouling in
intake risers and distribution piping causing clogging and loss in pressure head,

resulting in plant "tripping" and expensive manual cleaning costs. In addition, chlorine (as sodium hypochlorite) has a direct impact on the rate of organics and biofouling growth on the RO membranes.

It is understood in the membrane industry that thin film composite polyamide membranes have limited resistance to chlorine-based oxidants. Therefore, operators have relatively few options regarding chemicals, which can be safely used to disinfect RO-systems and prevent and mitigate biofouling on present on the membranes. In addition, to prevent any impact of chlorine the access of oxidizing capacity of chlorine is neutralized by adding, for example, sodium bisulphite. For desalination plants, mitigation strategies for excessive biofouling settlement and growth are necessary to guarantee plant operation and potable water delivery with minimal SWRO membrane flux reduction.

Small organic molecules are generally much more biodegradable than large organic molecules. Chlorine has been recognized as a proven oxidant to break down large organic molecules into smaller organic molecules. This is of particular importance when chlorination is used for drinking water plants. Feed waters that have high levels of organics and are chlorinated are particularly susceptible to the formation of disinfection by-products. It is acknowledged that these organic molecules can result in adverse health impacts. In addition, these small organic molecules are also far more biodegradable than the precursors, so chlorinating water increases the concentration of readily biodegradable organics in the water. Normally, the chlorine itself restricts the growth of biofilm, because of its toxicity. However, with RO membranes, the chlorine has to be neutralised so it doesn't oxidise the RO membrane, which means that the rate of biofouling is far greater if the feed water is chlorinated and dechlorinated than if it is not chlorinated.

Theoretically, Assimilable Organic Carbon (AOC) should be a very good measure of biofouling potential in SWRO systems, because it is a measure of the amount of organic carbon consumed by bacteria under standard conditions. However, the analysis is not straightforward, and there is massive variability in the values of AOC which can be measured by different laboratories. Consequently, it is difficult to use AOC as a standard measure of biofouling potential. The lack of a rapid, simple and consistent method of measuring biofouling potential in SWRO feed water makes research in this area much more complex and makes it very difficult to objectively compare results from different sites. It is, therefore, the current industry practice that biofouling potential to be measured based on the rate at which SWRO membranes foul (rate of Differential Pressure (DP) increase and Clean-in-Place (CIP) frequency).

Total Organic Carbon (TOC) is often characterized in molecule size like biopolymers, humics; building blocks; LMW (low molecular weight) organic acids and LMW neutrals. Small molecules are more easily degraded by micro-biological activity than larger molecules. This means that in case the ratio between relatively small molecules and large molecules is in favour of the small ones, the feed has a high micro-biofouling potential. Building blocks, sub-units of HS with molecular weights of 300–450 g/mol, are considered to be natural breakdown products of humics. They cannot be removed in flocculation processes. In Fig. 16.6 an example

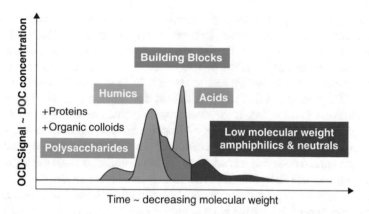

Fig. 16.6 Example of molecular size distribution of TOC. Reprodcued with permission from Ref. [25]; Copyright 2011 @ Intechopen; Creative Commons License

is given of such a molecule size distribution analysis. The results of these analyses will give information about the molecular size distribution of the available TOC.

Recently two SWRO plants applied an alternative dosing strategy using chlorine-based on the Ecodosing/Pulse-chlorination philosophy. This is based on tuning the required chlorine dosing for marine fouling treatment to a minimum and as a consequence reduce any potential effect of biofouling growth on the RO membranes. At the Al Fatah SWRO plant in Jubail KSA, the Ecodosing regime was implemented in September 2019 resulting in a clean seawater intake system and no increase in CIP frequency compared in the previous period wherein no chlorine dosing at the seawater intake took place [24]. It must be noted that it is of high importance to study the specific biological and seawater conditions at a site. These are important to find a right operational balance without being exposed to either marine biofouling in the intake pipes or increased membrane biofouling. At locations where algal blooms are prevalent, the operational approach for chlorination may also need to be altered accordingly to manage these events, when they occur.

(iii) Effect of chlorine on thermal desalination units

During operation of macrofouling mitigation using sodium hypochlorite in thermal desalination units, chlorine is suspected to cause corrosion of the copper alloy materials used in heat exchangers. Their combination of strength and corrosion resistance makes copper alloy materials one of today's most important engineering materials for highly stressed components in corrosive environments [26].

The corrosion resistance of copper alloys relies on the formation of a protective surface (oxide) film which forms quickly on exposure to seawater and continues to mature over a period of years. Lack of this passive layer (thickness approximately 3–10 nanometers) allows the alloy to a rather significant loss of material and shortens the lifespan of the heat exchanger. Initial exposure to clean seawater and good surface film formation is important to the long-term performance of these materials. The surface film is complex, multi-layered and can be brown, greenish-brown or brownish-black. It is predominantly comprised of cuprous oxide although it can contain nickel, iron oxide, cuprous hydroxychloride and cupric oxide. The initial film forms quite quickly over the first few days but takes about 2–3 months to mature in temperate waters. At low temperatures, the process is slower, but the film will form even in Arctic and Antarctic waters. In higher temperatures (water of around 27 °C common to inlets for Middle East desalination plants), the film can be expected to form in a few hrs. If the passive protective layer is not formed because of pollutants in the seawater, adding of iron(II)sulphate solution to the oxygen-containing water can enhance the formation of a protective layer on copper alloys. The solution should be added just upstream of the material to avoid the formation of trivalent iron because Fe^{3+} is ineffective and may even react adversely.

Copper alloys are in general resistant to chlorination at normal dosing levels used to control biofouling. Excessive chlorination, however, can be detrimental. Lewis [27] has reported that, in the presence of 0.25 ppm free chlorine, the corrosion of CuNi 90/10 increased during 30 days of exposure, but the effect of the chlorine weakened subsequently. Kirk et al. (1991) stated that according to general experience no negative effect of chlorine concentrations 0.2–0.5 ppm was indicated on the corrosion behaviour of copper-nickel alloy during many years in coastal power and process industries. Also, according to Tuthill [28] copper alloy tubing is resistant to chlorination at concentrations required to control biofouling with continuous chlorination at a 0.2–0.5 ppm residual chlorine. Francis [29] published the results of tests related to the effect of chlorine additions in the range between 0.3 ppm and 4.0 ppm on corrosion and jet impingement tests of Al brass, CuNi 90/10 and CuNi 70/30 exposed to natural seawater with and without the dosing of iron (II) sulphate solution.

The corrosion resistance of copper alloys used in thermal desalination units, both MSF and MED, to natural seawater and chlorine is considered to be able to withstand normal chlorine levels sufficient to control biological activity.

In seawater, bromide is normally present in a concentration of 65 mg/L. This bromide reacts with the hypochlorite to form bromine. Based on this reaction most of the available chlorine will be converted into bromine. It is known that bromine is much more aggressive towards metals than chlorine. From literature, it is known that this conversion from hypochlorite to bromine is depending on pH and temperature. It is stated that the lower the pH, the higher the conversion and the higher the temperature, the higher the conversion. Of course, the contact time also plays a role in this conversion.

Normally the pH of seawater is slightly caustic (pH 8.0–8.2) and the temperature is high (110–130 °C in the first stage). Bromine, just as chlorine is volatile, most of the bromine will evaporate in the first stages. Due to its volatility, a large amount of the bromine will not condense at the heat exchanger but will travel to the vent system. In the vent system, non-condensable gasses, such as N_2, O_2 and CO_2, are removed. In MSF, steam ejectors are used to remove the non-condensable gasses. Part of this steam is condensed. At a different location relatively high bromine concentrations have been measured in this condensate. Also, the MSF condensate has a lower pH (7.0), which is related to the absorption of CO_2 in the condensate.

From literature, it is well known that CO_2 play a critical role in corrosion observed in the vent system. To reduce the risks of CO_2, which also plays a role in bromine attack, the pH of the steam could be increased, by adding more ammonia. To what level is partly depending on the current pH of the condensate.

Based on the available information, it will be beneficial for the thermal desalination plants to reduce the chlorine dosage to a minimum. This will also reduce the corrosion and copper release to a minimum.

(b) Chlorine dioxide

Chlorine dioxide (ClO_2) is an effective disinfectant for biofouling control. It gained interest because of the reduction in the formation of some by-products [30]. Additionally, ClO_2 does not react with nitrogen compounds to form halogenated amines, as in the case of sodium hypochlorite. This can be advantageous in water systems having high loadings of certain compounds (e.g. ammonia or glycol). The concentrated gas is sensitive to pressure and temperature and has therefore to be generated on-site. Unlike gaseous chlorine, ClO_2 does not hydrolyse and exists as a dissolved gas in the pH range of 2–10 [31]. As such the disinfection rate of ClO_2 in once-through water systems is not affected by pH.

ClO_2 has high water solubility (up to 60 g/L at 20 °C) and a low Log Pow estimated to be negative and is therefore not expected to bioaccumulate or to partition to the sediment or soil and in the presence of hydroxyl ions is hydrolysed rapidly and undergoes rapid photolysis in water ($t\frac{1}{2} = 15$ min). It will volatilise readily from water at temperatures above 10 °C but then undergoes rapid photolysis in air ($t\frac{1}{2} = 2.4$ days). ClO_2 shows no absorption in the so-called atmospheric window (800–1200 nm) and therefore, is not considered to be a potential greenhouse gas.

ClO_2 will react in the environment to form chloride ions via the transient intermediate, chlorite. Organohalides or other organic derivatives do not appear to be formed during the reduction process.

Ecotoxicity studies show that ClO_2 is very toxic to fish (96 h LC50 = 0.021 mg/L), invertebrates (48 h EC50 = 0.063 mg/L) and algae (72 h EC50 = 1.096 mg/L) and is harmful to micro-organisms (3 h EC50 = 10.7 mg/L). There does not appear to be any significant difference in sensitivity between freshwater and marine data for fish invertebrates and algae. However, under normal operating conditions little if any ClO_2 is released in the process water. The direct degradation products of ClO_2, chlorite and chlorate, are more likely to be found at measurable concentrations. Since sodium chlorite is an inorganic substance it will not undergo biodegradation, i.e. microbial degradation to water and CO_2. With a high water solubility, low Log Pow and rapid degradation in water, there is little potential for bioaccumulation. Hence, sodium chlorite is of no concern with regards to secondary poisoning. Chlorate is relatively stable under aerobic aquatic conditions and does not hydrolyse or readily oxidize organic matter, but it will decompose rapidly under anaerobic conditions or in anaerobic niches. ClO_2 does not result in reactions that produce bromate from surface water containing bromide.

ClO_2 has also been considered as a potential disinfectant for RO membranes. ClO_2 is present as a dissolved gas in seawater. One advantage of ClO_2 is that it is a weaker oxidant than HOCl, HOBr and Ozone. A weaker oxidant is less damaging to the membrane, and apparently the ClO_2 gas can penetrate the biofilm better. However, study with polyamide RO membrane, exposed to high concentrations of ClO_2 showed excellent performance in terms of permeate flux and salt rejection [32].

In addition, since ClO_2 is a weaker oxidant in comparison to chlorine and not reacts with organic molecules, it is expected it will result in a much smaller rate of breakdown of large organic molecules into smaller organic molecules and act as precursors for biofouling growth rates on RO membranes.

(c) Chloramines

Monochloramine is produced on-site by means of mixing solutions of sodium hypochlorite and ammonia chloride. The optimum mass ratio chlorine/nitrogen must be 5/1 to produce monochloramine [33]. Chloramines have been used successfully for open cooling and wastewater systems where ammonia is present in the feed water stream and chlorine is added to obtain 1–2 ppm of chloramine. However, chloramines have been used with sporadic success, and failure, for surface waters where both ammonia and chlorine need to be added to the water to generate the chloramine. Special care should be taken that transition metals (Fe, Mn) are not present in the feedwater or deposited on the membrane, since these can accelerate oxidation reactions by 1–2 orders of magnitude. The use of chloramines are useful in situations where a toxic action is needed after several kilometer of pipelines as in the case of drinking water lines. Bromamines, however, are much more acute toxic and are effective in the same range as bromine.

(d) Ozone, Hydrogen peroxide and Peractic acid

Ozone (O_3) is an efficient biocide against the bacterial slime, but it produces deposits of manganese dioxide and the cost of ozone treatment is generally much higher than chlorine. Ozone is a very strong oxidant, more so than ClO_2, which in turn is a stronger oxidant than sodium hypochlorite. Ozone will react with all organic material present in the seawater. For this reason, ozone is difficult to use in other than very clean recirculating water systems, and it makes ozone unsuitable for larger once-through application

Hydrogen peroxide (H_2O_2) is sometimes applied as algaecide or biocide in small open and closed recirculating water systems. H_2O_2 disintegrates easily and reacts with some materials. High levels up to 15 ppm continuous dosing are necessary. H_2O_2 can also be used with peracetic acid as a disinfectant for RO membranes but the H_2O_2 concentration should not exceed 0.2% and temperature should not exceed 25 °C. Special care should be taken that transition metals (Fe, Mn) are not present in the feedwater or deposited on the membrane, since these can accelerate oxidation reactions by 1–2 orders of magnitude.

Peracetic acid (PAA), or peroxyacetic acid, is a weak acid which does not exist as a pure compound. It is a very powerful oxidant; the oxidation potential outranges that of chlorine and ClO_2. Verween [34] tested the toxicity of chlorine and PAA (15%) in the biofouling control of mussel *Mytilopsis leucophaeata* and *Dreissena polymorpha* embryos. The range of tested concentrations was chosen on the basis of preliminary research on mussel larvae, in which a dosage of 40 ppm PAA during 15 min per day was proposed. The 15% corresponds to 6 mg/L active PAA. Results show that both *M. leucophaeata* and *D. polymorpha* embryos have very low resistance to PAA, even at concentrations as low as 0.75 mg/L (active PAA) where tested exposure augmented 2 h. At 3 mg/L however, a 15 min-exposure is already lethal to 95% of all embryos for *D. polymorpha* and to more than 98% for embryonic *M. leucophaeata*. Because of the high costs of continuous dosage, initially an intermittent chemical dosage was also performed. This study indicates that a dosage as low as 3 mg/L active PAA during 15 min is as efficient as the proposed 6 mg/L active PAA in combating mussel fouling by *D. polymorpha* or *M. leucophaeata* [34]. The study of Verween [34] was applied with embryos which are more sensitive as larvea or adult mussels.

(e) Bromochlorodimethylhydantoin (BCDMH)

Hydantoin molecules have been used industrially for a number of years. One of the first applications of BCDMH to an industrial cooling system occurred in the late 1970s. A thorough study of the disinfection effectiveness of BCDMH showed that BCDMH was more effective than chlorine at pH 8.5 when dosed to the same residual against all organisms tested – *E. coli*, *E. aerogenes*, *P. aeruginosa*, and polybacteria [35]. BCDMH is mainly used to mitigate biological film forming bacteria in open recirculation systems with cooling towers. The product is used as a tablet which slowly releases the active products. However, for biofouling control at large-scale water intakes, it will result in a significantly higher cost in comparison with chlorine.

16.4.2.2 Non-oxidizing Biocides

Most of these chemicals were originally developed for bacterial disinfection and algae control in water treatment systems. They include organic film-forming anti-fouling compounds, gill membrane toxins, and non-organics. The proprietary formulations have a much higher per-volume cost than oxidizing chemicals but remain cost-effective due to lower use rates and rapid toxicity (Table 16.4). They often can provide better control of adult mussels due to the inability of mussels to detect them; because shells remain open, shorter exposures are required. Most are easy to apply and do not present corrosion problems for metal components. Although most compounds are biodegradable, detoxification or deactivation may be required to meet discharge requirements.

(a) Quaternary ammonium compounds (QAC's)

The application of quaternary ammonium compounds (QAC's) for once-through systems is very rare due to restrictions of discharging toxic concentrations in the outlet area which perforce detoxifying by fine clay dosing. Their spectrum of action is wide-ranging from micro-organisms to mollusks. Some products act very rapidly and can kill off an entire population within 48 h, others require a longer exposure [36]. QACs are surfactants which are absorbed on suspended matter in water or on colloids such as humic acids. To reduce the active compounds in the treated water before discharge, clay is added. This, in turn, forms a toxic sediment clay for benthic organisms which moves the problem from the water phase to the benthic community.

(b) Glutaraldehyde, formaldehyde and isothiazolin

A method for controlling the freshwater zebra mussels by using glutaraldehyde is documented. Glutaraldehyde is remarkably effective in controlling zebra mussels in industrial water systems. The effective dosage may vary between about 5–50 mg/L of glutaraldehyde added to the industrial process water infested with the zebra mussels. A usually effective dosage is between 10–25 ppm. The dosage may vary depending upon circumstances. Slug or continuous feed techniques may be used. Not suitable for once-through systems due to its long half-life. At pH 7.0 and 25 °C, glutaraldehyde has a half-life of 300 days. Glutaraldehyde has also been documented to remove biofilms.

For RO-membranes, addition of formaldehyde and/or gluteraldehyde (0.1–1.0%) has been reported to cause flux reduction of 10–50% to new RO-elements depending on specific membrane chemistry. To minimize the chances of flux reduction, membrane elements should be run for a minimum of 24-hrs prior to exposure to formaldehyde and or gluteraldehyde. However, this gives no guarantee for permanent flux loss. Also, isothiazolin (slug dose rate in the range of 50–100 ppm, with a contact time of about 4 h) used to target aerobic and anaerobic bacteria, fungi and algae are recommended for potable water systems by the supplier to use only off-line and dump the permeate during slug-dosing. 5-chloro-2-methyl-4-isothiazolin-3-one (CMIT) is another alternative. The bactericidal effect of CMIT can be studied by breaking the bond between bacteria and algae protein. When contacted with microbes, CMIT can quickly inhibit their growth, thus leading to death of these microbes. CMIT has strong inhibition and biocidal effects on ordinary bacteria, fungi and algae. It has good degradation kinetics resulting in no residual. The required dosing is 100–150 mg/L.

(c) 2,2-dibromo-3-nitrilopropionamide (DBNPA)

2,2-dibromo-3-nitrilopropionamide (DBNPA) is a fast-acting, non-oxidizing biocide which is very effective at low concentrations in controlling the growth of aerobic bacteria, anaerobic bacteria, fungi and algae. DBNPA is an advantageous disinfectant since it also quickly degrades to carbon dioxide, ammonia and bromide ion when in an aqueous environment. This allows the effluent to be safely discharged even in sensitive water bodies. It is degraded by reactions with water, nucleophiles, and UV light (rate is dependent on pH and temperature). The approximate half-lives are 24 h (pH 7), 2 h (pH 8), and 15 min (pH 9). The vast majority of microorganisms that come into contact with it are killed within 5–10 min. Most RO-chemical suppliers have a premixed private label version with varying solution concentrations of 5–20%. Recommended usage for RO systems is slug dosing 10–30 ppm of active ingredient for 30 min to 3 h, every 5 days (for waters less prone to biological fouling). During slug dosing, the permeate should be dumped to drain if product water is for potable use.

16.4.2.3 Coatings in Water Intake Structures and Conduits

The application of coatings in power plant cooling water conduits and other surfaces is well documented and there are a number of coatings specifically developed for this. In general, coatings are not widely used in desalination plants, while antifouling coatings and paints are able to play an important role in the reduction of fouling settlement and growth. They can indirectly reduce the amount of biocide needed for adequate control. For desalination plants it is important to protect the seawater intake and distribution piping system for marine biofouling settlement. In addition, the coating used cannot release any byproduct which can result in a negative impact on the process (pre-filtration and RO membranes) and on the produced water quality. Therefore the amount of coatings which can be used are very limited.

Generally, toxic and non-toxic coatings and paints are distinguished between. Non-toxic foul release coatings weaken the adhesive bond between the organism, film or deposit and the coating. The most promising product available for cooling water systems at this moment is the non-toxic, fouling release, silicone-based and fluoropolymer-based coatings. Although foul release coatings can be fouled by biological growth, the rate of fouling is significantly less and is normally removed by the cooling water flow (shear stress).

The silicone-based and fluoropolymer-based coatings should be applied to perfectly clean and dry surfaces, or to clean and almost dry (5% or less moisture) concrete over the appropriate epoxy primers. For this reason, the technique is more difficult to apply in existing situations, where dry conditions are more difficult to obtain than in newly built cooling water systems. In cooling water systems foul release coatings can be durable up to 10 years (including a re-coat of finish only after 5 years). Fluoropolymer-based coatings are more hard-wearing than the silicones-based ones. Attempts to toughen foul release paints have always resulted in diminished antifouling performance. The development of paint with both properties remains a challenge to the paint industry.

From comparison of application costs of foul release coatings to the alternative mitigation methods for cooling water systems, the foul release coatings could in principle be an interesting option. However, a silicone-based and fluoropolymer-based coatings should be applied to perfectly clean and dry surface, or to clean and almost dry (5% or less moisture) concrete over the appropriate epoxy primers. For this reason, the technique is more difficult to apply in existing situations, where dry conditions are very hard to obtain than in newly built systems.

Coatings, in general, can have a practical advantage of protecting specific parts of the seawater intake structure, e.g. intake screens, trash racks and intake walls structures. However, protecting intake pipes itself are difficult to maintain free of biofouling settlement using coatings due to the fact the coating needs replacement every couple of years. Also, the efficacy cannot be monitored, the antifouling efficiency is only traceable indirectly through performance indicators or by visual inspection. Coatings are at this moment not feasible as a prime antifouling method in desalination seawater intake systems.

16.5 Monitoring

For any applied biofouling control procedure, it is important to achieve insight into the efficiency of the treatment. Long-term monitoring at industrial provides insight into the type of biofouling and the abundance of populations settled in the seawater system. It will also provide insight into the arrival of new invasive species, which might require an adjustment of the applied biofouling control procedure.

16.5.1 Microfouling Monitoring

Biofilms are able to change the electrochemical characteristics of passivatable metals' surface, both in aerobic and/or anaerobic environment [37, 38]. To achieve real-time insight in the development of biofilm buildup several types of electrochemical probe systems have been developed, such as the BIOX™ and the BioGeorge™. The BIOX™ electrochemical sensors proved to be able to monitor early stages of biofilm growth and to optimize chlorination treatments [39]. The BIOX probe has been used at several Italian power plants. The BioGEORGE™ has shown to be effective for predicting biofilm activity and as a continuous measure of biocide effectiveness [40, 41]. The probe consists of a series of metallic discs comprising two nominally identical electrodes (Fig. 16.7).

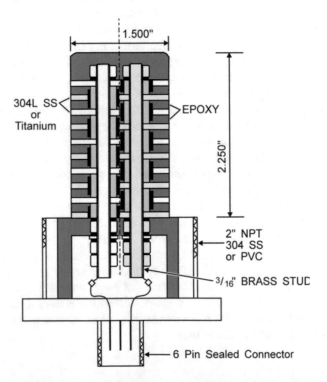

Fig. 16.7 The BIoGEORGE™ probe [40]. (Source: Structural Integrity Associates)

The electrodes are electrically isolated from each other and from the stainless-steel plug that serves as the body of the probe. An epoxy resin filled between the electrodes produces a right circular cylinder of metal discs and the insulating resin. One electrode (set of discs) is polarized relative to the other for a short period of time, typically 1 h, once each day. This applied polarization potential causes a current, designated as the "applied current", to flow between the electrodes. When a biofilm forms on the probe, it provides a more conductive path for the applied current than the general cooling water, increasing the current flow significantly over the baseline value. Metabolic processes in the biofilm, many of which involve oxidation/reduction reactions, also appear to enhance the applied current. The applied potential also produces slightly modified environments on the discs that are conducive to microbial activity and thus biofilm formation [42, 43, 44]. This will produce a biofilm on the probe sooner than on the general piping and therefore acts as an early warning system. In a study carried out by Bruijs et al. [40] it was shown the BioGeorge proved a helpful tool to tailor the biocide dosing regime to prevent any buildup of a biofilm. In Fig. 16.8 an example of this optimisation is presented.

The BIoGeorge™ system was found to be a helpful tool to optimize the chlorine pre-treatment at RO membranes [45].

16.5.2   Macrofouling Monitoring

To provide operators with a clear understanding of the levels of biofouling growth, a biomonitoring system can be installed. A biofouling monitoring system can provide the following insight:

- seasonality larvae production and settlement
- development fouling community, including densities and species composition
- efficacy of the antifouling treatment.

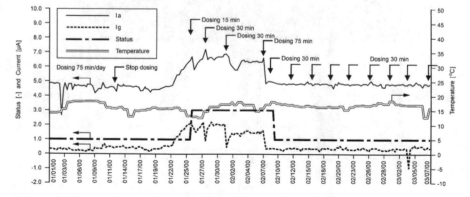

Fig. 16.8 Presentation of biofilm activity monitoring during sodium hypochlorite (TRO = 0.6 mg/L Cl₂) optimization. Reproduced with permission from Ref. [40] Copyright 2001 @ PowerPlant Chemistry

To achieve these monitoring objectives, a monitoring system should be applied to achieve real-time insight. This can be achieved by a visual check on e.g. shell settlement on intake screening and walls, however, this does not necessarily give insight in the actual fouling status inside a pipe or culvert. Head loss measurement can also be applied to achieve insight in increased wall roughness due to biofouling. Since the impact of biofouling can be irreversible (due to cementation of their shells to the wall surface) it is important to achieve insight at an early (larval) stage of settlement. To be able to achieve real-time insight in-plant, biomonitoring devices can be used, which are installed as a bypass to the seawater feed line. These monitors are preferably installed before the seawater pre-treatment for RO to obtain insight in the full spectrum of fouling species and fouling pressure. Several types of monitors has been developed and applied at different industrial sites. The Bio-box is a monitoring system specifically developed for Zebra Mussel settlement. It can however also be used for other types of macrofouling species. In Fig. 16.9 a schematic image of the BioBox is presented.

KEMA Nederland developed the KEMA Biofouling Monitor which was specifically designed to monitor the settlement of biofouling larvae. The monitor consists of four fouling coupons (standard roughened PVC plates) which are exposed to a fixed seawater flow. The coupons can be analysed according to a fixed time schedule which makes it possible to monitor both the rate of larval settlement and at the same time achieve insight in the cumulative growth and/or effect of the applied biofouling control procedure. In Fig. 16.10 the KEMA Biofouling Monitor is shown.

H2O Biofouling Solutions developed a monitor (Biovision Monitor™) which was designed to provide a representative, real-time image of the actual biofouling situation inside the different parts of the seawater intake and culverts. The Biovision monitor consists of five fouling coupons and is designed in a way that the coupons are exposed to a range of velocities (0.2–2 m/s) and turbulences. This will provide

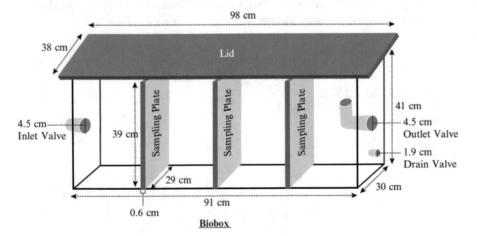

Fig. 16.9 Schematic representation of side-stream sampler commonly known as a Bio-box. Reproduced with permission from Ref [15]; Copyright 2012 @ Springer

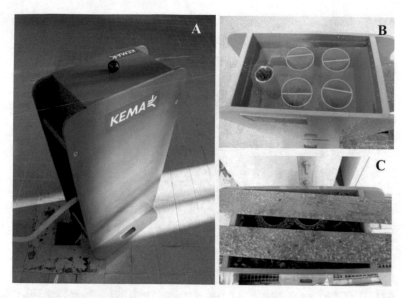

Fig. 16.10 KEMA biofouling monitor (**A**); Top view with the four fouling coupons (**B**); Fouled coupons with marine biofouling (**C**). (Source: KEMA Nederland B.V)

an overall insight of the potential risk for biofouling settlement in the different spots in the intake and culverts where generally different seawater flows and/or turbulence is present. In addition, since different type of biofouling species have their own preference for certain flows to settle and grow, the Biovision monitor creates optimal conditions for all the different biofouling species. In Fig. 16.11 the Biovision monitor is presented.

In-plant monitoring systems provide data which offers real-time insight in the potential fouling risk in seawater intake and culverts. It helps the plant to observe any potential fouling risk at an early stage and take counter actions to prevent high numbers of fouling which could put the plants operational output at risk. Effectiveness of a certain coating can be easily tested using the fouling coupons of a biofouling monitor. This makes it easy to monitor the efficiency and wear of the coatingduring operation of the seawater system.

16.6 Conclusions and Outlook

Biofouling settlement is a common phenomenon for desalination plants using an open sea water intake system. Operators of large-scale seawater intake systems need to rely on both the efficiency of the water treatment procedure as well as the monitoring equipment. For this reason, proven technologies with over 10 years proven industry practice are the prime choice. Of all available biofouling mitigation

Fig. 16.11 H2O biovision monitor™ (**A**); Fouled coupons with marine biofouling (**B**). (Source: H2O Biofouling Solutions B.V)

methods, chlorination has shown to be an established practice in the past decades to be efficient in preventing biofouling in seawater intakes and culverts at a relatively low costs.

The desalination industry is well aware of the benefits and disadvantages of chlorination for macrofouling settlement prevention in their seawater feed system. Especially for SWRO there are strict limitations in respect to the amount of chlorine used. However, the applied chlorination regimes are often "standard" practice and not tailored to local requirements or potential negative side effects. This often results in an inefficient control of marine biofouling species in the intake piping or increased fouling rates on the RO membranes. The uniqueness of each seawater, intake system and desalination system make it a challenge to determine how to maximize the efficacy of chlorination on the intake while minimizing biofouling on the SWRO membranes.

The table below provides a concise evaluation of technologies in terms of applicability. Only chlorination can be considered proven-technology (>10 years field experience). For desalination facilities the use of alternative non-oxidative chemicals is generally not feasible, because of the fact that many chemicals can only be used for off-line SWRO membrane cleaning and often need strict discharge regulations. The use of coatings could provide some additional protection at specific parts in the seawater intake, e.g. intake bar screens or trash racks. However, coatings are at this moment in time not yet suitable to use as a biofouling control method in the seawater intake and piping systems of desalination plants.

Further work is required in this regard to understand how best to monitor and control biological and organic fouling resulting from the chlorination process. Other promising technologies that have shown efficient mitigation of macrofouling

Table 16.5 Feasibility for macrofouling mitigation and control in seawater desalination systems

Options	Feasibility	Remark
Oxidizing biocides	Yes	Safety requirements and regulatory issues (discharge)
Non-oxidizing biocides	No	Can only be used in off-line SWRO membrane cleaning and often need strict discharge regulations.
Heat treatment	No	No (partial) recirculation or steam injection possible.
Increase in water velocity	Yes (for distribution piping)	Increase in water velocity can prevent settlement in intake conduits
	No (for intake risers and screening compartments)	Increase in seawater velocity has no impact on macrofouling settlement.
Oxygen depletion	Yes (for intake risers and screening compartments)	If intake risers and screening compartments can be separately closed along with availability of additional redundant intake pipe.
	No (for intake chamber)	Cannot be separately closed
Coatings	No (not as prime technology)	Only applicable to areas with low flow/turbulence, provides some additional protection.
UV light	No	Limited industry practice

are yet determined as fit-for-purpose (available but little field experience) or are non-applicable for desalination plants. The approaches to macrofouling mitigation and control in seawater desalination systems, the following feasibilities are assessed as summarized in Table 16.5.

References

1. IDA global data, Global Water Intelligence, DesalData 2016 Plant Inventory Report, International Desalination Association, Oxford, 2016
2. M. Kurihara, H. Takeuchi, SWRO-PRO system in 'Mega-ton water system, for energy reduction and low environmental impact. Water **10**, 48 (2018). https://doi.org/10.3390/w10010048
3. J.W. Whitehouse, M. Khalanski, M.G. Saroglia, H.A. Jenner, The control of biofouling in marine and estuarine power stations: A collaborative research working group report for use by station designers and station managers. CEGB NW Region 191-9-85, CEGB (England), EdF (France), ENEL (Italy), KEMA (The Netherlands), (1985), pp. 1–48.
4. S. Abarzua, S. Jakubowski, Biotechnological investigation for the prevention of biofouling. I. Biological and biochemical principles for the prevention of biofouling. Mar. Ecol. Prog. Ser. **123**, 301–312 (1995)
5. H.A. Jenner, S. Rajagopal, G. Van der Velde, M.S. Daud, Perforation of ABS pipes by boring bivalve Martesia striata: A case study. Intl. Biodeter. Biodegrad. **52**, 229–232 (2003)
6. I.E. Idelchik, *Handbook of Hydraulic Resistance*, 3rd edn. (Begell House Inc., New York, 1996)
7. L.W. Hall, *Power Plant Chlorination*, 1st edn. (Electric Power Research Institute, Palo Alto, 1981)
8. Woods Hole Oceanographic Institution, *Marine Fouling and its Prevention*, 1st edn. (US Naval Institute, Annapolis, 1952)

9. H.J.G. Polman, F. Verhaart, M.C.M. Bruijs, Impact of biofouling in intake pipes on the hydraulics and efficiency of pumping capacity. Desalin. Water Treat. **51**, 997–1003 (2013)
10. J. Ritchie, Report on prevention of growth of mussels in sub-marine shafts and tunnels at Westbank electric station, Portobello. Trans. R. Scot. Soc. Arts **19**, 1–20 (1927)
11. M.C.M. Bruijs, H.A. Jenner, Cooling water system design in relation to fouling pressure, in *Operational and Environmental Consequences of Large Industrial Cooling*, ed. by W. Systems, S. Rajagopal, H. A. Jenner, V. P. Venugopalan, (Springer, 2012). https://doi.org/10.1007/978-1-4614-1698-2_19.
12. J.D. Ackerman, C.M. Cotrell, C. Ross Ethier, D. Grant Allen, J.K. Spelt, A wall jet to measure the attachment strength of zebra mussels. Can. J. Fish. Aquat. Sci. **52**, 126–135 (1995)
13. M. Kawabe, Sea level variations at the Izu Islands and typical stable paths of the Kuroshio. J. Oceanogr. Soc. Jpn **41**, 307–326 (1985)
14. H.A. Jenner, Control of mussel fouling in the Netherlands: Experimental and existing methods, Symposium on condensor macrofouling control technologies, The State-of-the-art. EPRI. Hyannis. Massachusetts. June 1983.
15. R. Claudi, H.A. Jenner, G.L. Mackie, Monitoring: The underestimated need in macrofouling control, in *Operational and Environmental Consequences of Large Industrial Cooling Water Systems*, ed. by S. Rajagopal, H. A. Jenner, V. P. Venugopalan, (Springer, 2012). https://doi.org/10.1007/978-1-4614-1698-2_19.
16. D.J. Bitter, Advanced ultraviolet disinfection unit provides first cost-effective uses for UV against biofouling. Wateronline, July 11, 2013, https://www.wateronline.com/doc/advanced-ultraviolet-disinfection-unit-provides-first-cost-effective-uses-for-uv-against-biofouling-0001
17. S.F. Pucherelli, R. Claudi, T. Prescott, Control of biofouling in hydropower cooling systems using HOD ultraviolet light. Manag. Biol. Invasion. **9**, 451–461 (2018). https://doi.org/10.3391/mbi.2018.9.4.08
18. A. De Zwaan, M. Mathieu, Cellular biochemistry and endocrinology, in *The Mussel Mytilus: Ecology, Physiology, Genetics and Culture*, ed. by E. Gosling, (Elsevier, Amsterdam, 1992), pp. 223–307
19. R.J. Diaz, R. Rosenberg, Marine benthic hypoxia: A review of its ecological effects and the behavioural responses of benthic macrofauna. Oceanogr. Mar. Biol. Ann. Rev. **33**, 245–303 (1995)
20. M. Khalanski, Organic products generated by the chlorination of cooling water at marine power stations. Journées d'Etudes du Cebedeau, Tribune de l'Eau No 619-620-621, (2002), 24–39.
21. C.J. Taylor, The effects of biological fouling control at coastal and estuarine power stations. Mar. Pollut. Bull. **53**, 30–48 (2006)
22. H.J.G. Polman, H.A. Jenner, Pulse-chlorination®, the best available technique in macrofouling mitigation using chlorine. PowerPlant Chem. **4**, 93–97 (2002)
23. H.J.G. Polman, M.C.M. Bruijs, L.P. Venhuis, S.A. van Dijk, More than 10 year experience with pulse-chlorination® dosing regime against macrofouling, in *Heat Exchanger Fouling – Mitigation and Cleaning Technologies*, 2nd edn., (Handbook, Publico, 2011), pp. 240–251. ISBN 3-934736-20-3
24. H.J.G. Polman, M. Kanavoutsos, T. Attenborough, H. Kamal, Finding the biofouling control balance for SWRO plants. International Desalination Association conference Dubai 2019, In press.
25. K. Gaid, A large review of the pre treatment, in *Expanding Issues in Desalination*, ed. by R. Y. Ning, (IntechOpen, 2011). https://doi.org/10.5772/19680
26. CDA, Aluminium bronze alloys for industry. Copper Development Association, CDA Publication No 83. (1986)
27. R.O. Lewis, The influence of biofouling counter measures on corrosion of heat exchanger materials in seawater. Mater. Perfom. **22**, 31 (1981)
28. A.H. Tuthill, Guidelines for the Use of Copper Alloys in Seawater, NiDI, CDA. (1987), https://www.nickelinstitute.org/media/1688/guidelinesfortheuseofcopperalloysinseawater_12003_.pdf

29. R. Francis, H. Campbell, Chlorine review of BNF studies of the effect of chlorine and pollutants on the corrosion of copper alloy condenser tubes, EFC Marine Corrosion Workshop, Eurocorr 2008
30. B.W. Lykins, M.H. Griese, Using chlorine dioxide for trihalomethane control. J. AWWA **78**(6), 88–93 (1986)
31. E.M. Aieta, J.D. Berg, A review of chlorine dioxide in drinking water treatment. J. AWWA **78**, 62–72 (1986)
32. J.H. Koh, A. Jang, Effect of chlorine dioxide (ClO_2) on polyamide-based RO membrane for seawater desalination process: Exposure to high concentration of ClO_2. Desalin. Water Treat. **80**, 11–17 (2017). https://doi.org/10.5004/dwt.2017.20679
33. L. Duvivier, Formation et elimination des organo-halogenes lors de la desinfection des eaux. These de doctorat es-sciences. Universite catholique de Louvain-la-Neuve, 1993
34. A.M. Verween, M. Vincx, S. Degraer, Comparative toxicity of chlorine and peracetic acid in the biofouling control of Mytilopsis leucophaeata and Dreissena polymorpha embryos (Mollusca, Bivalvia), *Intl. Biodeter.* Biodegradation **63**, 523–528 (2009)
35. Z. Zhang, J.V. Matson, Organic Halogen Stabilizers: Mechanisms and Disinfection Efficiencies, Paper TP89–05, (Cooling Tower Institute, Houston, 1989)
36. J.C. Petrille, M.W. Werner, (1993). *A Combined Treatment Approach Using a Non-oxidizing Molluscicide and Heat to Control Zebra Mussels*. 3rd International Zebra Mussel Conference, Toronto, February 23–26 1993.
37. P. Chandrasekaran, S.C. Dexter, *Factor Contributing to Ennoblement of Passive Metals Due to Biofilm in Seawater*, Proc. 12th International Corrosion Congress, NACE International, 1993, 3696–3707.
38. A. Mollica, Biofilm and corrosion on active-passive alloys in seawater. Intl. Biodeter. Biodegrad. **29**, 213–229 (1992)
39. A. Mollica, P. Cristiani, On-line biofilm monitoring by "BIOX" electrochemical probe. Water Sci. Technol. **47**, 45–49 (2003)
40. M.C.M. Bruijs, L.P. Venhuis, H.A. Jenner, D.G. Daniels, G.J. Licina, Biocide optimisation using an on-line biofilm monitor. J. Power Plant Chem. **3**, 400–405 (2001)
41. W.E. Garrett, G.J. Licina, *Eighth EPRI Service Water System Reliability Improvement Seminar* (Electric Power Research Institute, Palo Alto, 1995)
42. J. Guezennec, N.J. Dowling, M. Conte, E. Antoine, L. Fiksdal, Cathodic protection in marine sediments and the aerated seawater column, in *Microbially Influenced Corrosion and Biodeterioration*, ed. by N. J. Dowling, M. W. Mittleman, J. C. Danko, (NACE, Washington, 1991), pp. 6.43–6.50
43. G. Nekoksa, B. Gutherman, Cathodic protection criteria for controlling microbially influenced corrosion in power plants. Electric Power Research Institute, Palo Alto, EPRI NP-7312.
44. G.J. Licina, G. Nekoksa, An electrochemical method for on-line monitoring of biofilm activity, Paper No. 403, Corrosion 93, (NACE International, Houston, 1993).
45. D. Brumfield, J. Licina, Increasing the effectiveness of RO membranes by improved pretreatment, Ultrapure Water, October, 21–24 2009.

Chapter 17
Recent Strategies in Designing Antifouling Desalination Membranes

Mohamed Afizal Mohamed Amin, Pei Sean Goh, Ahmad Fauzi Ismail, and Dayang Norafizan Awang Chee

17.1 Introduction

Many recent and important improvements in membrane-based desalination technologies are mainly focused on improving existing processes such as development of fouling-resistant membranes. Membrane technology is an emerging domain in the desalination application due to its capability that can remove a lot of unwanted substances in the produced water. In addition, the membrane can be tailored so that it can possess a specific property that act as a shielding in protecting it from different types of foulants. Fouling is a severe bottleneck for membranes specifically in a desalination application such as reverse osmosis (RO) [58], forward osmosis (FO) [29], membrane distillation (MD) [19], ultrafiltration (UF) [6], microfiltration (MF) [41], nanofiltration (NF) [46] and pressure retarded osmosis (PRO) [64].

M. A. M. Amin
Advanced Membrane Technology Research Centre, School of Chemical and Energy Engineering, Faculty of Engineering, Universiti Teknologi Malaysia, Johor, Johor Bahru, Malaysia

Department of Chemical Engineering and Sustainability Energy, Faculty of Engineering, Universiti Malaysia Sarawak, Kota Samarahan, Sarawak, Malaysia

P. S. Goh · A. F. Ismail (✉)
Advanced Membrane Technology Research Centre, School of Chemical and Energy Engineering, Faculty of Engineering, Universiti Teknologi Malaysia, Johor, Johor Bahru, Malaysia
e-mail: afauzi@utm.my

D. N. A. Chee
Advanced Membrane Technology Research Centre, School of Chemical and Energy Engineering, Faculty of Engineering, Universiti Teknologi Malaysia, Johor, Johor Bahru, Malaysia

Faculty of Resource Science and Technology, Universiti Malaysia Sarawak, Kota Samarahan, Sarawak, Malaysia

© Springer Nature Switzerland AG 2020
V. S. Saji et al. (eds.), *Corrosion and Fouling Control in Desalination Industry*,
https://doi.org/10.1007/978-3-030-34284-5_17

Fouling is defined as an accumulation, deposition, and/or adsorption of foulants on the membrane surface (external fouling) or within the membranes pores (internal fouling), which seriously hampers the application of membrane technologies, typically by significantly deteriorating the filtration rate, permeate flow, solute removal efficiency, and pressure differential across the membranes [18]. Fouling mechanism varies for different types of desalination membranes. For instance, in microporous type-membranes such as that used in UF and MF, fouling usually occur by pore blocking, solute adsorption, and cake/gel layer formation. Meanwhile, for salt rejecting types like NF and RO membranes, fouling is mainly governed by the adsorption of contaminants on the surface, and scaling by divalent ions.

Without proper mitigation strategy, fouling could highly risk the whole desalination operation plant by reducing the membrane life-expectancy and increases the operation and maintenance costs and in certain scenario, it could lead to temporary plant shutdowns [17]. Thus, research and development efforts are directed towards preventing and/ or mitigating the membrane fouling specifically in desalination field. This is currently being done by developing a better understanding of the membrane fouling such as the characteristics of foulants, foulants interactions as well as developing membranes with specific anti-fouling properties.

The present chapter aims to discuss novel strategies in designing more efficient desalination membranes. The types of fouling encountered, the membrane surface properties affecting fouling and the novel approaches implemented in designing antifouling membranes are explained.

17.2   Types of Fouling

Foulants can be identified into four main categories which are inorganic fouling (or scaling), adsorbed organic fouling, colloidal fouling and biofouling [42, 47].

17.2.1   Inorganic Fouling

Inorganic fouling resulted from the increment concentration of single or combined inorganic salts particularly calcium sulfate, calcium phosphate and calcium carbonate beyond their solubility limits and their ultimate precipitation on the membrane surfaces [7, 51]. Analysis on raw water for desalination plant revealed that the constituent in that water body comprise of both cations (Al^{3+}, Na^+, K^+, Ca^{2+} etc.) and anions (Cl^-, Br^-, SO_4^{2-}, HCO_3^-) where it has a high possibility to form unwanted scale. Butt et al. revealed that high concentrations of Ca^{2+}, SO_4^{2-}, and HCO_3^- under certain temperature conditions, can give rise to scale-forming salts for example $CaCO_3$ and $CaSO_4$ [5].

Thus, the mitigation strategies for this type of foulant for most of the membrane processes like seawater desalination and water softening are often focused on the pre-treatment of feed water or optimization of operational conditions [71]. For instance, calcium carbonate scaling can be controlled by suitable antiscalants or lowering the feed water pH. calcium phosphate scaling during wastewater desalination is more difficult to cope with, primarily because of a lack of efficient antiscalants [51].

17.2.2   Organic Fouling

Organic fouling is caused by dissolved natural organic matter (NOM) such as proteins, humic substances, and polysaccharides in the source waters. This foulant might cause the permeate flux to decline due to strong adsorption on the membrane surface and have also been known to cause performance deterioration. NOM is ubiquitous in natural waters, and its removal is not well elucidated. It can be categorized via molecular weight (M_W) or different wetting properties [16].

With respect to M_W, Lee et al. reported that medium to low M_W component of NOM (300–1000 Da) is responsible for the initiation of fouling, where bulk of the fouling observed is due to very high M_W 'colloidal' NOM (>50,000 Da) [32]. Zularisam et al. studied the behavior of membrane fouling with respect to fractionated NOMs in term of wetting property which are hydrophobic, transphilic and hydrophilic. Transphilic is the intermediate or transitional polarity between hydrophobic and hydrophilic properties. Based on the result, they found out that each fraction of NOM exhibited different primary fouling mechanisms. The mechanisms accounting for the fouling by hydrophobic fraction, hydrophilic part and transphilic components were concentration polarization (low concentration of the solute in the permeate than concentration in the bulk, due to solute retention), adsorptive fouling (attraction of foulant on the membrane body due to electrostatic interaction) and cake layer deposition (compaction of the foulant on the membrane surface), respectively. The responsible foulant that affected the membrane performance was the hydrophobic fraction which was found to cause the highest flux decline during the membrane filtration [74].

17.2.3   Colloidal Fouling

Colloidal fouling occurs due to the deposition of colloids particles and it is pervasive in natural waters. Colloid comprise of clay minerals and suspended particles whereby it covers wide size range from a few nanometers to a few micrometers. Most colloids carry negative surface charge in a wide pH range and size range (from nanometers to a micrometers). Colloidal fouling of pressure-driven membranes can be categorized into two phenomena. First, colloidal fouling caused by the accumulation of particles on the membrane surface also known as cake layer. This type of fouling usually occurs on RO, NF and UF membranes. This cake layer provides an additional hydraulic resistance to water flow through the membrane and, thus, reduces the product water flux. Second, colloidal fouling caused by initially accumulated particles on the membrane surface followed by pore plugging and that occurs most probably on MF membranes [72].

17.2.4   Biofouling

Biofouling is an inevitable aspect of membrane filtration. Biofouling initiates through the adhesion, accumulation and growth of microorganisms on the surface of membrane including bacterial cells, such as Gram negative *Escherichia coli* (*E. coli*) and Gram positive *Staphylococcus aureus* (*S. aureus*), and algae, starting the development of a biofilm [25, 30]. The formation of biofilm occurs in five stages; (1) bacterial cell attachment; (2) cell to cell adhesion; (3) proliferation; (4) maturation, and (5) dispersion of planktonic bacteria. In this process, once the biofilm is stabilized on the membrane surface (through weak hydrogen bonding and van der Waal's, or electrostatic interactions) continuous film growth happens [21]. In this stage metabolic activities release extracellular polymeric substances (EPS) which provide an anchoring effect to the biofilm structure [47]. Finally, after over growth, the microorganisms are detached from the membrane surface from where they are dispersed to new sites to reinitialize the biofilm formation and that slowly hinder or block the water pathway, leading to undesired water flux decline, overall higher energy consumption and in worst case, physical degradation of certain types of membranes.

There are numerous factors influencing the microorganism attachment on the membrane surface. Those factors can be categorised into three major groups which are membrane surface properties (roughness, wetting, charge and pore dimension), raw water parameters (temperature, pH, ionic strength, nutrients, osmotic pressure and velocity) and microbial properties (size, cell surface wetting property and charge) [8]. The exploitation of antifouling membranes is mostly concentrated on alleviating fouling of organic foulants and biofoulants [71].

17.3   Membrane Surface Properties Affecting Fouling

The performance of a membrane depends largely on its structural properties, such as pore size and morphology, and surface properties such as charge and roughness. The skin layer or top surface layer plays a vital role on the overall performance. The enhanced membrane antifouling resistance was mainly due to the improved membrane surface properties that minimize deposition and adsorption of foulants on the membrane surface.

It is known that the interaction between membrane surface and foulants is strongly related to the surface properties (wetting, charge, and roughness). A mem-

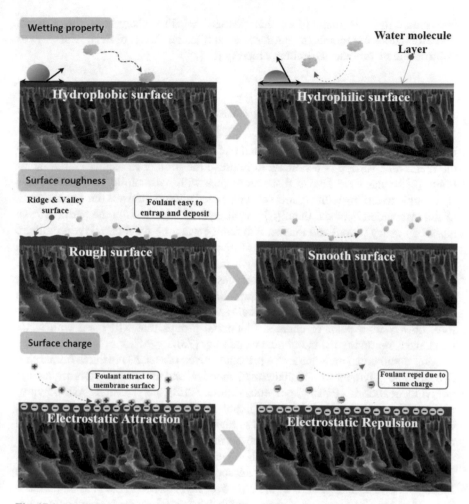

Fig. 17.1 Membrane surface behavior towards foulants

brane with a smooth and hydrophilic surface of similar charge to foulants seems to possess good characteristic in reducing the fouling (Fig. 17.1) [20, 50, 53]. Membrane with high surface wetting property (hydrophobic) will tend to attract more hydrophobic organic molecules compared to the membrane with less wetting property (hydrophilic). As the hydrophilicity increase, membrane surface chemistry will prefer to bind more water molecules and thus form a water barrier between hydrophobic foulant and the membrane surface. Rougher membranes have greater affinity towards foulants than the smoother membrane. Membrane surface with rough topography usually exhibit high number of ridges–valley structure and tend to trap and accumulate more foulants.

Apart from that, foulants accumulate on the membrane surfaces due to stronger electrostatic force of attraction between the oppositely charged surface and the foulants. For instance, positively charged foulants such as cationic surfactants will not attract towards membranes that possess negative charges on its surface. Therefore, tailoring membrane via surface modification is one of the useful ways to improve the membrane antifouling property [59].

17.3.1 Wettability Properties of Membrane Surfaces

Numerous researches have demonstrated that the presence of a hydrophilic layer on the membrane surface is beneficial to improve the fouling resistance of the membrane [2]. It has been known that membranes with hydrophilic surfaces are less susceptible to fouling with organic substances and microorganisms due to a decrease of the interaction between them [57] while hydrophobic membrane surfaces have higher tendency to foul and thus tend to reduce water permeability. Hydrophilicity refers primarily to hydrogen bonding of membrane with water molecules. Hence, having a high percentage of atoms which can form a hydrogen bond, either in the main chain or in pendant groups, contributes to the water-loving nature of a material. Membranes including grafted hydrophilic polymers are characterized as having oxygen or nitrogen atoms (as well as halogens) in their backbone structure; therefore, they contain polar or charged functional groups that will bond with water molecules, improving the membrane wettability [73].

Most commercial membranes for pressure-driven processes are made from rather hydrophobic polymers with high thermal, chemical and mechanical stability such as polyvilidenefluoride (PVDF), polyethersulfone (PES), polysulfone (PS), polypropylene (PP), polyacrylonitrile (PAN), polyamide (PA) and polyethylene (PE). Usually these types of materials possess high wetting values ($90° \leq \theta < 180°$) and tend to adsorb various kind of solutes from feed streams [28].

The utmost reason which caused the hydrophobic membrane to become easily fouled with organic compounds and microorganisms is that there are almost no hydrogen bonding interactions in the boundary layer between the membrane interface and water. The repulsion of water molecules away from the hydrophobic membrane surface is a spontaneous process with increasing entropy and therefore foulant molecules tend to adsorb onto the membrane surface and dominate the boundary layer [28]. On the other hand, membrane skin with a hydrophilic layer tend to possess a high surface tension and might be able to form hydrogen bonds with the surrounding water molecules to reconstruct a thin water boundary between the membrane and bulk solutions. Hydrophilic interfaces form a pure water layer, which reduces binding sites for foulants to adsorb and eventually deposit on the membrane surface. Hence, it can automatically prevent or reduce many undesirable adsorption or adhesion of hydrophobic foulants on the membrane surface [39]. In addition, it is generally accepted that an increase in hydrophilicity offers better fouling resistance because protein and many other foulants are hydrophobic in nature [50]. A well-known measurement for the hydrophilicity of a surface is the contact angle (θ) of a

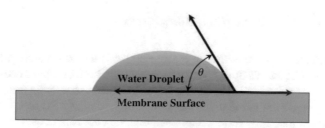

Fig. 17.2 Schematic diagram of a droplet on a membrane surface with the contact angle θ, and the surface tension of the interfaces

water droplet on the surface, as illustrated in Fig. 17.2. The membrane is considered hydrophilic when the water contact angle is smaller than 90° and if the angle is beyond 90°, the membrane is considered hydrophobic. For super-hydrophobic case the value of contact angles can reach greater than 150°.

17.3.2   Membrane Surface Charge

The membrane surface for the desalination process often carries a negative charge [55]. This charge is usually caused by sulfonic or carboxylic acid groups in the top of membrane layer, which may be deprotonated in feed solution. The negative surface charge of the membranes can be affected by the pH of feed solutions due to an increase in dissociation of carboxylic or sulphonic functional groups. Yoon et al. studied the effect of pH and salt conductivity towards the rejection of perchlorate anion ($ClO_4^?$) and effective diffusion coefficient of NF and UF membranes [66]. They revealed that $ClO_4^?$ rejection by negatively charged NF and UF membranes was greater due to electrostatic exclusion especially with increasing pH. The variation of membrane surface charge with increase in pH and conductivity was found to reduce the effective diffusion coefficient (hindered diffusion) as the more negative membrane surface charge enhanced electrostatic repulsion between the perchlorate anion and the negatively charged surface.

The membrane charge is particularly important in mitigating fouling if the existing foulants are in the form of ions. Existence of similar charges on the membrane and the foulant might result in electrostatic repulsion between them and hence prevent the foulants deposition. For instance, a negative surface charge of the membrane may have a beneficial effect to reduce membrane fouling during protein filtration at neutral pH, because most of the proteins have negative charge at such conditions [28]. An electrically charged membrane can significantly reduce the surface scale-formation [50]. The membrane charge can be measured by using a zeta potential analyzer.

17.3.3   Membrane Surface Roughness

An approximate linear relationship existed between the RO membrane skin layer surface roughness and its flux. A correlation between both of it have been studied by Hirose and co-workers in 1996. In this study, RO membranes with rough skin layer surface exhibited high saline water flux due to high effective RO area without sacrificing salt rejection. They further explained that the unevenness of cross-linked aromatic polyamide on the skin layer of RO membranes enlarge the effective area of the membrane. Thus, the contact area between water molecule and membrane skin increase and induce more water transport. In other words, flux is able to be manipulated by controlling the unevenness of the skin layer surface structures of cross-linked aromatic polyamide composite RO membranes [22]. However, no correlation study have been made between membrane surface roughness and fouling in this research.

Recent study in antifouling desalination membrane showed a strong correlation between the fouling and the membrane surface roughness where an increase in surface roughness will tend to enhance the interaction of foulants towards membrane surface [72]. In view of this, most of the researchers are focusing on tailoring smooth membrane surfaces [50]. As membrane surfaces become smooth, the number of "valleys" that can act as adsorption sites for foulants is reduced and the membrane is less prone to fouling. Whilst, for rougher membrane surface, fouling becomes more severe due to highly entrapment of foulants [39]. Since 90's, atomic force microscopy (AFM) has been used as a tool to study the effect of surface roughness.

17.4   Membrane Surface Modification Strategies

Membrane modification is a common strategy used to reduce fouling tendency. The surface modification approaches described in this chapter are schematically shown in (Fig. 17.3).

17.4.1   Surface Modification

Surface modification of a polymer thin film composite (TFC) RO membrane can be employed via physical or chemical methods. In physical modification, the materials interact with Polyamide (PA) layer of RO membrane and attach it through van der Waals attraction, electrostatic interaction or hydrogen bonding. In contrast, in chemical modification, the materials bonds to the membrane surface by covalent bonds and have better chemical and structural stabilities. Several membrane surface modification techniques have been reported ranging from direct chemical treatment,

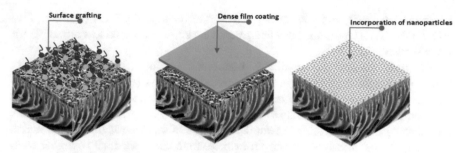

Fig. 17.3 Illustration of membrane surface modification strategies

known as hydrophilization treatment, to surface coating and surface grafting (via either 'grafting to' or 'grafting from' mechanisms) [12]. The 'grafting to' method involves the use of a backbone chain with functional groups that are distributed randomly along the chain. The formation of the graft copolymer originates from the coupling reaction between the functional backbone and the end-groups of the branches that are reactive. Meanwhile, in the 'grafting from' method, the macromolecular backbone is modified in order to introduce active sites capable of initiating functionality [4].

Apart from that, plenty of studies have been made in the development of desalination membranes via the incorporation of nanoparticles onto the membrane skin. Significant breakthroughs have been achieved through incorporation of conventional nanomaterials such as metal oxides, silica, zeolites as well as the emerging carbon-based nanomaterials such as carbon nanotubes and graphene [45].

17.4.1.1   Surface Coatings

Surface coating approach was known to be less complicated with higher throughput procedure in modifying membrane surface with antifouling properties [12]. Coating will act as a defensive layer to reduce or eliminate the adsorption and deposition of foulants onto membrane surface. Surface coating can be segregated into two distinct techniques known as 'coating-to' and 'coating-from'. For the coating-to technique, membranes are post-modified with antifouling polymeric materials or inorganic nanomaterials via spray coating, dip coating or spin coating. For the 'coating-from' technique, the coatings are in situ generated over the membrane surface [71].

Nikolaeve and co-workers applied hydrophilic material on the membrane surface via spray coating. In this study, they successfully bonded a hydrophilic hyperbranched poly (amido amine) onto the skin layer of a membrane by spraying it after the interfacial polymerization onto the surface [44]. Result showed that bovine serum albumin (BSA) adsorption on modified membrane surfaces reduced significantly (from 6.05% for unmodified to 2.86% for modified membrane). Surface roughness reduced from 38 nm to 35 nm after modification. The laboratory membrane possesses a more hydrophilic surface, as outlined by the lower water contact angle (43° vs. 33°), and a lower negative surface charge at pH 7.4 (-35.5 mV vs -18.5 mV). As a result, the modified membrane surface become less attractive for protein adsorption.

Yu and co-workers modified membrane surface by using poly-(N-isopropylacrylamide) and poly-(acrylic acid) copolymers (P(NIPAm-co-AAc)) by simple surface coating procedure [67]. Membrane topography studies showed that the modified membrane appears to comprise a more nodular structure and exhibits an unevenly distributed surface feature indicative of surface roughness increment. Interestingly, the membrane surface roughness found decreased with a higher coating solution concentration. Membrane surface hydrophilicity decreased from approximately 68° to 48°. The modified membranes acquired a more negative charge at pH 9.0 (-48.13 mV) than the neat membrane (-36.76 mV). The results of the fouling experiments with BSA aqueous solution revealed that the fouling resistance was improved significantly when compared to the unmodified membrane. The water fluxes and salt rejections of a 720 h testing period for the modified membrane exhibited good durability and high long-term performance stability.

The use of plasma for surface coating is regarded as one of the most promising technologies due to the absence of solvent or any hazardous liquid in the procedure [18, 73]. Plasma polymerization can be assumed to be one of the less damaging methods of membrane modification and this offers great advantage for membrane treatment. It consists of electrical ionization of a monomer, which results in the generation of reactive monomer fragments. These fragments recombine on membrane surface to form cross-linked structures. The advantages of plasma polymers over conventional polymers are that they show a much higher degree of cross-linking and adhere strongly to the substrate, the coatings are uniform and do not require the use of harsh solvents that may damage the substrate. Depending upon the monomer used, plasma polymerization can be used to chemically modify the surface of substrates as well as the monomers and combination of monomers with desired functional groups [73].

Zou et al. deposited triethylene glycol dimethyl ether (triglyme) on top of the commercial TFC RO membranes specifically to reduce organic fouling. Water flux study using a protein solution were conducted to evaluate the change of hydrophilicity and anti-fouling properties of the membrane. The result shows that, after 210 min of filtration process, no flux decline was evident for the modified membranes, whereas a 27% reduction of the initial flux was observed for the uncoated membrane. After the membrane surface was clean with water, flux recovery reaches up to 99.5% for the modified membranes and 91.0% for the untreated. Overall, it was found that surface hydrophilic modification of RO membranes by plasma polymerization of hydrophilic polymer triglyme effectively reduced the organic fouling without sacrificing salt rejections [73].

Polydopamine (PDA) is a novel bio-inspired polymer sharing similar properties to the adhesive secretions of mussels. This polymer can be applied to many surfaces without additional preparation and only requires that there exists good contact between the dopamine coating solution and the surface to be coated. PDA has been used to tailor fouling resistance property on desalination membranes. Previous investigations revealed that, the PDA could increase hydrophilicity when it was applied at the selective layer of the membrane. This resulted in reduced adhesion to the surface by proteins and other foulants. There are few studies on the use of PDA as surface coater in modifying the rejection layer of TFC membranes. For instance, Liu et al. deposited PDA on both side of the FO membrane surfaces [35]. Resultant membrane exhibited the increment in surface hydrophilicity with improved fouling resistance and fouling reversibility. The wetting property for both sides was enhanced compared to unmodified membrane; top (82.4° vs 80.5°) and bottom (82.2° vs 37.4°). Modified membrane successfully achieved 93.0% of water flux recovery after combined fouling (alginate + silica), demonstrating much superior performance compared with the unmodified double-skinned (81.7%) and single-skinned (61.7%) under same testing conditions. The surface charge for the bottom side became slightly less negative (from −21.3 mV for neat to −16.4 mV for modified membrane) at the testing pH of 6.5; while the surface charge for top side become slightly more negative (−20.7 mV vs −20.9 mV). Gou et al. applied a highly selective PDA coating on a TFC FO membrane and investigated the effects of coating on FO mass transport and antifouling behavior. It is showed that the PDA coating significantly improved membrane surface hydrophilicity (~ 42° of control membrane to ~ 25–29° of PDA coated membranes) as well as reduced membrane surface roughness (48.9 nm vs 36.1 nm). The coated membrane presented an improved antifouling performance compared to the control membrane (alginate was used as a model foulant) because of significantly reduced foulant adhesion [20].

Several studies have been conducted in proposing alternative monomers to produce PDA-alike surface modification. Chwatko et al. tailored membrane surface using polynorepinephrine on top of a TFC membrane. This strategy has been shown to produce smoother and thinner coating. In addition, due to the presence additional aliphatic hydroxyl groups, the polynorepinephrine coating can be further modified to tailor specific functionalities while maintaining its attachment to the coated surface [9].

Although surface coating can be easily applied for altering antifouling properties with high coverage density, the coating layer can be easily detached during long-term operation or aggressive cleaning procedures, because of the relatively weak noncovalent interactions with the membrane surface [71].

17.4.1.2 Surface Grafting

Surface grafting is another common modification strategy. Surface grafting refers to the addition of polymer chains onto a membrane surface. Usually, for desalination membranes, the hydrophilic or charged functional groups are introduced into the membrane surface for changing the surface chemistry and topology. As a result, foulant adsorption will be reduced resulting in excellent membrane antifouling property [69]. The surface graft polymerization to modify the thin film polymeric membranes can be performed by different techniques such as ultraviolet-induced [43, 49], free-radical graft [69] and redox-initiated graft polymerization [23] methods.

Hong and co-workers applied redox-initiated graft polymerization of acrylic acid (AA) onto the surface of PA TFC membranes. The experimental results indicated that the membrane surfaces became more hydrophilic and smoother after grafting. The modified membranes possess better separation performance with significant enhancement of flux at excellent retention. The fouling resistance of the modified membrane obviously increased with a higher maintained flux ratio and a lower irreversible fouling factor [23]. Zhang and co-workers modified a commercial aromatic PA TFC membrane through free-radical graft polymerization using 3-allyl-5,5-dimethylhydantoin (ADMH) and N,N′-methylenebis(acrylamide) (MBA). After grafting, hydrophilicity and salt rejection of ADMH + MBA membrane increased providing high antimicrobial and anti-biofouling abilities [69].

Densely grafted hydrophilic polymer brushes with suitable wettability and anti-fouling properties, especially anti-biofouling can be prepared by the surface-initiated controlled radical polymerization with specifically designed hydrophilic functional groups [27]. For example, a study on surface-initiated atom transfer radical polymerization (ATRP) on the membrane surface has been conducted by Yang and co-workers. The researchers grafted poly cysteine methacrylate (pCysMA) brush on polysulfone membrane and studied their antifouling properties. The results displayed that the pCysMA showed better hydrophilicity and effectively resisted the adsorption of BSA protein [63].

Zwitterionic polymers with both positively and negatively charged units have been developed as a kind of new generation antifouling materials to control the membrane fouling, attributed to their strong capability of forming a hydration layer via electrostatic interaction and hydrogen bond [33]. Various innovative methods have been developed to modify TFC membrane with zwitterionic materials, including initiated chemical vapor deposition (iCVD) [54], ATRP [36], solution polymerization and solvent evaporation [68]. Zhang et al. grafted a zwitterionic poly(sulfobetaine mathacrylate) onto the membrane surface using surface-initiated ATRP. Results proved that zwitterionic-coated membrane was highly resistant to protein adhesion [70]. Studies showed superior anti-fouling properties for membrane surface deposited with the zwitterionic thin film through iCVD process [54, 62]. On the other hand, Meng and co-workers employed surface initiated redox graft polymerization method to graft zwitterionic poly (4-(2-sulfoethyl)-1-(4-vinylbenzyl) pyridinium betaine) (PSVBP). The prepared membrane was demonstrated to have superior anti-fouling ability [40]. While, Azari and co-workers used zwitterionic amino acid 3-(3,4-dihydroxyphenyl)-L-alanine to modify a PA TFC membrane surface [3]. Recently, Mahdavi and co-workers introduced zwitterionic polymers using poly(2-(methacryloyloxy)ethyl dimethyl-(3-sulfopropyl)ammonium hydroxide) (PMSA)-grafted graphene oxide (GO) (PMSA-g-GO) as an antifouling additive [37].

Wu and co-workers grafted a commercial RO membrane with the positively charged and hydrophilic polyvinylamine (PVAm). The membrane surface became more hydrophilic with reduced surface roughness after the modification. More importantly, the membrane surface charge climbed from negative to positive values

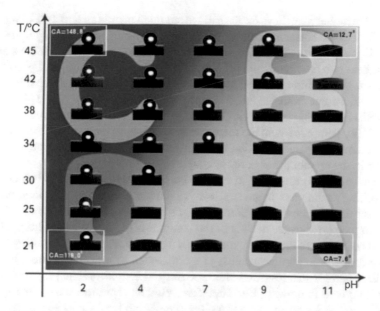

Fig. 17.4 Changes of contact angles at different temperature and pH for a modified substrate. Reproduced with permission from Ref. [60]; Copyright 2006 @ John Wiley and Sons

when PVAm grafted with still higher concentrations. Fouling behaviors of the membrane were investigated with two different proteins, one polysaccharide, one surfactant, and one colloid model foulant. The results revealed better antifouling properties for PVAm-grafted membrane when compared to a commercial RO membranes with PVA coating layer [59].

Another interesting strategy in membrane surface modification is by tuning the surface chemistry of a polymer surface with reversible surface wettability (between hydrophilicity and hydrophobicity) at different condition (changes in temperature and pH) [15]. The first attempt on this method was introduced in 2006 by Xia and co-workers. In this study, a thin double stimuli-responsive layer was grafted from random copolymer of poly-(N-isopropylacrylamide) and poly-(acrylic acid) on both a flat and a roughly etched silicon substrate to obtain a smart surface upon varying both a temperature and pH (Fig. 17.4) [60].

Micropatterned PA TFC was developed by using two microfabrication methods, combined processes of vapor- and non-solvent-induced phase separation micromolding, as well as micro- imprinting lithography [11]. Recently, ElSherbiny's group successfully modified membrane surface by combining "surface micropatterning" and "double stimuli responsivity" (surface that can switch reversibly between hydrophilicity and hydrophobicity upon varying both the temperature and the pH value) approaches onto aromatic PA TFC RO membranes [12]. The modified membrane showed significant switchable wettability, without adversely influencing the separation performance. The surface modified micropatterned PA TFC membranes exhibited various degrees of grafting between 0.2–0.5 mg cm^{-2}, which were higher than those for flat surface modified membranes, 0.1–0.16 mg cm^{-2}. The ability of the surface modified membranes to switch between very hydrophilic and hydrophobic properties was assessed via the measurement of static water contact angle at two environmental pH values, pH 3 (<pKa of PAAc) and pH 7 (>pKa of PAAc), and within a temperature range of 25–55 °C. A large transition in the water contact angle from 82° (at 55 °C and pH = 3) to 28° (at 25 °C and pH = 7) was accomplished by this membrane [12]. These new membranes showed a superior permeability compared with the reference membrane, without a detrimental impact on the membrane selectivity due to enhanced active surface area and significant increase in the membrane surface roughness.

According to the results of studies tabulated in Table 17.1, it can be concluded that grafting techniques in general can significantly enhance the antibiofouling characteristics of membranes. Somehow, grafting at certain degrees may also contribute to the performance decline of the membrane (e.g. pure water flux) [61]. For some cases, although the surface modification of the membrane via grafting method led to a decrease in pure water flux, properties such as foulant rejection, total flux loss, irreversible flux loss and flux recovery (i.e. antifouling property) of the membrane were improved [49]. Therefore, vigilant measures during the grafting process need to be taken to improve the membrane performance, particularly in term of flux and solute rejection.

17.4.2   Addition of Nanoparticles in the Membrane Matrix

Incorporating nanomaterials into the thin film of RO membrane has become one of the most attractive strategy in combating foulants in desalination. The immobilization of nanomaterials on the membrane surface imparts the anti-fouling property by improving the hydrophilicity so that the foulants are repelled and washed away from the membrane surface during the filtration process without deteriorating the membrane surface (Fig. 17.5). For instance, Emadzadeh *et.al* modified PA layer of TFC membranes using nanoporous titanate (mNTs). Results of the study showed that the modified membrane demonstrated a lower degree of flux decline compared to the control membrane. The membrane contact angle decreases from 73.1° for the control membrane to around 41.3° for the modified membrane indicating a remarkable improvement in membrane surface hydrophilicity. The improved membrane hydrophilicity resulting from embedding mNTs in the PA selective layer was mainly due to the super-hydrophilicity of the mNTs. The decreased in contact angle, tend to increase surface energy from 93.96 mJ/m^2 for neat membrane to 127.5 mJ/m^2 for the modified membrane. In addition, zeta potential of membranes become more negative as the amount of mNTs increased [13].

Studies showed that addition of GO can offer a number of significant advantages such as improved anti-microbial and anti-fouling capability, high chemical stability and increased permeability of the membranes [38, 56]. The capability of GO can be

Table 17.1 Membrane surface grafting approaches reported

Strategy of grafting	Membrane performance	Ref.
Free-radical grafted polymerization of ADMH onto the surfaces of PA TFC membranes and then crosslinked by MBA	The membrane hydrophilicity increased 47%. Membrane chlorine resistances was enhanced with better anti-microbial efficiencies with reduction ratio of bacteria cells above 90%.	[69]
UV photo-grafting between acrylic and amino monomer on top PES polymer	The contact angle measurements were significantly increased from 64° to ~ 50°. The protein rejection was improved from 91% for neat membrane to 99% for modified membrane.	[49]
Polymerization of AA grafted onto the surfaces of PA TFC membranes via UV photo-induced method	The antifouling of the membrane improved and evidenced by the higher maintained flux ratios and the lower irreversible fouling factors during the filtration of feed solutions containing humic acid, dye and BSA.	[24]
Polymerization of AA grafted onto the surface of PA TFC via redox-initiated	The modified membrane surfaces become more hydrophilic with a strongly reduced water contact angles (24°), with lower surface roughness (33.5 nm). The antifouling property significantly improved with the higher maintained flux ratio and the lower irreversible fouling factor compared to the unmodified.	[23]
Poly(2-hydroxyethyl methacrylate) (pHEMA) and poly[poly(ethylene glycol)methacrylate] (pPEG) grafted via ATRP onto the surface of PA TFC membrane	Surface roughness for pHEMA and pPEG grafted membrane was smoother as compared with neat membrane with hydrophilicity increased by approximately 4.6% and 7.6% respectively after 60 min polymerization time. The bacterial adhesion decreased to 4.8% and 0.1% for pHEMA and pPEG respectively. Dynamic biofouling filtration test showed that modified membrane exhibited smaller decline in water permeability and larger final permeability after 20 h of filtration as compared with the pristine PA TFC membrane.	[65]
Poly(sulfobetaine)-grafted PVDF hollow fiber MF membrane	As the grafting amount reached 513 $\mu g/cm^2$, the value of contact angle dropped to 22.1° and the amount of protein adsorption decreased to zero compared to the nascent PVDF membrane.	[33]
Conversion of copolymer coatings poly(4-vinylpyridine-co-ethylene glycol diacrylate) synthesized via iCVD to zwitterionic structures containing polycarboxy-betaine acrylic acetate units via a post-deposition quaternizing reaction with 3-bromopropionic acid	The modified membrane showed three improvements; firstly coating resulted in higher hydrophilicity (±50%), secondly, it significantly enhanced resistance against bacterial adhesion (reduction >95%) thirdly, it resulted in smoother surface roughness (7.1 nm).	[54]

(continued)

Table 17.1 (continued)

Strategy of grafting	Membrane performance	Ref.
Polydopamine- [2-(methacryloyloxy)-ethyl] dimethyl- (3-sulfopropyl)ammonium hydroxide zwitterionic polymer brush grafted via ATRP	The modified membrane exhibited reduced surface roughness (> 50%), enhanced hydrophilicity (> 70%), and lower surface charge. The excellent fouling resistance was demonstrated by significantly reduced adsorption of proteins (zero adsorption) and bacteria (reduction of colony-forming units (CFU) >80%).	[36]
Poly(sulfobetaine methacrylate) (pSBMA) zwitterionic polymer grafted on the surface of PA TFC membrane	The roughness of pSBMA-coated membrane surface decreased from 17.5 nm to 12.8 nm with relative protein adsorption 6.7% when compared with untreated membrane (100%) after 60 min reaction time.	[70]
Poly[2-(dimethylamino)ethyl methacrylate-co-ethylene glycol dimethacry late] (PDE) with 1,3-propane sultone grafted on to TFC-PA membrane	The iCVD zwitterionic coating on RO membrane is highly smooth (<3 nm). The adhesion of bacteria reduced more than 99% when antifouling test was conducted.	[62]
PSVBP grafted onto a PA TFC membrane by employing surface-initiated free radical polymerization	The dynamic water contact angle significantly decreased to only 7.2° after PSVBP grafting. A protein fouling test indicates superior antifouling property in the short term but lost the advantage for long-term operation. The modified membrane can restore 90% of the initial flux after rinsing with brine.	[40]
Amino acid 3-(3,4-dihydroxyphenyl)-l- alanine (l-DOPA) grafted onto commercial PA TFC membrane	The hydrophilicity of the modified membranes was significantly improved (nearly 50% reduction contact angle to ≈20°). The amount of BSA adsorbed was reduced by almost 50%.	[3]
PMSA-grafted GO introduced onto PA TFC membrane	Hydrophilicity improved from 63° to 19° and surface roughness reduced from 37.2 nm to 10.9 nm after modification. Anti-fouling ability of the RO membrane was improved with irreversible fouling ratio decreased from 20.7% for the neat membrane to 5.9% for the modified membrane.	[37]
Positively charged PVAm grafted onto PA TFC membrane	The membrane surface became more hydrophilic and smoother after the PVAm grafting. The PVAm-grafted membrane showed lower flux decline after organic fouling. The flux recovery of the modified membrane after rinsing was higher than that of unmodified membrane.	[59]

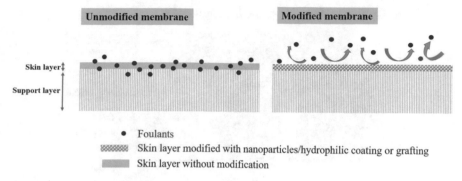

Foulants

Skin layer modified with nanoparticles/hydrophilic coating or grafting

Skin layer without modification

Fig. 17.5 Illustration showing the anti-fouling phenomena

increased by functionalizing the GO with other compounds such as zwitterionic polymer. Through this modification, it can protect GO nanosheets against non-specific interactions, improve the dispersion ability and introduce useful functional groups for enhancing specific properties. Several researches have demonstrated that the zwitterionic functionalized GO (ZGO) exhibited remarkable anti-biofouling performance, which could be attributed to the more negative zeta potential, more surface smoothness and good hydrophilicity [26, 37].

Efforts have been made in modifying and optimizing membrane surface using different two-dimensional (2D) materials such as nano clays. 2D nano clay has a unique charge property and hydrophilic nature. Membrane incorporated with both negatively and positively charged clay nanomaterials showed an increased hydrophilicity and improved desalination performance. Dong and co-workers reported that the incorporation of a cationic and an anionic nanoclays respectively resulted in a more and less negatively charged membrane surface, and thus exhibited different electrostatic repulsion effects and improved antifouling performances towards protein, cationic surfactant, and natural organic matter [10]. Figure 17.6 illustrates the structure of Na-montmorillonite (MMT), the cationic clay and Mg/Al-layered double hydroxide (LDH), the anionic clay employed in this work.

2D nanomaterials can be projected as next generation materials for desalination membranes [31]. Several works on 2D nanomaterials are available where researchers focused in fabricating novel 2D materials such as molybdenum disulfide (MoS_2) [14], h-boron nitride nanosheets [48] and titanium carbide ($Ti_3C_2T_x$; T represents terminating functional groups (O, OH, and/or F), and x the number of terminating groups) [52]. For instance, MoS_2 which is made up of hexagonal layers of Mo and S atoms possess highly negatively charged properties with the hydrophilic sites provided by the S/Mo atoms. This material is very useful to modify the physicochemical properties of TFC RO membrane surface, including hydrophilicity, roughness and electrostatic charge. Li et al. prepared a few-layered MoS_2 sheets via liquid-phase exfoliation from bulk MoS_2 crystals and subsequently incorporated into TFN membrane via interfacial polymerization. It was observed that MoS_2 has successfully decreased the membrane selective layer thickness, improved surface hydrophilicity

A. Cationic clay Na-montmorillonite

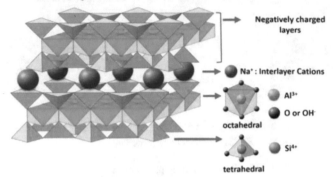

B. Anionic clay Mg/Al-layered double hydroxide

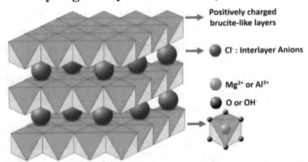

Fig. 17.6 (**A**) Structure of Na-montmorillonite (MMT), the cationic clay and (**B**) Mg/Al-layered double hydroxide (LDH), the anionic clay. Reproduced with permission from Ref. [10]; Copyright 2015 @ Elsevier

(90° for neat and 71° for 0.01 wt% MoS_2-TFN) and increased the surface roughness (58 nm for neat and 80.6 nm for 0.01 wt% MoS_2-TFN). Fouling test demonstrated that 91% of the normalized water flux was maintained for 0.01 wt% MoS_2-TFN membrane using 100 ppm BSA as the protein foulant [34].

Another approach investigated is incorporation of metal organic frameworks (MOFs). MOFs are crystalline porous compounds that belong to organic–inorganic hybrid materials constructed from metal ions coordinated to organic linkers. This type of material possesses a narrow pore size distribution with tunable pore size and surface chemistry and large surface area. MOFs have been used as a filler in RO membrane with a capability to treat wastewater rich with dye. Among MOFs, zeolitic imidazolate framework (ZIF) nanoparticles have attracted great interest due to its high stability in water and other solvents, high thermal and chemical stability as well as fouling propensity. Aljundi investigated the addition of ZIF-8 to the membrane selective layer and its effect on the improvement of membrane fouling-resistance [1]. Scanning electron microscopy (SEM) micrographs showed a relatively smoother membrane surface which may be due to displacement of 'valley' structure with incorporation of ZIF-8 nanoparticles. The contact angle of the best performing membrane was much lower compared to the control membrane. The work shown that, the increased hydrophilicity may be due to a higher concentration of exposed carboxylic groups on the surface. The results also evidenced that, ZIF-8 nanoparticles can significantly enhance the PA fouling-resistance and the fouling was remarkably reduced by more than 75% under BSA condition.

17.5   Conclusions and Outlook

In commercial or industrial operations, to ensure that desalination membrane can last for a longer period, membrane should possess a good fouling propensity with high water flux recovery even after long term operation. In this chapter, the recent strategies to mitigate antifouling via membrane surface modifications were described. Basically, there are three main factors that could affect the membrane behavior towards foulant rejection: surface wetting behavior, surface charge and surface roughness. The antifouling modification layer on the membrane skin has to be mechanically strong with good adhesion to the surface. Compared to the physical modification approaches, which is usually unstable for long term operation, the membrane grafting is a better option. Membrane grafting offers superior performance due to less leaching and stripping issues, attributed to the stronger chemical bonding. However, it is often difficult to control the polymerization parameters in the grafting process, therefore it is hard to maintain the excellent antifouling properties of the modified membrane. Another promising technique to overcome the membrane fouling is the incorporation of nanoparticles into the thin film RO membrane. Yet, agglomeration and dispersion issues of nanoparticles has become significant challenges in this modification. Although to totally eradicate the membrane fouling is impossible, the continuous efforts in studying more effective techniques as well as their combinations might be able to mitigate membrane fouling and improve fouling reversibility.

References

1. I.H. Aljundi, Desalination characteristics of TFN-RO membrane incorporated with ZIF-8 nanoparticles. Desalination **420**, 12–20 (2017)
2. M. Asadollahi, D. Bastani, S.A. Musavi, Enhancement of surface properties and performance of reverse osmosis membranes after surface modification: A review. Desalination **420**, 330–383 (2017)
3. S. Azari, L. Zou, Using zwitterionic amino acid l-DOPA to modify the surface of thin film composite polyamide reverse osmosis membranes to increase their fouling resistance. J. Membr. Sci. **401–402**, 68–75 (2012)
4. S. Berger, A. Synytska, L. Ionov, K.J. Eichhorn, M. Stamm, Stimuli-responsive bicomponent polymer janus particles by "grafting from"/"grafting to" approaches. Macromolecules **41**, 9669–9676 (2008)

5. F.H. Butt, F. Rahman, U. Baduruthamal, Characterization of foulants by autopsy of RO desalination membranes. Desalination **114**, 51–64 (1997)
6. H. Chang, T. Li, B. Liu, C. Chen, Q. He, J.C. Crittenden, Smart ultrafiltration membrane fouling control as desalination pretreatment of shale gas fracturing wastewater: The effects of backwash water. Environ. Intl. **130**, 104869 (2019)
7. S.C. Chen, G.L. Amy, T.S. Chung, Membrane fouling and anti-fouling strategies using RO retentate from a municipal water recycling plant as the feed for osmotic power generation. Water Res. **88**, 144–155 (2016)
8. Y. Chun, D. Mulcahy, L. Zou, I.S. Kim, A short review of membrane fouling in forward osmosis processes. Membranes **7**, 1–23 (2017)
9. M. Chwatko, J.T. Arena, J.R. McCutcheon, Norepinephrine modified thin film composite membranes for forward osmosis. Desalination **423**, 157–164 (2017)
10. H. Dong, L. Wu, L. Zhang, H. Chen, C. Gao, Clay nanosheets as charged filler materials for high-performance and fouling-resistant thin film nanocomposite membranes. J. Membr. Sci. **494**, 92–103 (2015)
11. I.M.A. ElSherbiny, A.S.G. Khalil, M. Ulbricht, Surface micro-patterning as a promising platform towards novel polyamide thin-film composite membranes of superior performance. J. Membr. Sci. **529**, 11–22 (2017)
12. I.M.A. ElSherbiny, A.S.G. Khalil, M. Ulbricht, Tailoring surface characteristics of polyamide thin-film composite membranes toward pronounced switchable wettability. Adv. Mater. Interface **6**, 1–12 (2019)
13. D. Emadzadeh, M. Ghanbari, W.J. Lau, M. Rahbari-Sisakht, D. Rana, T. Matsuura, B. Kruczek, A.F. Ismail, Surface modification of thin film composite membrane by nanoporous titanate nanoparticles for improving combined organic and inorganic antifouling properties. Mater. Sci. Eng. C **75**, 463–470 (2017)
14. J. Feng, M. Graf, K. Liu, D. Ovchinnikov, D. Dumcenco, M. Heiranian, V. Nandigan, N.R. Aluru, A. Kis, A. Radenovic, Single-layer MoS$_2$ nanopores as nanopower generators. Nature **536**, 197–200 (2016)
15. S. Feng, Y. Xing, S. Deng, W. Shang, D. Li, M. Zhang, Y. Hou, Y. Zheng, An integrative mesh with dual wettable on–off switch of water/oil. Adv. Mater. Interface **5**, 1–6 (2018)
16. W. Gao, H. Liang, J. Ma, M. Han, Z.l. Chen, Z.s. Han, G. Li, Membrane fouling control in ultrafiltration technology for drinking water production: A review. Desalination **272**, 1–8 (2011)
17. N. Ghaffour, T.M. Missimer, G.L. Amy, Technical review and evaluation of the economics of water desalination: Current and future challenges for better water supply sustainability. Desalination **309**, 197–207 (2013)
18. P.S. Goh, W.J. Lau, M.H.D. Othman, A.F. Ismail, Membrane fouling in desalination and its mitigation strategies. Desalination **425**, 130–155 (2018)
19. D. González, J. Amigo, F. Suárez, Membrane distillation: Perspectives for sustainable and improved desalination. Renew. Sust. Energy Rev. **80**, 238–259 (2017)
20. H. Guo, Z. Yao, J. Wang, Z. Yang, X. Ma, C.Y. Tang, Polydopamine coating on a thin film composite forward osmosis membrane for enhanced mass transport and antifouling performance. J. Membr. Sci. **551**, 234–242 (2018)
21. H.M. Hegab, A. ElMekawy, T.G. Barclay, A. Michelmore, L. Zou, C.P. Saint, M. Ginic-Markovic, Fine-tuning the surface of forward osmosis membranes via grafting graphene oxide: Performance patterns and biofouling propensity. ACS Appl. Mater. Interfaces **7**, 18004–18016 (2015)
22. M. Hirose, H. Ito, Y. Kamiyama, Effect of skin layer surface structures on the flux behaviour of RO membranes. J. Membr. Sci. **121**, 209–215 (1996)
23. T. Hong Anh Ngo, K. Dinh Do, D. Thi Tran, Surface modification of polyamide TFC membranes via redox-initiated graft polymerization of acrylic acid. J. Appl. Polym. Sci. **134**, 1–8 (2017)

24. T. Hong Anh Ngo, D.T. Tran, C. Hung Dinh, Surface photochemical graft polymerization of acrylic acid onto polyamide thin film composite membranes. J. Appl. Polym. Sci. **134**, 1–9 (2017)
25. K.Y. Jee, D.H. Shin, Y.T. Lee, Surface modification of polyamide RO membrane for improved fouling resistance. Desalination **394**, 131–137 (2016)
26. S. Jin, N. Zhou, D. Xu, J. Shen, Synthesis and characterization of poly(2-methacryloyloxyethyl phosphorylcholine) onto graphene oxide. Polym. Adv. Technol. **24**, 685–691 (2013)
27. M. Kobayashi, Y. Terayama, H. Yamaguchi, M. Terada, D. Murakami, K. Ishihara, A. Takahara, Wettability and antifouling behavior on the surfaces of superhydrophilic polymer brushes. Langmuir **28**, 7212–7222 (2012)
28. V. Kochkodan, N. Hilal, A comprehensive review on surface modified polymer membranes for biofouling mitigation. Desalination **356**, 187–207 (2015)
29. S.E. Kwan, E. Bar-Zeev, M. Elimelech, Biofouling in forward osmosis and reverse osmosis: Measurements and mechanisms. J. Membr. Sci. **493**, 703–708 (2015)
30. J. Landaburu-Aguirre, R. García-Pacheco, S. Molina, L. Rodríguez-Sáez, J. Rabadán, E. García-Calvo, Fouling prevention, preparing for re-use and membrane recycling. Towards circular economy in RO desalination. Desalination **393**, 16–30 (2016)
31. J. Lawler, Incorporation of graphene-related carbon nanosheets in membrane fabrication for water treatment: A review. Membranes **6**, 57 (2016)
32. E.K. Lee, V. Chen, A.G. Fane, Natural organic matter (NOM) fouling in low pressure membrane filtration - Effect of membranes and operation modes. Desalination **218**, 257–270 (2008)
33. Q. Li, J. Imbrogno, G. Belfort, X.L. Wang, Making polymeric membranes antifouling via "grafting from" polymerization of zwitterions. J. Appl. Polym. Sci. **132**, 1–12 (2015)
34. Y. Li, S. Yang, K. Zhang, B. Van der Bruggen, Thin film nanocomposite reverse osmosis membrane modified by two dimensional laminar MoS_2 with improved desalination performance and fouling-resistant characteristics. Desalination **454**, 48–58 (2019)
35. X. Liu, S.L. Ong, H.Y. Ng, Fabrication of mesh-embedded double-skinned substrate membrane and enhancement of its surface hydrophilicity to improve anti-fouling performance of resultant thin-film composite forward osmosis membrane. J. Membr. Sci. **511**, 40–53 (2016)
36. C. Liu, J. Lee, J. Ma, M. Elimelech, Antifouling thin-film composite membranes by controlled architecture of zwitterionic polymer brush layer. Environ. Sci. Technol. **51**, 2161–2169 (2017)
37. H. Mahdavi, A. Rahimi, Zwitterion functionalized graphene oxide/polyamide thin film nanocomposite membrane: Towards improved anti-fouling performance for reverse osmosis. Desalination **433**, 94–107 (2018)
38. K.A. Mahmoud, B. Mansoor, A. Mansour, M. Khraisheh, Functional graphene nanosheets: The next generation membranes for water desalination. Desalination **356**, 208–225 (2015)
39. B.S. Mbuli, E.N. Nxumalo, S.D. Mhlanga, R.W. Krause, V.L. Pillay, Y. Oren, C. Linder, B.B. Mamba, Development of antifouling polyamide thin-film composite membranes modified with amino-cyclodextrins and diethylamino-cyclodextrins for water treatment. J. Appl. Polym. Sci. **131**, 1–10 (2014)
40. J. Meng, Z. Cao, L. Ni, Y. Zhang, X. Wang, X. Zhang, E. Liu, A novel salt-responsive TFC RO membrane having superior antifouling and easy-cleaning properties. J. Membr. Sci. **461**, 123–129 (2014)
41. S. Meng, W. Fan, X. Li, Y. Liu, D. Liang, X. Liu, Intermolecular interactions of polysaccharides in membrane fouling during microfiltration. Water Res. **143**, 38–46 (2018)
42. S.S. Mitra, A.R. Thomas, G.T. Gang, Evaluation and characterization of seawater RO membrane fouling. Desalination **247**, 94–107 (2009)
43. L.Y. Ng, A. Ahmad, A.W. Mohammad, Alteration of polyethersulphone membranes through UV-induced modification using various materials: A brief review. Arab. J. Chem. **10**, S1821–S1834 (2017)
44. D. Nikolaeva, C. Langner, A. Ghanem, M.A. Rehim, B. Voit, J. Meier-Haack, Hydrogel surface modification of reverse osmosis membranes. J. Membr. Sci. **476**, 264–276 (2015)

45. C.S. Ong, P.S. Goh, W.J. Lau, N. Misdan, A.F. Ismail, Nanomaterials for biofouling and scaling mitigation of thin film composite membrane: A review. Desalination **393**, 2–15 (2016)

46. J. Park, S. Lee, J. You, S. Park, Y. Ahn, W. Jung, K.H. Cho, Evaluation of fouling in nanofiltration for desalination using a resistance-in-series model and optical coherence tomography. Sci. Total Environ. **642**, 349–355 (2018)

47. M. Qasim, M. Badrelzaman, N.N. Darwish, N.A. Darwish, N. Hilal, Reverse osmosis desalination: A state-of-the-art review. Desalination **459**, 59–104 (2019)

48. S. Qin, D. Liu, G. Wang, D. Portehault, C.J. Garvey, Y. Gogotsi, W. Lei, Y. Chen, High and stable ionic conductivity in 2D nanofluidicion channels between boron nitride layers. J. Am. Chem. Soc. **139**, 6314–6320 (2017)

49. A. Rahimpour, UV photo-grafting of hydrophilic monomers onto the surface of nano-porous PES membranes for improving surface properties. Desalination **265**, 93–101 (2011)

50. D. Rana, T. Matsuura, Surface modifications for antifouling membranes. Chem. Rev. **110**, 2448–2471 (2010)

51. K. Rathinam, Y. Oren, W. Petry, D. Schwahn, R. Kasher, Calcium phosphate scaling during wastewater desalination on oligoamide surfaces mimicking reverse osmosis and nanofiltration membranes. Water Res. **128**, 217–225 (2018)

52. C.E. Ren, K.B. Hatzell, M. Alhabeb, Z. Ling, K.A. Mahmoud, Y. Gogotsi, Charge- and size-selective ion sieving through $Ti_3C_2T_x$ MXene membranes. J. Phys. Chem. Lett. **6**, 4026–4031 (2015)

53. H. Salehi, M. Rastgar, A. Shakeri, Anti-fouling and high water permeable forward osmosis membrane fabricated via layer by layer assembly of chitosan/graphene oxide. Appl. Surf. Sci. **413**, 99–108 (2017)

54. H.Z. Shafi, Z. Khan, R. Yang, K.K. Gleason, Surface modification of reverse osmosis membranes with zwitterionic coating for improved resistance to fouling. Desalination **362**, 93–103 (2015)

55. Y. Shim, H.J. Lee, S. Lee, S.H. Moon, J. Cho, Effects of natural organic matter and ionic species on membrane surface charge. Environ. Sci. Technol. **36**, 3864–3871 (2002)

56. J. Wang, Y. Wang, J. Zhu, Y. Zhang, J. Liu, B. Van der Bruggen, Construction of TiO_2@graphene oxide incorporated antifouling nanofiltration membrane with elevated filtration performance. J. Membr. Sci. **533**, 279–288 (2017)

57. H. Wang, W. Wang, L. Wang, B. Zhao, Z. Zhang, X. Xia, H. Yang, Y. Xue, N. Chang, Enhancement of hydrophilicity and the resistance for irreversible fouling of polysulfone (PSF) membrane immobilized with graphene oxide (GO) through chloromethylated and quaternized reaction. Chem. Eng. J. **334**, 2068–2078 (2018)

58. J. Wu, Z. Wang, W. Yan, Y. Wang, J. Wang, S. Wang, Improving the hydrophilicity and fouling resistance of RO membranes by surface immobilization of PVP based on a metal-polyphenol precursor layer. J. Membr. Sci. **496**, 58–69 (2015)

59. J. Wu, Z. Wang, Y. Wang, W. Yan, J. Wang, S. Wang, Polyvinylamine-grafted polyamide reverse osmosis membrane with improved antifouling property. J. Membr. Sci. **495**, 1–13 (2015)

60. F. Xia, L. Feng, S. Wang, T. Sun, W. Song, W. Jiang, L. Jiang, Dual-responsive surfaces that switch between superhydrophilicity and superhydrophobicity. Adv. Mater. **18**, 432–436 (2006)

61. J. Xu, Z. Wang, L. Yu, J. Wang, S. Wang, A novel reverse osmosis membrane with regenerable anti-biofouling and chlorine resistant properties. J. Membr. Sci. **435**, 80–91 (2013)

62. R. Yang, J. Xu, G. Ozaydin-Ince, S.Y. Wong, K.K. Gleason, Surface-tethered zwitterionic ultrathin antifouling coatings on reverse osmosis membranes by initiated chemical vapor deposition. Chem. Mater. **23**, 1263–1272 (2011)

63. L. Yang, S.-T. Ma, H. Xia, Y. Bu, J.-J. Huang, S.-J. Gu, Grafting of poly (cysteine methacrylate) brush from polysulfone membrane via surface-initiated ATRP and their anti-protein fouling property. J. Fiber Bioeng. Inform. **10**, 231–237 (2017)

64. T. Yang, C.F. Wan, J.Y. Xiong, T.S. Chung, Pre-treatment of wastewater retentate to mitigate fouling on the pressure retarded osmosis (PRO) process. Sep. Purif. Technol. **215**, 390–397 (2019)
65. Z. Yang, D. Saeki, H. Wu, T. Yoshioka, H. Matsuyama, Effect of polymer structure modified on RO membrane surfaces via surface-initiated ATRP on dynamic biofouling behavior. J. Membr. Sci. **582**, 111–119 (2019)
66. Y. Yoon, G. Amy, J. Cho, N. Her, J. Pellegrino, Transport of perchlorate (ClO_4^-) through NF and UF membranes. Desalination **147**, 11–17 (2002)
67. S. Yu, Z. Lü, Z. Chen, X. Liu, M. Liu, C. Gao, Surface modification of thin-film composite polyamide reverse osmosis membranes by coating N-isopropylacrylamide-co-acrylic acid copolymers for improved membrane properties. J. Membr. Sci. **371**, 293–306 (2011)
68. S.F. Zhang, P. Rolfe, G. Wright, W. Lian, A.J. Milling, S. Tanaka, K. Ishihara, Physical and biological properties of compound membranes incorporating a copolymer with a phosphorylcholine head group. Biomaterials **19**, 691–700 (1998)
69. Z. Zhang, Z. Wang, J. Wang, S. Wang, Enhancing chlorine resistances and anti-biofouling properties of commercial aromatic polyamide reverse osmosis membranes by grafting 3-allyl-5,5-dimethylhydantoin and N,N'-Methylenebis(acrylamide). Desalination **309**, 187–196 (2013)
70. Y. Zhang, Z. Wang, W. Lin, H. Sun, L. Wu, S. Chen, A facile method for polyamide membrane modification by poly(sulfobetaine methacrylate) to improve fouling resistance. J. Membr. Sci. **446**, 164–170 (2013)
71. R. Zhang, Y. Liu, M. He, Y. Su, X. Zhao, M. Elimelech, Z. Jiang, Antifouling membranes for sustainable water purification: Strategies and mechanisms. Chem. Soc. Rev. **45**, 5888–5924 (2016)
72. X. Zhu, M. Elimelech, Colloidal fouling of reverse osmosis membranes: Measurements and fouling mechanisms. Environ. Sci. Technol. **31**, 3654–3662 (1997)
73. L. Zou, I. Vidalis, D. Steele, A. Michelmore, S.P. Low, J.Q.J.C. Verberk, Surface hydrophilic modification of RO membranes by plasma polymerization for low organic fouling. J. Membr. Sci. **369**, 420–428 (2011)
74. A.W. Zularisam, A.F. Ismail, M.R. Salim, M. Sakinah, H. Ozaki, The effects of natural organic matter (NOM) fractions on fouling characteristics and flux recovery of ultrafiltration membranes. Desalination **212**, 191–208 (2007)

Correction to: Biofouling in RO Desalination Membranes

Nawrin Anwar, Liuqing Yang, Wen Ma, Haamid Sani Usman, and Md. Saifur Rahaman

Correction to:
Chapter 13 in: V. S. Saji et al. (eds.),
Corrosion and Fouling Control in Desalination Industry,
https://doi.org/10.1007/978-3-030-34284-5_13

The original version of the book was inadvertently published without the copyright information for Fig. 13.2 in Chap. 13. The correct figure caption has been updated in the chapter as given below:

Figure 13.2 Formation of irreversible fouling on polyamide membrane. (Reproduced the inset image with permission from Dr. Florian Beyer; Copyright @ Dr. Florian Beyer)

The updated online version of this chapter can be found at
https://doi.org/10.1007/978-3-030-34284-5_13

Index

© Springer Nature Switzerland AG 2020
V. S. Saji et al. (eds.), *Corrosion and Fouling Control in Desalination Industry*,
https://doi.org/10.1007/978-3-030-34284-5

Printed in the United States
by Baker & Taylor Publisher Services